D1573032

S E M S M E S

EMS Textbooks in Mathematics

EMS Textbooks in Mathematics is a series of books aimed at students or professional mathematicians seeking an introduction into a particular field. The individual volumes are intended not only to provide relevant techniques, results, and applications, but also to afford insight into the motivations and ideas behind the theory. Suitably designed exercises help to master the subject and prepare the reader for the study of more advanced and specialized literature.

Jørn Justesen and Tom Høholdt, *A Course In Error-Correcting Codes*
Markus Stroppel, *Locally Compact Groups*
Peter Kunkel and Volker Mehrmann, *Differential-Algebraic Equations*
Dorothee D. Haroske and Hans Triebel, *Distributions, Sobolev Spaces, Elliptic Equations*
Thomas Timmermann, *An Invitation to Quantum Groups and Duality*
Oleg Bogopolski, *Introduction to Group Theory*
Marek Jarnicki and Peter Pflug, *First Steps in Several Complex Variables: Reinhardt Domains*
Tammo tom Dieck, *Algebraic Topology*
Mauro C. Beltrametti et al., *Lectures on Curves, Surfaces and Projective Varieties*
Wolfgang Woess, *Denumerable Markov Chains*
Eduard Zehnder, *Lectures on Dynamical Systems. Hamiltonian Vector Fields and Symplectic Capacities*
Andrzej Skowroński and Kunio Yamagata, *Frobenius Algebras I. Basic Representation Theory*
Piotr W. Nowak and Guoliang Yu, *Large Scale Geometry*
Joaquim Bruna and Juliá Cufí, *Complex Analysis*
Eduardo Casas-Alvero, *Analytic Projective Geometry*

Fabrice Baudoin

Diffusion Processes and Stochastic Calculus

BOWLING GREEN STATE
UNIVERSITY LIBRARIES

European Mathematical Society

Author:

Fabrice Baudoin
Department of Mathematics
Purdue University
150 N. University Street
West Lafayette, IN 47907
USA

E-mail: fbaudoin@purdue.edu

2010 Mathematics Subject Classification: 60-01, 60G07, 60J60, 60J65, 60H05, 60H07

Key words: Brownian motion, diffusion processes, Malliavin calculus, rough paths theory, semigroup theory, stochastic calculus, stochastic processes

ISBN 978-3-03719-133-0

The Swiss National Library lists this publication in The Swiss Book, the Swiss national bibliography, and the detailed bibliographic data are available on the Internet at http://www.helveticat.ch.

This work is subject to copyright. All rights are reserved, whether the whole or part of the material is concerned, specifically the rights of translation, reprinting, re-use of illustrations, recitation, broadcasting, reproduction on microfilms or in other ways, and storage in data banks. For any kind of use permission of the copyright owner must be obtained.

© European Mathematical Society 2014

Contact address:

European Mathematical Society Publishing House
Seminar for Applied Mathematics
ETH-Zentrum SEW A27
CH-8092 Zürich
Switzerland

Phone: +41 (0)44 632 34 36
Email: info@ems-ph.org
Homepage: www.ems-ph.org

Typeset using the author's TEX files: I. Zimmermann, Freiburg
Printing and binding: Beltz Bad Langensalza GmbH, Bad Langensalza, Germany
∞ Printed on acid free paper
9 8 7 6 5 4 3 2 1

BOWLING GREEN STATE
UNIVERSITY LIBRARIES

Preface

This book originates from a set of lecture notes for graduate classes I delivered since 2005, first at the Paul Sabatier University in Toulouse, and then at Purdue University in West Lafayette. The lecture notes benefited much from the input and criticism from several students and have been modified and expanded numerous times before reaching the final form of this book.

My motivation is to present at the graduate level and in a concise but complete way the most important tools and ideas of the classical theory of continuous time processes and at the same time introduce the readers to more advanced theories: the theory of Dirichlet forms, the Malliavin calculus and the Lyons rough paths theory.

Several exercises of various levels are distributed throughout the text in order to test the understanding of the reader. Results proved in these exercises are sometimes used in later parts of the text so I really encourage the reader to have a dynamic approach in his reading and to try to solve the exercises.

I included at the end of each chapter a short section listing references for the reader wishing to complement his reading or looking for more advanced theories and topics.

Chapters 1, 2, 5 and 6 are essentially independent from Chapters 3 and 4. I often used the materials in Chapters 1, 2, 5 and 6 as a graduate course on stochastic calculus and Chapters 3 and 4 on their own as a course on Markov processes and Markov semigroups assuming some of the basic results of Chapter 1. Chapter 7 is almost entirely independent from the other chapters. It is an introduction to Lyons rough paths theory which offers a deterministic approach to understand differential equations driven by very irregular processes including as a special case Brownian motion.

To conclude, I would like to express my gratitude to the students who pointed out typos and inaccuracies in various versions of the lecture notes leading to this book and to my colleague Cheng Ouyang for a detailed reading of an early draft of the manuscript. Of course, all the remaining typos and mistakes are my own responsibility and a list of corrections will be kept online on my personal blog. Finally, I thank Igor Kortchmeski for letting me use his nice picture of a random stable tree on the cover of the book.

West Lafayette, May 2014 Fabrice Baudoin

Contents

Preface v

Conventions and frequently used notations xi

Introduction 1

1 Stochastic processes 6
- 1.1 Measure theory in function spaces . . . 6
- 1.2 Stochastic processes . . . 8
- 1.3 Filtrations . . . 9
- 1.4 The Daniell–Kolmogorov extension theorem . . . 11
- 1.5 The Kolmogorov continuity theorem . . . 16
- 1.6 Stopping times . . . 18
- 1.7 Martingales . . . 20
- 1.8 Martingale inequalities . . . 32
- Notes and comments . . . 34

2 Brownian motion 35
- 2.1 Definition and basic properties . . . 35
- 2.2 Basic properties . . . 39
- 2.3 The law of iterated logarithm . . . 42
- 2.4 Symmetric random walks . . . 45
- 2.5 Donsker theorem . . . 55
- Notes and comments . . . 62

3 Markov processes 63
- 3.1 Markov processes . . . 63
- 3.2 Strong Markov processes . . . 72
- 3.3 Feller–Dynkin diffusions . . . 75
- 3.4 Lévy processes . . . 87
- Notes and comments . . . 96

4 Symmetric diffusion semigroups 97
- 4.1 Essential self-adjointness, spectral theorem . . . 97
- 4.2 Existence and regularity of the heat kernel . . . 106
- 4.3 The sub-Markov property . . . 109
- 4.4 L^p-theory: The interpolation method . . . 112

4.5 L^p-theory: The Hille–Yosida method 114
4.6 Diffusion semigroups as solutions of a parabolic Cauchy problem 125
4.7 The Dirichlet semigroup 127
4.8 The Neumann semigroup 131
4.9 Symmetric diffusion processes 132
Notes and comments 135

5 Itô calculus 138
5.1 Variation of the Brownian paths 138
5.2 Itô integral 140
5.3 Square integrable martingales and quadratic variations 146
5.4 Local martingales, semimartingales and integrators 154
5.5 Döblin–Itô formula 161
5.6 Recurrence and transience of the Brownian motion in higher dimensions 166
5.7 Itô representation theorem 168
5.8 Time changed martingales and planar Brownian motion 171
5.9 Burkholder–Davis–Gundy inequalities 175
5.10 Girsanov theorem 178
Notes and comments 183

6 Stochastic differential equations and Malliavin calculus 185
6.1 Existence and uniqueness of solutions 185
6.2 Continuity and differentiability of stochastic flows 190
6.3 The Feynman–Kac formula 193
6.4 The strong Markov property of solutions 199
6.5 Stratonovitch stochastic differential equations and the language of vector fields 201
6.6 Malliavin calculus 205
6.7 Existence of a smooth density 216
Notes and comments 219

7 An introduction to Lyons' rough paths theory 221
7.1 Continuous paths with finite p-variation 221
7.2 The signature of a bounded variation path 226
7.3 Estimating iterated integrals 228
7.4 Rough differential equations 239
7.5 The Brownian motion as a rough path 245
Notes and comments 254

Appendix A Unbounded operators 257

Appendix B Regularity theory 262

References 269

Index 275

Conventions and frequently used notations

Unless specified otherwise, the Borel measures we consider on $\mathbb{R}^n$ are assumed to be Radon measures, that is, are finite on compact sets.

$\mathbb{R}_{\geq 0}$	$[0, +\infty)$
$\mathbb{R}^{n \times p}$	set of $n \times p$ matrices
${}^cA$ or A^c	complement of the set A
$[x]$	integer part of x
$x \wedge y$	minimum between x and y
$\Gamma(x)$	$\Gamma(x) = \int_0^{+\infty} t^{x-1} e^{-t} dt$
$\mathcal{A}(A, B)$	set of functions $A \to B$
$\mathcal{C}(A, B)$	set of continuous functions $A \to B$
$\mathcal{C}^k(A, B)$	functions $A \to B$, k-times continuously differentiable
$\mathcal{C}_c(A, B)$	functions $A \to B$, smooth and compactly supported inside A
$\mathcal{C}_0(A, B)$	continuous functions $A \to B$ whose limit at ∞ is 0
$\mathcal{C}^{k,l}(A \times B, C)$	functions $A \times B \to C$ which are k-times continuously differentiable in the first variable and l times in the second
$\mathcal{T}(A, B)$	σ-field on $\mathcal{A}(A, B)$ generated by cylinders
$\mathcal{B}(A, B)$	σ-field on $\mathcal{C}(A, B)$ generated by cylinders
$\mathcal{B}(A)$	Borel σ-field on A
$L^p_\mu(A, B)$	L^p space of functions $A \to B$ for the measure μ
$L^p(\mathcal{F}, \mathbb{P})$	real L^p space of $\mathcal{F}$ measurable random variables
$\Delta_n[0, t]$	$\{0 = t_0^n \leq t_1^n \leq \dots \leq t_n^n = t\}$
$\int H_s dM_s$	Itô integral
$\int H_s \circ dM_s$	Stratonovitch integral
$\mathcal{H}_s(\mathbb{R}^n)$	Sobolev space of order s
$\mathcal{H}_s^0(\Omega)$	closure of $\mathcal{C}_c(\Omega, \mathbb{C})$ in $\mathcal{H}_s(\mathbb{R}^n)$
$\boldsymbol{D}$	Malliavin derivative
δ	divergence operator
$\mathbb{D}^{k,p}$	domain of $\boldsymbol{D}^k$ in $L^p(\mathcal{F}, \mathbb{P})$
$\Vert \cdot \Vert_{p\text{-var},[s,t]}$	p-variation norm on $[s, t]$
$\Vert \cdot \Vert_{\infty,[s,t]}$	supremum norm on $[s, t]$
$C^{p\text{-var}}([s, t], \mathbb{R}^d)$	continuous paths $[s, t] \to \mathbb{R}^d$ with bounded p-variation
$\int_{\Delta^k[s,t]} dx^I$	$\int_{s \leq t_1 \leq t_2 \leq \dots \leq t_k \leq t} dx^{i_1}(t_1) \dots dx^{i_k}(t_k)$
$\Omega^p([0, T], \mathbb{R}^d)$	space of p-rough paths

Introduction

The first stochastic process that has been extensively studied is the Brownian motion, named in honor of the botanist Robert Brown, who observed and described in 1828 the random movement of particles suspended in a liquid or gas. One of the first mathematical studies of this process goes back to the mathematician Louis Bachelier, in 1900, who presented in his thesis [2] a stochastic modelling of the stock and option markets. But, mainly due to the lack of rigorous foundations of probability theory at that time, the seminal work of Bachelier has been ignored for a long time by mathematicians. However, in his 1905 paper, Albert Einstein brought this stochastic process to the attention of physicists by presenting it as a way to indirectly confirm the existence of atoms and molecules. The rigorous mathematical study of stochastic processes really began with the mathematician Andrei Kolmogorov. His monograph [46] published in Russian in 1933 built up probability theory in a rigorous way from fundamental axioms in a way comparable to Euclid's treatment of geometry. From this axiomatic, Kolmogorov gives a precise definition of stochastic processes. His point of view stresses the fact that a stochastic process is nothing else but a random variable valued in a space of functions (or a space of curves). For instance, if an economist reads a financial newspaper because he is interested in the prices of barrel of oil for last year, then he will focus on the curve of these prices. According to Kolmogorov's point of view, saying that these prices form a stochastic process is then equivalent to saying that the curve that is seen is the realization of a random variable defined on a suitable probability space. This point of view is mathematically quite deep and provides existence results for stochastic processes as well as pathwise regularity results.

Joseph Doob writes in the introduction to his famous book "Stochastic processes" [19]:

A stochastic process is any process running along in time and controlled by probability laws...more precisely any family of random variables where a random variable ... is simply a measurable function ...

Doob's point of view, which is consistent with Kolmogorov's and built on the work by Paul Lévy, is nowadays commonly given as a definition of a stochastic process. Relying on this point of view that emphasizes the role of time, Doob's work, developed during the 1940s and the 1950s has quickly become one of the most powerful tools available to study stochastic processes.

Let us now describe the seminal considerations of Bachelier. Let X_t denote the price at time t of a given asset on a financial market (Bachelier considered a given quantity of wheat). We will assume that $X_0 = 0$ (otherwise, we work with $X_t - X_0$). The first observation is that the price X_t can not be predicted with absolute certainty.

It seems therefore reasonable to assume that X_t is a random variable defined on some probability space. One of the initial problems of Bachelier was to understand the distribution of prices at given times, that is, the distribution of the random variable $(X_{t_1}, \dots, X_{t_n})$, where $t_1, \dots, t_n$ are fixed.

The two following fundamental observations of Bachelier were based on empirical observations:

- If τ is very small then, in absolute value, the price variation $X_{t+\tau} - X_t$ is of order $\sigma\sqrt{\tau}$, where σ is a positive parameter (nowadays called the volatility of the asset).
- The expectation of a speculator is always zero[1] (nowadays, a generalization of this principle is called the absence of arbitrage).

Next, Bachelier assumes that for every $t > 0$, X_t has a density with respect to the Lebesgue measure, let us say $p(t,x)$. It means that if $[x-\varepsilon, x+\varepsilon]$ is a small interval around x, then

$$\mathbb{P}(X_t \in [x-\varepsilon, x+\varepsilon]) \simeq 2\varepsilon p(t,x).$$

The two above observations imply that for τ small,

$$p(t+\tau, x) \simeq \frac{1}{2} p(t, x - \sigma\sqrt{\tau}) + \frac{1}{2} p(t, x + \sigma\sqrt{\tau}).$$

Indeed, due to the first observation, if the price is x at time $t+\tau$, it means that at time t the price was equal to $x - \sigma\sqrt{\tau}$ or to $x + \sigma\sqrt{\tau}$. According to the second observation, each of these cases occurs with probability $\frac{1}{2}$.

Now Bachelier assumes that $p(t,x)$ is regular enough and uses the following approximations coming from a Taylor expansion:

$$\begin{aligned}
p(t+\tau, x) &\simeq p(t,x) + \tau \frac{\partial p}{\partial t}(t,x), \\
p(t, x - \sigma\sqrt{\tau}) &\simeq p(t,x) - \sigma\sqrt{\tau} \frac{\partial p}{\partial x}(t,x) + \frac{1}{2}\sigma^2 \tau \frac{\partial^2 p}{\partial x^2}(t,x), \\
p(t, x + \sigma\sqrt{\tau}) &\simeq p(t,x) + \sigma\sqrt{\tau} \frac{\partial p}{\partial x}(t,x) + \frac{1}{2}\sigma^2 \tau \frac{\partial^2 p}{\partial x^2}(t,x).
\end{aligned}$$

This yields the equation

$$\frac{\partial p}{\partial t} = \frac{1}{2}\sigma^2 \frac{\partial^2 p}{\partial x^2}(t,x).$$

This is the so-called heat equation, which is the primary example of a diffusion equation. Explicit solutions to this equation are obtained using the Fourier transform, and by using the fact that at time 0, p is the Dirac distribution at 0, it is

[1] Quoted and translated from the French: *It seems that the market, the aggregate of speculators, can believe in neither a market rise nor a market fall, since, for each quoted price, there are as many buyers as sellers.*

computed that

$$p(t,x) = \frac{e^{-\frac{x^2}{2\sigma^2 t}}}{\sigma\sqrt{2\pi t}}.$$

It means that X_t has a Gaussian distribution with mean 0 and variance σ^2. Let now $0 < t_1 < \dots < t_n$ be fixed times and $I_1, \dots, I_n$ be fixed intervals. In order to compute $\mathbb{P}(X_{t_1} \in I_1, \dots, X_{t_n} \in I_n)$ the next step is to assume that the above analysis did not depend on the origin of time, or, more precisely, that the best information available at time t is given by the price X_t. This leads to the following computation:

$$\begin{aligned}
\mathbb{P}(X_{t_1} \in I_1, X_{t_2} \in I_2) &= \int_{I_1} \mathbb{P}(X_{t_2} \in I_2 | X_{t_1} = x_1) p(t_1, x_1) dx_1 \\
&= \int_{I_1} \mathbb{P}(X_{t_2 - t_1} + x_1 \in I_2 | X_{t_1} = x_1) p(t_1, x_1) dx_1 \\
&= \int_{I_1 \times I_2} p(t_2 - t_1, x_2 - x_1) p(t_1, x_1) dx_1 dx_2,
\end{aligned}$$

which is easily generalized to

$$\begin{aligned}
&\mathbb{P}(X_{t_1} \in I_1, \dots, X_{t_n} \in I_n) \\
&= \int\limits_{I_1 \times \dots \times I_n} p(t_n - t_{n-1}, x_n - x_{n-1}) \dots p(t_2 - t_1, x_2 - x_1) p(t_1, x_1) dx_1 dx_2 \dots dx_n.
\end{aligned} \tag{0.1}$$

In many regards, the previous computations were not rigorous but heuristic. One of our first motivations is to provide a rigorous construction of this object X_t on which Bachelier worked and which is called a Brownian motion.

From a rigorous point of view the question is: Does there exist a sequence of random variables $\{X_t, t \geq 0\}$ such that $t \to X_t$ is continuous and such that the property (0.1) is satisfied? Chapter 1 will give a positive answer to this question. We will see how to define and construct processes. In particular we will prove the existence of Brownian motion and then study several of its properties.

Chapter 1 sets the foundations. It deals with the basic definitions and results that are required to rigorously deal with stochastic processes. We introduce the relevant σ-fields and prove the fundamental Daniell–Kolmogorov theorem which may be seen as an infinite-dimensional version of the Carathéodory extension of measure theorem. It is the basic theorem to prove the existence of a stochastic process. However, despite its importance and usefulness, the Daniell–Kolmogorov result relies on the axiom of choice and as such is non-constructive and does not give any information or insight about the stochastic process that has been proved to exist. The

Kolmogorov continuity theorem fills one of these gaps and gives a useful criterion to ensure that we can work with a process whose sample paths are continuous. Chapter 1 also includes a thorough study of continuous martingales. We focus on Doob's theorems: The stopping theorem, the regularization result and the maximal inequalities. Martingale techniques are essential to study stochastic processes. They give the tools to handle stopping times which are naturally associated to processes and provide the inequalities which are the cornerstones of the theory of stochastic integration which is presented in Chapter 5.

Chapter 2 is devoted to the study of the most important stochastic process: The Brownian motion. As a consequence of the Daniell–Kolmogorov and Kolmogorov continuity theorems, we prove the existence of this process and then study some of its most fundamental properties. From many point of views, Brownian motion can be seen as the continuous random walk in continuous time. This is made precise at the end of the chapter where we give an alternative proof of the existence of the Brownian motion as a limit of suitably rescaled random walks.

Chapter 3 is devoted to the study of Markov processes. Our goal is to emphasize the role of the theory of semigroups when studying Markov processes. More precisely, we wish to understand how one can construct a Markov process from a semigroup and then see what are the properties inherited from the semigroup to the sample path properties of the process. We will particularly focus on the class of Feller–Dynkin Markov processes which are a class of Markov processes enjoying several nice properties, like the existence of regular versions and the strong Markov property. We finish the chapter with a study of the Lévy processes which form an important subset of the class of Feller–Dynkin Markov processes.

Chapter 4 can be thought an introduction to the theory of symmetric Dirichlet forms. As we will see, this theory and the set of tools attached to it belong much more to functional analysis than to probability theory. The basic problem is the construction of a Markov semigroup and of a Markov process only from the generator. More precisely, the question is: Given a diffusion operator L, does L generate a Markov semigroup $\boldsymbol{P}_t$ and if yes, is this semigroup the transition semigroup of a continuous Markov process? We will answer positively this question in quite a general framework under the basic assumption that L is elliptic and essentially self-adjoint with respect to some Borel measure.

Chapter 5 is about stochastic calculus and its applications to the study of Brownian motion. Stochastic calculus is an integral and differential calculus with respect to Brownian motion or more generally with respect to martingales. It allows one to give a meaningful sense to integrals with respect to Brownian motion paths and to define differential equations driven by such paths. It is not a straightforward extension of the usual calculus because Brownian motion paths are not regular enough, they are only γ-Hölder continuous with $\gamma < 1/2$. The range of applications of the stochastic calculus is huge and still growing today. We mention in particular that the applications to mathematical finance have drawn a lot of attention on this calculus. Actually, most of the modern pricing theories for derivatives on financial markets are based on Itô–Döblin's formula which is the chain rule for stochastic calculus.

Chapter 6 deals with the theory of stochastic differential equations. Stochastic differential equations are the differential equations that correspond to Itô's integration theory. They give a very powerful tool to construct a Markov process from its generator. We will prove the basic existence and uniqueness results for such equations and quickly turn to the basic properties of the solution. One of the problems we are mostly interested in is the existence of a smooth density for the solution of a stochastic differential equation. This problem gave birth to the so-called Malliavin calculus which is the study of the Sobolev regularity of Brownian functionals and that we study in some detail.

Chapter 7 is an introduction to the theory of rough paths that was developed by Lyons in the 1990s. The theory is deterministic. It allows one to give a sense to solutions of differential equations driven by very irregular paths. Stochastic differential equations driven by Brownian motions are then seen as a very special case of rough differential equations. The advantage of the theory is that it goes much beyond the scope of Itô calculus when dealing with differential systems driven by random noises and comes with powerful estimates.

Chapter 1
Stochastic processes

In this chapter we set the rigorous foundations on which the theory of general stochastic processes is built. Our first task will be to build a measure theory in infinite dimension and then to study sample path properties of stochastic processes. The second part of the chapter is devoted to the study of martingales. Martingale techniques will be used in almost every part of this book.

1.1 Measure theory in function spaces

Stochastic processes can be seen as random variables taking their values in a function space. It is therefore important to understand the naturally associated σ-algebras.

Let $\mathcal{A}(\mathbb{R}_{\geq 0}, \mathbb{R}^d)$, $d \geq 1$, be the set of functions $\mathbb{R}_{\geq 0} \to \mathbb{R}^d$. We denote by $\mathcal{T}(\mathbb{R}_{\geq 0}, \mathbb{R}^d)$ the σ-algebra generated by the so-called cylindrical sets

$$\{f \in \mathcal{A}(\mathbb{R}_{\geq 0}, \mathbb{R}^d),\ f(t_1) \in I_1, \dots, f(t_n) \in I_n\},$$

where

$$t_1, \dots, t_n \in \mathbb{R}_{\geq 0}$$

and where $I_1, \dots, I_n$ are products of intervals: $I_i = \Pi_{k=1}^d (a_i^k, b_i^k]$.

Remark 1.1. As a σ-algebra, $\mathcal{T}(\mathbb{R}_{\geq 0}, \mathbb{R}^d)$ is also generated by the following families:

$$\{f \in \mathcal{A}(\mathbb{R}_{\geq 0}, \mathbb{R}^d), f(t_1) \in B_1, \dots, f(t_n) \in B_n\},$$

where $t_1, \dots, t_n \in \mathbb{R}_{\geq 0}$ and where $B_1, \dots, B_n$ are Borel sets in $\mathbb{R}^d$;

$$\{f \in \mathcal{A}(\mathbb{R}_{\geq 0}, \mathbb{R}^d), (f(t_1), \dots, f(t_n)) \in B\},$$

where $t_1, \dots, t_n \in \mathbb{R}_{\geq 0}$ and where B is a Borel set in $(\mathbb{R}^d)^{\otimes n}$.

Exercise 1.2. Show that the following sets are not in $\mathcal{T}([0,1], \mathbb{R})$:

(1)
$$\{f \in \mathcal{A}([0,1], \mathbb{R}),\ \sup_{t \in [0,1]} f(t) < 1\},$$

(2)
$$\{f \in \mathcal{A}([0,1], \mathbb{R}),\ \exists t \in [0,1] f(t) = 0\}.$$

The above exercise shows that the σ-algebra $\mathcal{T}(\mathbb{R}_{\geq 0}, \mathbb{R}^d)$ is not rich enough to include *natural* events; this is due to the fact that the space $\mathcal{A}(\mathbb{R}_{\geq 0}, \mathbb{R}^d)$ is by far too big.

In this book, we shall mainly be interested in processes with continuous paths. In this case, we use the space of continuous functions $\mathcal{C}(\mathbb{R}_{\geq 0}, \mathbb{R}^d)$ endowed with the σ-algebra $\mathcal{B}(\mathbb{R}_{\geq 0}, \mathbb{R}^d)$ generated by

$$\{f \in \mathcal{C}(\mathbb{R}_{\geq 0}, \mathbb{R}^d),\ f(t_1) \in I_1, \dots, f(t_n) \in I_n\}$$

where $t_1, \dots, t_n \in \mathbb{R}_{\geq 0}$ and where $I_1, \dots, I_n$ are products of intervals $\Pi_{k=1}^d (a_i^k, b_i^k]$. This σ-algebra enjoys nice properties. It is generated by the open sets of the (metric) topology of uniform convergence on compact sets.

Proposition 1.3. *The σ-algebra $\mathcal{B}(\mathbb{R}_{\geq 0}, \mathbb{R}^d)$ is generated by the open sets of the topology of uniform convergence on compact sets.*

Proof. We make the proof when the dimension $d = 1$ and let the reader adapt it in higher dimension. Let us first recall that, on $\mathcal{C}(\mathbb{R}_{\geq 0}, \mathbb{R})$ the topology of uniform convergence on compact sets is given by the distance

$$d(f, g) = \sum_{n=1}^{+\infty} \frac{1}{2^n} \min\Big(\sup_{0 \leq t \leq n} |f(t) - g(t)|, 1\Big).$$

This distance endows $\mathcal{C}(\mathbb{R}_{\geq 0}, \mathbb{R})$ with the structure of a complete, separable, metric space. Let us denote by $\mathcal{O}$ the σ-field generated by the open sets of this metric space. First, it is clear that the cylinders

$$\{f \in \mathcal{C}(\mathbb{R}_{\geq 0}, \mathbb{R}),\ f(t_1) < a_1, \dots, f(t_n) < a_n\}$$

are open sets that generate $\mathcal{B}(\mathbb{R}_{\geq 0}, \mathbb{R})$. Thus, we have $\mathcal{B}(\mathbb{R}_{\geq 0}, \mathbb{R}) \subset \mathcal{O}$. On the other hand, since for every $g \in \mathcal{C}(\mathbb{R}_{\geq 0}, \mathbb{R})$, $n \in \mathbb{N}$, $n \geq 1$ and $\rho > 0$

$$\begin{aligned}
&\{f \in \mathcal{C}(\mathbb{R}_{\geq 0}, \mathbb{R}),\ \sup_{0 \leq t \leq n} |f(t) - g(t)| \leq \rho\} \\
&\quad = \bigcap_{t \in \mathbb{Q}, 0 \leq t \leq n} \{f \in \mathcal{C}(\mathbb{R}_{\geq 0}, \mathbb{R}), |f(t) - g(t)| \leq \rho\},
\end{aligned}$$

we deduce that

$$\{f \in \mathcal{C}(\mathbb{R}_{\geq 0}, \mathbb{R}),\ \sup_{0 \leq t \leq n} |f(t) - g(t)| \leq \rho\} \in \mathcal{B}(\mathbb{R}_{\geq 0}, \mathbb{R}).$$

Since $\mathcal{O}$ is generated by the above sets, this implies $\mathcal{O} \subset \mathcal{B}(\mathbb{R}_{\geq 0}, \mathbb{R})$. □

Exercise 1.4. Show that the following sets are in $\mathcal{B}([0, 1], \mathbb{R})$:

(1)
$$\{f \in \mathcal{C}([0,1],\mathbb{R}),\ \sup_{t\in[0,1]} f(t) < 1\}.$$

(2)
$$\{f \in \mathcal{C}([0,1],\mathbb{R}),\ \exists t \in [0,1] f(t) = 0\}.$$

1.2 Stochastic processes

Let $(\Omega, \mathcal{F}, \mathbb{P})$ be a probability space.

Definition 1.5. On $(\Omega, \mathcal{F}, \mathbb{P})$, a ($d$-dimensional) *stochastic process* is a sequence $(X_t)_{t\geq 0}$ of $\mathbb{R}^d$-valued random variables that are measurable with respect to $\mathcal{F}$.

A process $(X_t)_{t\geq 0}$ can also be seen as a mapping

$$X(\omega) \in \mathcal{A}(\mathbb{R}_{\geq 0}, \mathbb{R}^d), \quad t \to X_t(\omega).$$

The mappings $t \to X_t(\omega)$ are called the *paths* of the process. The mapping $X\colon (\Omega, \mathcal{F}) \to (\mathcal{A}(\mathbb{R}_{\geq 0}, \mathbb{R}^d), \mathcal{T}(\mathbb{R}_{\geq 0}, \mathbb{R}^d))$ is measurable. The probability measure defined by

$$\mu(A) = \mathbb{P}(X^{-1}(A)), A \in \mathcal{T}(\mathbb{R}_{\geq 0}, \mathbb{R}^d)$$

is then called the *law* (or *distribution*) of $(X_t)_{t\geq 0}$.

For $t \geq 0$, we denote by π_t the mapping that transforms $f \in \mathcal{A}(\mathbb{R}_{\geq 0}, \mathbb{R}^d)$ into $f(t)$: $\pi_t\colon f \to f(t)$. The stochastic process $(\pi_t)_{t\in \mathbb{R}_{\geq 0}}$ which is thus defined on the probability space $(\mathcal{A}(\mathbb{R}_{\geq 0}, \mathbb{R}), \mathcal{T}(\mathbb{R}_{\geq 0}, \mathbb{R}^d), \mu)$ is called the *canonical process associated to X*. It is a process with distribution μ.

Definition 1.6. A process $(X_t)_{t\geq 0}$ is said to be *measurable* if the mapping

$$(t, \omega) \to X_t(\omega)$$

is measurable with respect to the σ-algebra $\mathcal{B}(\mathbb{R}_{\geq 0}) \otimes \mathcal{F}$, that is, if

$$\forall A \in \mathcal{B}(\mathbb{R}^d),\ \{(t,\omega), X_t(\omega) \in A\} \in \mathcal{B}(\mathbb{R}_{\geq 0}) \otimes \mathcal{F}.$$

The paths of a measurable process are, of course, measurable functions $\mathbb{R}_{\geq 0} \to \mathbb{R}^d$.

Definition 1.7. If a process X takes its values in $\mathcal{C}(\mathbb{R}_{\geq 0}, \mathbb{R}^d)$, that is, if the paths of X are continuous functions, then we say that X is a *continuous process*.

If $(X_t)_{t\geq 0}$ is a continuous process then the mapping

$$X: (\Omega, \mathcal{F}) \to (\mathcal{C}(\mathbb{R}_{\geq 0}, \mathbb{R}^d), \mathcal{B}(\mathbb{R}_{\geq 0}, \mathbb{R}))$$

is measurable and the distribution of X is a probability measure on the space $(\mathcal{C}(\mathbb{R}_{\geq 0}, \mathbb{R}^d), \mathcal{B}(\mathbb{R}_{\geq 0}, \mathbb{R}^d))$. Moreover, a continuous process is measurable in the sense of Definition 1.6:

Proposition 1.8. *A continuous stochastic process is measurable.*

Proof. Let $(X_t)_{t\geq 0}$ be a continuous process. Let us first prove that if A is a Borel set in $\mathbb{R}$, then

$$\{(t,\omega) \in [0,1] \times \Omega,\ X_t(\omega) \in A\} \in \mathcal{B}(\mathbb{R}_{\geq 0}) \otimes \mathcal{F}.$$

For $n \in \mathbb{N}$, let

$$X_t^n = X_{\frac{[2^n t]}{2^n}}, t \in [0,1],$$

where $[x]$ denotes the integer part of x. Since the paths of X^n are piecewise constant we have

$$\{(t,\omega) \in [0,1] \times \Omega, X_t^n(\omega) \in A\} \in \mathcal{B}(\mathbb{R}_{\geq 0}) \otimes \mathcal{F}.$$

Moreover, for all $t \in [0,1]$, $\omega \in \Omega$ we have

$$\lim_{n\to+\infty} X_t^n(\omega) = X_t(\omega),$$

which implies

$$\{(t,\omega) \in [0,1] \times \Omega, X_t(\omega) \in A\} \in \mathcal{B}(\mathbb{R}_{\geq 0}) \otimes \mathcal{F}.$$

In the same way we obtain that for all $k \in \mathbb{N}$,

$$\{(t,\omega) \in [k,k+1] \times \Omega, X_t(\omega) \in A\} \in \mathcal{B}(\mathbb{R}_{\geq 0}) \otimes \mathcal{F}.$$

Observing

$$\{(t,\omega) \in \mathbb{R} \times \Omega, X_t(\omega) \in A\} = \bigcup_{k\in\mathbb{N}} \{(t,\omega) \in [k,k+1] \times \Omega, X_t(\omega) \in A\}$$

yields the sought of conclusion. □

1.3 Filtrations

A stochastic process $(X_t)_{t\geq 0}$ may also be seen as a random system evolving in time. This system carries some information. More precisely, if one observes the paths of a stochastic process up to a time $t > 0$, one is able to decide if an event

$$A \in \sigma(X_s, s \leq t)$$

has occurred (here and in the sequel $\sigma(X_s, s \leq t)$ denotes the smallest σ-field that makes all the random variables $\{(X_{t_1}, \dots, X_{t_n}), 0 \leq t_1 \leq \dots \leq t_n \leq t\}$ measurable). This notion of information carried by a stochastic process is modeled by filtrations.

Definition 1.9. Let $(\Omega, \mathcal{F}, \mathbb{P})$ be a probability space. A *filtration* $(\mathcal{F}_t)_{t\geq 0}$ is a non-decreasing family of sub-σ-algebras of $\mathcal{F}$.

As a basic example, if $(X_t)_{t\geq 0}$ is a stochastic process defined on $(\Omega, \mathcal{F}, \mathbb{P})$,then

$$\mathcal{F}_t = \sigma(X_s, s \leq t)$$

is a filtration. This filtration is called the natural filtration of the process X and often denoted by $(\mathcal{F}_t^X)_{t\geq 0}$.

Definition 1.10. A stochastic process $(X_t)_{t\geq 0}$ is said to be *adapted to a filtration* $(\mathcal{F}_t)_{t\geq 0}$ if for every $t \geq 0$ the random variable X_t is measurable with respect to $\mathcal{F}_t$.

Of course, a stochastic process is always adapted with respect to its natural filtration. We may observe that if a stochastic process $(X_t)_{t\geq 0}$ is adapted to a filtration $(\mathcal{F}_t)_{t\geq 0}$ and if $\mathcal{F}_0$ contains all the subsets of $\mathcal{F}$ that have a zero probability, then every process $(\tilde{X}_t)_{t\geq 0}$ that satisfies

$$\mathbb{P}(\tilde{X}_t = X_t) = 1, \quad t \geq 0,$$

is still adapted to the filtration $(\mathcal{F}_t)_{t\geq 0}$.

We previously defined the notion of measurability for a stochastic process. In order to take into account the dynamic aspect associated to a filtration, the notion of progressive measurability is needed.

Definition 1.11. A stochastic process $(X_t)_{t\geq 0}$ which is adapted to a filtration $(\mathcal{F}_t)_{t\geq 0}$ is said to be *progressively measurable* with respect to the filtration $(\mathcal{F}_t)_{t\geq 0}$ if for every $t \geq 0$,

$$\forall A \in \mathcal{B}(\mathbb{R}), \ \{(s, \omega) \in [0, t] \times \Omega, X_s(\omega) \in A\} \in \mathcal{B}([0, t]) \otimes \mathcal{F}_t.$$

By using the diagonal method, it is possible to construct adapted but not progressively measurable processes. However, the next proposition whose proof is left as an exercise to the reader shows that an adapted and continuous stochastic process is automatically progressively measurable.

Proposition 1.12. *A continuous stochastic process $(X_t)_{t\geq 0}$, that is adapted with respect to a filtration $(\mathcal{F}_t)_{t\geq 0}$, is also progressively measurable with respect to it.*

1.4 The Daniell–Kolmogorov extension theorem

The Daniell–Kolmogorov extension theorem is one of the first deep theorems of the theory of stochastic processes. It provides existence results for nice probability measures on path (function) spaces. It is however non-constructive and relies on the axiom of choice. In what follows, in order to avoid heavy notations we restrict the presentation to the one-dimensional case $d = 1$. The multi-dimensional extension is straightforward.

Definition 1.13. Let $(X_t)_{t\geq 0}$ be a stochastic process. For $t_1, \dots, t_n \in \mathbb{R}_{\geq 0}$ we denote by $\mu_{t_1,\dots,t_n}$ the probability distribution of the random variable

$$(X_{t_1}, \dots, X_{t_n}).$$

It is therefore a probability measure on $\mathbb{R}^n$. This probability measure is called a *finite-dimensional distribution of the process* $(X_t)_{t\geq 0}$.

If two processes have the same finite-dimensional distributions, then it is clear that the two processes induce the same distribution on the path space $\mathcal{A}(\mathbb{R}_{\geq 0}, \mathbb{R})$ because cylinders generate the σ-algebra $\mathcal{T}(\mathbb{R}_{\geq 0}, \mathbb{R})$.

The finite-dimensional distributions of a given process satisfy the two following properties: If $t_1, \dots, t_n \in \mathbb{R}_{\geq 0}$ and if τ is a permutation of the set $\{1, \dots, n\}$, then

(1) $\mu_{t_1,\dots,t_n}(A_1 \times \dots \times A_n) = \mu_{t_{\tau(1)},\dots,t_{\tau(n)}}(A_{\tau(1)} \times \dots \times A_{\tau(n)})$, $A_i \in \mathcal{B}(\mathbb{R})$,

(2) $\mu_{t_1,\dots,t_n}(A_1 \times \dots \times A_{n-1} \times \mathbb{R}) = \mu_{t_1,\dots,t_{n-1}}(A_1 \times \dots \times A_{n-1})$, $A_i \in \mathcal{B}(\mathbb{R})$.

Conversely, we have

Theorem 1.14 (Daniell–Kolmogorov theorem). *Assume that we are given for every* $t_1, \dots, t_n \in \mathbb{R}_{\geq 0}$ *a probability measure* $\mu_{t_1,\dots,t_n}$ *on* $\mathbb{R}^n$. *Let us assume that these probability measures satisfy*

(1) $\mu_{t_1,\dots,t_n}(A_1 \times \dots \times A_n) = \mu_{t_{\tau(1)},\dots,t_{\tau(n)}}(A_{\tau(1)} \times \dots \times A_{\tau(n)})$, $A_i \in \mathcal{B}(\mathbb{R})$,

and

(2) $\mu_{t_1,\dots,t_n}(A_1 \times \dots \times A_{n-1} \times \mathbb{R}) = \mu_{t_1,\dots,t_{n-1}}(A_1 \times \dots \times A_{n-1})$, $A_i \in \mathcal{B}(\mathbb{R})$.

Then there is a unique probability measure μ *on* $(\mathcal{A}(\mathbb{R}_+, \mathbb{R}), \mathcal{T}(\mathbb{R}_+, \mathbb{R}))$ *such that for* $t_1, \dots, t_n \in \mathbb{R}_{\geq 0}$, $A_1, \dots, A_n \in \mathcal{B}(\mathbb{R})$*:*

$$\mu(\pi_{t_1} \in A_1, \dots, \pi_{t_n} \in A_n) = \mu_{t_1,\dots,t_n}(A_1 \times \dots \times A_n).$$

The Daniell–Kolmogorov theorem is often used to construct processes thanks to the following corollary:

Corollary 1.15. *Assume given for every $t_1, \dots, t_n \in \mathbb{R}_{\geq 0}$ a probability measure $\mu_{t_1,\dots,t_n}$ on $\mathbb{R}^n$. Let us further assume that these measures satisfy the assumptions of the Daniell–Kolmogorov theorem. Then there exists a probability space $(\Omega, \mathcal{F}, \mathbb{P})$ as well as a process $(X_t)_{t\geq 0}$ defined on this space such that the finite-dimensional distributions of $(X_t)_{t\geq 0}$ are given by the $\mu_{t_1,\dots,t_n}$'s.*

Proof. As a probability space we chose

$$(\Omega, \mathcal{F}, \mathbb{P}) = (\mathcal{A}(\mathbb{R}_{\geq 0}, \mathbb{R}), \mathcal{T}(\mathbb{R}_{\geq 0}, \mathbb{R}), \mu)$$

where μ is the probability measure given by the Daniell–Kolmogorov theorem. The canonical process $(\pi_t)_{t\geq 0}$ defined on $\mathcal{A}(\mathbb{R}_{\geq 0}, \mathbb{R})$ by $\pi_t(f) = f(t)$ satisfies the required property. □

We now turn to the proof of the Daniell–Kolmogorov theorem. This proof proceeds in several steps.

As a first step, let us recall the Carathéodory extension theorem that is often useful for the effective construction of measures (for instance the construction of the Lebesgue measure on $\mathbb{R}$).

Theorem 1.16 (Carathéodory theorem). *Let Ω be a non-empty set and let $\mathcal{A}$ be a family of subsets that satisfy*

(1) $\Omega \in \mathcal{A}$*;*

(2) *if* $A, B \in \mathcal{A}$, $A \cup B \in \mathcal{A}$*;*

(3) *if* $A \in \mathcal{A}$, $\Omega \backslash A \in \mathcal{A}$.

Let $\sigma(\mathcal{A})$ be the σ-algebra generated by $\mathcal{A}$. If μ_0 is σ-additive measure on $(\Omega, \mathcal{A})$, then there exists a unique σ-additive measure μ on $(\Omega, \sigma(\mathcal{A}))$ such that for $A \in \mathcal{A}$,

$$\mu_0(A) = \mu(A).$$

As a second step, we prove the following fact:

Lemma 1.17. *Let $B_n \subset \mathbb{R}^n$, $n \in \mathbb{N}$ be a sequence of Borel sets that satisfy*

$$B_{n+1} \subset B_n \times \mathbb{R}.$$

Let us assume that for every $n \in \mathbb{N}$ a probability measure μ_n is given on $(\mathbb{R}^n, \mathcal{B}(\mathbb{R}^n))$ and that these probability measures are compatible in the sense that

$$\mu_n(A_1 \times \dots \times A_{n-1} \times \mathbb{R}) = \mu_{n-1}(A_1 \times \dots \times A_{n-1}), \quad A_i \in \mathcal{B}(\mathbb{R}).$$

and satisfy

$$\mu_n(B_n) > \varepsilon,$$

where $0 < \varepsilon < 1$. There exists a sequence of compact sets $K_n \subset \mathbb{R}^n$, $n \in \mathbb{N}$, such that

- $K_n \subset B_n$,
- $K_{n+1} \subset K_n \times \mathbb{R}$,
- $\mu_n(K_n) \geq \frac{\varepsilon}{2}$.

Proof. For every n, we can find a compact set $K_n^* \subset \mathbb{R}^n$ such that

$$K_n^* \subset B_n$$

and

$$\mu_n(B_n \backslash K_n^*) \leq \frac{\varepsilon}{2^{n+1}}.$$

Let us consider

$$K_n = (K_1^* \times \mathbb{R}^{n-1}) \cap \cdots \cap (K_{n-1}^* \times \mathbb{R}) \cap K_n^*.$$

It is easily checked that

- $K_n \subset B_n$,
- $K_{n+1} \subset K_n \times \mathbb{R}$.

Moreover, we have

$$\begin{aligned}
\mu_n(K_n) &= \mu_n(B_n) - \mu_n(B_n \backslash K_n) \\
&= \mu_n(B_n) - \mu_n(B_n \backslash ((K_1^* \times \mathbb{R}^{n-1}) \cap \cdots \cap (K_{n-1}^* \times \mathbb{R}) \cap K_n^*)) \\
&\geq \mu_n(B_n) - \mu_n(B_n \backslash ((K_1^* \times \mathbb{R}^{n-1}))) - \cdots - \mu_n(B_n \backslash (K_{n-1}^* \times \mathbb{R})) \\
&\quad - \mu_n(B_n \backslash K_n^*) \\
&\geq \mu_n(B_n) - \mu_1(B_1 \backslash K_1^*) - \cdots - \mu_n(B_n \backslash K_n^*) \\
&\geq \varepsilon - \frac{\varepsilon}{4} - \cdots - \frac{\varepsilon}{2^{n+1}} \\
&\geq \frac{\varepsilon}{2}.
\end{aligned}$$

□

With this in hands, we can now turn to the proof of the Daniell–Kolmogorov theorem.

Proof. For the cylinder

$$\mathcal{C}_{t_1,\ldots,t_n}(B) = \{f \in \mathcal{A}(\mathbb{R}_+, \mathbb{R}), (f(t_1), \ldots, f(t_n)) \in B\}$$

where

$$t_1, \ldots, t_n \in \mathbb{R}_{\geq 0}$$

and where B is a Borel subset of $\mathbb{R}^n$, we define

$$\mu(\mathcal{C}_{t_1,\ldots,t_n}(B)) = \mu_{t_1,\ldots,t_n}(B).$$

Thanks to the assumptions on the $\mu_{t_1,\dots,t_n}$'s, it is seen that such a μ is well defined and satisfies

$$\mu(\mathcal{A}(\mathbb{R}_{\geq 0}, \mathbb{R})) = 1.$$

The set $\mathcal{A}$ of all the possible cylinders $\mathcal{C}_{t_1,\dots,t_n}(B)$ satisfies the assumption of Carathéodory theorem. Therefore, in order to conclude, we have to show that μ is σ-additive, that is, if $(C_n)_{n\in\mathbb{N}}$ is a sequence of pairwise disjoint cylinders and if $C = \bigcup_{n\in\mathbb{N}} C_n$ is a cylinder then

$$\mu(C) = \sum_{n=0}^{+\infty} \mu(C_n).$$

This is the difficult part of the theorem. Since for $N \in \mathbb{N}$,

$$\mu(C) = \mu(C \setminus \bigcup_{n=0}^{N} C_n) + \mu(\bigcup_{n=0}^{N} C_n),$$

we just have to show that

$$\lim_{N\to+\infty} \mu(D_N) = 0,$$

where $D_N = C \setminus \bigcup_{n=0}^{N} C_n$.

The sequence $(\mu(D_N))_{N\in\mathbb{N}}$ is positive decreasing and therefore converges. Let assume that it converges toward $\varepsilon > 0$. We shall prove that in this case

$$\bigcap_{N\in\mathbb{N}} D_N \neq \emptyset,$$

which is clearly absurd.

Since D_N is a cylinder, the event $\bigcup_{N\in\mathbb{N}} D_N$ only involves a countable sequence of times $t_1 < \dots < t_n < \dots$ and we may assume (otherwise we can add other convenient sets in the sequence of the D_N's) that every D_N can be described as follows:

$$D_N = \{f \in \mathcal{A}(\mathbb{R}_{\geq 0}, \mathbb{R}), (f(t_1), \dots, f(t_N)) \in B_N\},$$

where $B_n \subset \mathbb{R}^n$, $n \in \mathbb{N}$, is a sequence of Borel sets such that

$$B_{n+1} \subset B_n \times \mathbb{R}.$$

Since we assumed $\mu(D_N) \geq \varepsilon$, we can use the previous lemma to construct a sequence of compact sets $K_n \subset \mathbb{R}^n$, $n \in \mathbb{N}$, such that

- $K_n \subset B_n$,
- $K_{n+1} \subset K_n \times \mathbb{R}$,
- $\mu_{t_1,\dots,t_n}(K_n) \geq \frac{\varepsilon}{2}$.

Since K_n is non-empty, we pick

$$(x_1^n, \dots, x_n^n) \in K_n.$$

The sequence $(x_1^n)_{n\in\mathbb{N}}$ has a convergent subsequence $(x_1^{j_1(n)})_{n\in\mathbb{N}}$ that converges toward $x_1 \in K_1$. The sequence $((x_1^{j_1(n)}, x_2^{j_1(n)})_{n\in\mathbb{N}}$ has a convergent subsequence that converges toward $(x_1, x_2) \in K_2$. By pursuing this process[1] we obtain a sequence $(x_n)_{n\in\mathbb{N}}$ such that for every n,

$$(x_1, \dots, x_n) \in K_n.$$

The event

$$\{f \in \mathcal{A}(\mathbb{R}_+, \mathbb{R}), (f(t_1), \dots, f(t_N)) = (x_1, \dots, x_N)\}$$

is in D_N, and this leads to the expected contradiction. Therefore, the sequence $(\mu(D_N))_{N\in\mathbb{N}}$ converges toward 0, which implies the σ-additivity of μ. □

The Daniell–Kolmogorov theorem is the basic tool to prove the existence of a stochastic process with given finite-dimensional distributions. As an example, let us illustrate how it may be used to prove the existence of the so-called Gaussian processes.

Definition 1.18. A real-valued stochastic process $(X_t)_{t\geq 0}$ defined on $(\Omega, \mathcal{F}, \mathbb{P})$ is said to be a *Gaussian process* if all the finite-dimensional distributions of X are Gaussian random variables.

If $(X_t)_{t\geq 0}$ is a Gaussian process, its finite-dimensional distributions can be characterized, through Fourier transform, by its mean function

$$m(t) = \mathbb{E}(X_t)$$

and its covariance function

$$R(s,t) = \mathbb{E}((X_t - m(t))(X_s - m(s))).$$

We can observe that the covariance function $R(s,t)$ is symmetric ($R(s,t) = R(t,s)$) and positive, that is, for $a_1, \dots, a_n \in \mathbb{R}$ and $t_1, \dots, t_n \in \mathbb{R}_{\geq 0}$,

$$\begin{aligned}\sum_{1\leq i,j\leq n} a_i a_j R(t_i, t_j) &= \sum_{1\leq i,j\leq n} a_i a_j \mathbb{E}((X_{t_i} - m(t_i))(X_{t_j} - m(t_j))) \\ &= \mathbb{E}\Big(\Big(\sum_{i=1}^n (X_{t_i} - m(t_i))\Big)^2\Big) \\ &\geq 0.\end{aligned}$$

Conversely, as an application of the Daniell–Kolmogorov theorem, we let the reader prove as an exercise the following proposition.

[1]It is here that the axiom of choice is needed.

Proposition 1.19. *Let $m\colon \mathbb{R}_{\geq 0} \to \mathbb{R}$ and let $R\colon \mathbb{R}_{\geq 0} \times \mathbb{R}_{\geq 0} \to \mathbb{R}$ be a symmetric and positive function. There exists a probability space $(\Omega, \mathcal{F}, \mathbb{P})$ and a Gaussian process $(X_t)_{t\geq 0}$ defined on it whose mean function is m and whose covariance function is R.*

Exercise 1.20. Let $(X_t)_{0\leq t\leq T}$ be a continuous Gaussian process. Show that the random variable $\int_0^T X_s ds$ is a Gaussian random variable.

1.5 The Kolmogorov continuity theorem

The Daniell–Kolmogorov theorem is a very useful tool since it provides existence results for stochastic processes. However this theorem does not say anything about the paths of this process. The following theorem, due to Kolmogorov, makes precise that, under mild conditions, we can work with processes whose paths are quite regular.

Definition 1.21. A function $f\colon \mathbb{R}_{\geq 0} \to \mathbb{R}^d$ is said to be *Hölder with exponent* $\alpha > 0$ or *α-Hölder* if there exists a constant $C > 0$ such that

$$\|f(t) - f(s)\| \leq C|t-s|^\alpha$$

for $s, t \in \mathbb{R}_{\geq 0}$.

Hölder functions are in particular continuous.

Definition 1.22. A stochastic process $(\widetilde{X}_t)_{t\geq 0}$ is called a *modification of the process* $(X_t)_{t\geq 0}$ if for $t \geq 0$,

$$\mathbb{P}(X_t = \widetilde{X}_t) = 1.$$

Remark 1.23. We observe that if $(\widetilde{X}_t)_{t\geq 0}$ is a modification of $(X_t)_{t\geq 0}$ then $(\widetilde{X}_t)_{t\geq 0}$ has the same distribution as $(X_t)_{t\geq 0}$.

Theorem 1.24 (Kolmogorov continuity theorem). *Let $\alpha, \varepsilon, c > 0$. If a d-dimensional process $(X_t)_{t\in[0,1]}$ defined on a probability space $(\Omega, \mathcal{F}, \mathbb{P})$ satisfies*

$$\mathbb{E}(\|X_t - X_s\|^\alpha) \leq c|t-s|^{1+\varepsilon}$$

for $s, t \in [0, 1]$, then there exists a modification of the process $(X_t)_{t\in[0,1]}$ that is a continuous process and whose paths are γ-Hölder for every $\gamma \in [0, \frac{\varepsilon}{\alpha})$.

Proof. We make the proof for $d = 1$ and let the reader extend it as an exercise to the case $d \geq 2$. For $n \in \mathbb{N}$, we write

$$\mathcal{D}_n = \left\{\frac{k}{2^n},\ k = 0, \ldots, 2^n\right\}$$

and

$$\mathcal{D} = \bigcup_{n \in \mathbb{N}} \mathcal{D}_n.$$

Let $\gamma \in [0, \frac{\varepsilon}{\alpha})$. From Chebychev's inequality,

$$\begin{aligned}
\mathbb{P}(\max_{1 \le k \le 2^n} |X_{\frac{k}{2^n}} - X_{\frac{k-1}{2^n}}| \ge 2^{-\gamma n}) &= \mathbb{P}(\bigcup_{1 \le k \le 2^n} |X_{\frac{k}{2^n}} - X_{\frac{k-1}{2^n}}| \ge 2^{-\gamma n}) \\
&\le \sum_{k=1}^{2^n} \mathbb{P}(|X_{\frac{k}{2^n}} - X_{\frac{k-1}{2^n}}| \ge 2^{-\gamma n}) \\
&\le \sum_{k=1}^{2^n} \frac{\mathbb{E}(|X_{\frac{k}{2^n}} - X_{\frac{k-1}{2^n}}|^{\alpha})}{2^{-\gamma \alpha n}} \\
&\le c 2^{-n(\varepsilon - \gamma \alpha)}.
\end{aligned}$$

Therefore, since $\gamma\alpha < \varepsilon$, we deduce

$$\sum_{n=1}^{+\infty} \mathbb{P}(\max_{1 \le k \le 2^n} |X_{\frac{k}{2^n}} - X_{\frac{k-1}{2^n}}| \ge 2^{-\gamma n}) < +\infty.$$

From the Borel–Cantelli lemma, we can thus find a set $\Omega^* \in \mathcal{F}$ such that $\mathbb{P}(\Omega^*) = 1$ and such that for $\omega \in \Omega^*$, there exists $N(\omega)$ such that for $n \ge N(\omega)$,

$$\max_{1 \le k \le 2^n} |X_{\frac{k}{2^n}}(\omega) - X_{\frac{k-1}{2^n}}(\omega)| \le 2^{-\gamma n}.$$

In particular, there exists an almost surely finite random variable C such that for every $n \ge 0$,

$$\max_{1 \le k \le 2^n} |X_{\frac{k}{2^n}}(\omega) - X_{\frac{k-1}{2^n}}(\omega)| \le C 2^{-\gamma n}.$$

We now claim that the paths of the restricted process $X_{/\Omega^*}$ are consequently γ-Hölder on $\mathcal{D}$. Indeed, let $s, t \in \mathcal{D}$. We can find n such that

$$|s - t| \le \frac{1}{2^n}.$$

We now pick an increasing and stationary sequence $(s_k)_{k \ge n}$ converging toward s such that $s_k \in \mathcal{D}_k$ and

$$|s_{k+1} - s_k| = 2^{-(k+1)} \quad \text{or} \quad 0.$$

In the same way, we can find an analogue sequence $(t_k)_{k \ge n}$ that converges toward t and such that s_n and t_n are neighbors in $\mathcal{D}_n$. We then have:

$$X_t - X_s = \sum_{i=n}^{+\infty} (X_{s_{i+1}} - X_{s_i}) + (X_{s_n} - X_{t_n}) + \sum_{i=n}^{+\infty} (X_{t_i} - X_{t_{i+1}}),$$

where the above sums are actually finite.

Therefore,

$$\begin{aligned}|X_t - X_s| &\le C2^{-\gamma n} + 2\sum_{k=n}^{+\infty} C2^{-\gamma(k+1)} \\ &\le 2C \sum_{k=n}^{+\infty} 2^{-\gamma k} \\ &\le \frac{2C}{1-2^{-\gamma}} 2^{-\gamma n}.\end{aligned}$$

Hence the paths of $X_{/\Omega^*}$ are γ-Hölder on the set $\mathcal{D}$. For $\omega \in \Omega^*$, let $t \to \widetilde{X}_t(\omega)$ be the unique continuous function that agrees with $t \to X_t(\omega)$ on $\mathcal{D}$. For $\omega \notin \Omega^*$, we set $\widetilde{X}_t(\omega) = 0$. The process $(\widetilde{X}_t)_{t\in[0,1]}$ is the desired modification of $(X_t)_{t\in[0,1]}$. □

Exercise 1.25. Let $\alpha, \varepsilon, c > 0$. Let $(X_t)_{t\in[0,1]}$ be a continuous Gaussian process such that for $s, t \in [0,1]$,

$$\mathbb{E}(\|X_t - X_s\|^\alpha) \le c|t-s|^{1+\varepsilon}.$$

Show that for every $\gamma \in [0, \varepsilon/\alpha)$, there is a positive random variable η such that $\mathbb{E}(\eta^p) < \infty$, for every $p \ge 1$ and such that for every $s, t \in [0,1]$,

$$\|X_t - X_s\| \le \eta |t-s|^\gamma, \quad \text{a.s.}$$

Hint. You may use the Garsia–Rodemich–Rumsey inequality which is stated in Theorem 7.34, Chapter 7, in greater generality: Let $p \ge 1$ and $\alpha > p^{-1}$, then there exists a constant $C_{\alpha,p} > 0$ such that for any continuous function f on $[0,T]$, and for all $t, s \in [0,T]$ one has

$$\|f(t) - f(s)\|^p \le C_{\alpha,p}|t-s|^{\alpha p - 1} \int_0^T \int_0^T \frac{\|f(x)-f(y)\|^p}{|x-y|^{\alpha p+1}} dx dy.$$

1.6 Stopping times

In the study of a stochastic process it is often useful to consider some properties of the process that hold up to a random time. A natural question is for instance: For how long is the process less than a given constant?

Definition 1.26. Let $(\mathcal{F}_t)_{t\ge 0}$ be a filtration on a probability space $(\Omega, \mathcal{F}, \mathbb{P})$. Let T be a random variable, measurable with respect to $\mathcal{F}$ and valued in $\mathbb{R}_{\ge 0} \cup \{+\infty\}$. We say that T is a *stopping time* of the filtration $(\mathcal{F}_t)_{t\ge 0}$ if for $t \ge 0$,

$$\{T \le t\} \in \mathcal{F}_t.$$

Often, a stopping time will be the time during which a stochastic process adapted to the filtration $(\mathcal{F}_t)_{t\geq 0}$ satisfies a given property. The above definition means that for any $t \geq 0$, at time t, one is able to decide if this property is satisfied or not.

Among the most important examples of stopping times are the (first) hitting times of a closed set by a continuous stochastic process.

Exercise 1.27 (First hitting time of a closed set by a continuous stochastic process). Let $(X_t)_{t\geq 0}$ be a continuous process adapted to a filtration $(\mathcal{F}_t)_{t\geq 0}$. Let

$$T = \inf\{t \geq 0, X_t \in F\},$$

where F is a closed subset of $\mathbb{R}$. Show that T is a stopping time of the filtration $(\mathcal{F}_t)_{t\geq 0}$.

Given a stopping time T, we may define the σ-algebra of events that occur before the time T:

Proposition 1.28. *Let T be a stopping time of the filtration $(\mathcal{F}_t)_{t\geq 0}$. Let*

$$\mathcal{F}_T = \{A \in \mathcal{F}, \forall t \geq 0, A \cap \{T \leq t\} \in \mathcal{F}_t\}.$$

Then $\mathcal{F}_T$ is a σ-algebra.

Proof. Since for every $t \geq 0$, $\emptyset \in \mathcal{F}_t$, we have that $\emptyset \in \mathcal{F}_T$. Let us now consider $A \in \mathcal{F}_T$. We have

$$^cA \cap \{T \leq t\} = \{T \leq t\} \backslash (A \cap \{T \leq t\}) \in \mathcal{F}_t,$$

and thus $^cA \in \mathcal{F}_T$. Finally, if $(A_n)_{n\in\mathbb{N}}$ is a sequence of subsets of $\mathcal{F}_T$,

$$\Big(\bigcap_{n\in\mathbb{N}} A_n\Big) \cap \{T \leq t\} = \bigcap_{n\in\mathbb{N}} (A_n \cap \{T \leq t\}) \in \mathcal{F}_t. \qquad \square$$

If T is a stopping time of a filtration with respect to which a given process is adapted, then it is possible to stop this process in a natural way at the time T. We let the proof of the corresponding proposition as an exercise to the reader.

Proposition 1.29. *Let $(\mathcal{F}_t)_{t\geq 0}$ be a filtration on a probability space $(\Omega, \mathcal{F}, \mathbb{P})$ and let T be an almost surely finite stopping time of the filtration $(\mathcal{F}_t)_{t\geq 0}$. Let $(X_t)_{t\geq 0}$ be a stochastic process that is adapted and progressively measurable with respect to the filtration $(\mathcal{F}_t)_{t\geq 0}$. The stopped stochastic process $(X_{t\wedge T})_{t\geq 0}$ is progressively measurable with respect to the filtration $(\mathcal{F}_{t\wedge T})_{t\geq 0}$.*

BOWLING GREEN STATE
UNIVERSITY LIBRARIES

1.7 Martingales

We introduce and study in this section martingales in continuous time. Such processes were first introduced and extensively studied by Joseph Doob. Together with Markov processes they are among the most important class of stochastic processes and lie at the hearth of the theory of stochastic integration.

Definition 1.30. Let $(\mathcal{F}_t)_{t\geq 0}$ be a filtration defined on a probability space $(\Omega, \mathcal{F}, \mathbb{P})$. A process $(M_t)_{t\geq 0}$ that is adapted to $(\mathcal{F}_t)_{t\geq 0}$ is called a *submartingale* with respect to this filtration if

(1) for every $t \geq 0$, $\mathbb{E}(|M_t|) < +\infty$;

(2) for every $t \geq s \geq 0$,
$$\mathbb{E}(M_t \mid \mathcal{F}_s) \geq M_s.$$

A stochastic process $(M_t)_{t\geq 0}$ which is adapted to $(\mathcal{F}_t)_{t\geq 0}$ and such that $(-M_t)_{t\geq 0}$ is a submartingale is called a *supermartingale*.

Finally, a stochastic process $(M_t)_{t\geq 0}$ which is adapted to $(\mathcal{F}_t)_{t\geq 0}$ and which is at the same time both a submartingale and a supermartingale is called a *martingale*.

The following exercises provide some first properties of these processes.

Exercise 1.31. Let $(\mathcal{F}_t)_{t\geq 0}$ be a filtration defined on a probability space $(\Omega, \mathcal{F}, \mathbb{P})$ and let X be an integrable and $\mathcal{F}$-measurable random variable. Show that the process $(\mathbb{E}(X \mid \mathcal{F}_t))_{t\geq 0}$ is a martingale with respect to the filtration $(\mathcal{F}_t)_{t\geq 0}$.

Exercise 1.32. Let $(\mathcal{F}_t)_{t\geq 0}$ be a filtration defined on a probability space $(\Omega, \mathcal{F}, \mathbb{P})$ and let $(M_t)_{t\geq 0}$ be a submartingale with respect to the filtration $(\mathcal{F}_t)_{t\geq 0}$. Show that the function $t \to \mathbb{E}(M_t)$ is non-decreasing.

Exercise 1.33. Let $(\mathcal{F}_t)_{t\geq 0}$ be a filtration defined on a probability space $(\Omega, \mathcal{F}, \mathbb{P})$ and let $(M_t)_{t\geq 0}$ be a martingale with respect to the filtration $(\mathcal{F}_t)_{t\geq 0}$. Let now $\psi\colon \mathbb{R} \to \mathbb{R}$ be a convex function such that for $t \geq 0$, $\mathbb{E}(|\psi(M_t)|) < +\infty$. Show that the process $(\psi(M_t))_{t\geq 0}$ is a submartingale.

The following theorem, which is due to Doob, turns out to be extremely useful. It shows that martingales behave in a very nice way with respect to stopping times.

Proposition 1.34 (Doob stopping theorem). *Let $(\mathcal{F}_t)_{t\geq 0}$ be a filtration defined on a probability space $(\Omega, \mathcal{F}, \mathbb{P})$ and let $(M_t)_{t\geq 0}$ be a continuous stochastic process that is adapted to the filtration $(\mathcal{F}_t)_{t\geq 0}$. The following properties are equivalent:*

(1) *$(M_t)_{t\geq 0}$ is a martingale with respect to the filtration $(\mathcal{F}_t)_{t\geq 0}$;*

(2) *For any almost surely bounded stopping time T of the filtration $(\mathcal{F}_t)_{t\geq 0}$ such that $\mathbb{E}(|M_T|) < +\infty$, we have*
$$\mathbb{E}(M_T) = \mathbb{E}(M_0).$$

BOWLING GREEN STATE
UNIVERSITY LIBRARIES

Proof. Let us assume that $(M_t)_{t\geq 0}$ is a martingale with respect to the filtration $(\mathcal{F}_t)_{t\geq 0}$. Let now T be a stopping time of the filtration $(\mathcal{F}_t)_{t\geq 0}$ that is almost surely bounded by $K > 0$. Let us first assume that T takes its values in a finite set:

$$0 \leq t_1 < \cdots < t_n \leq K.$$

Thanks to the martingale property we have

$$\begin{aligned}
\mathbb{E}(M_T) &= \mathbb{E}\Big(\sum_{i=1}^{n} M_T 1_{T=t_i}\Big) \\
&= \sum_{i=1}^{n} \mathbb{E}(M_{t_i} 1_{T=t_i}) \\
&= \sum_{i=1}^{n} \mathbb{E}(M_{t_n} 1_{T=t_i}) \\
&= \mathbb{E}(M_{t_n}) \\
&= \mathbb{E}(M_0).
\end{aligned}$$

The theorem is therefore proved whenever T takes its values in a finite set. If T takes an infinite number of values, we approximate T by the following sequence of stopping times:

$$\tau_n = \sum_{k=1}^{2^n} \frac{kK}{2^n} 1_{\{\frac{(k-1)K}{2^n} \leq T < \frac{kK}{2^n}\}}.$$

The stopping time τ_n takes its values in a finite set and when $n \to +\infty$, $\tau_n \to T$. To conclude the proof of the first part of the proposition, we therefore have to show that

$$\lim_{n\to+\infty} \mathbb{E}(M_{\tau_n}) = \mathbb{E}(M_T).$$

For this, we are going to show that the family $(M_{\tau_n})_{n\in\mathbb{N}}$ is uniformly integrable.

Let $A \geq 0$. Since τ_n takes its values in a finite set, by using the martingale property and Jensen's inequality, it is easily checked that

$$\mathbb{E}(|M_K| 1_{M_{\tau_n}\geq A}) \geq \mathbb{E}(|M_{\tau_n}| 1_{M_{\tau_n}\geq A}).$$

Therefore,

$$\mathbb{E}(M_{\tau_n} 1_{M_{\tau_n}\geq A}) \leq \mathbb{E}(M_K 1_{\sup_{0\leq s\leq K} M_s \geq A}) \xrightarrow[A\to+\infty]{} 0.$$

By uniform integrability, we deduce that

$$\lim_{n\to+\infty} \mathbb{E}(M_{\tau_n}) = \mathbb{E}(M_T),$$

from which it is concluded that

$$\mathbb{E}(M_T) = \mathbb{E}(M_0).$$

Conversely, let us now assume that for any almost surely bounded stopping time T of the filtration $(\mathcal{F}_t)_{t\geq 0}$ such that $\mathbb{E}(|M_T|) < +\infty$, we have

$$\mathbb{E}(M_T) = \mathbb{E}(M_0).$$

Let $0 \leq s \leq t$ and $A \in \mathcal{F}_s$. By using the stopping time

$$T = s1_A + t1_{^cA},$$

we are led to

$$\mathbb{E}((M_t - M_s)1_A) = 0,$$

which implies the martingale property for $(M_t)_{t\geq 0}$. □

The hypothesis that the paths of $(M_t)_{t\geq 0}$ be continuous is actually not strictly necessary; however, the hypothesis that the stopping time T be almost surely bounded is essential, as is shown in the following exercise.

Exercise 1.35. Let $(\mathcal{F}_t)_{t\geq 0}$ be a filtration defined on a probability space $(\Omega, \mathcal{F}, \mathbb{P})$ and let $(M_t)_{t\geq 0}$ be a continuous martingale with respect to the filtration $(\mathcal{F}_t)_{t\geq 0}$ such that $M_0 = 0$ almost surely. For $a > 0$, we write $T_a = \inf\{t > 0, M_t = a\}$. Show that T_a is a stopping time of the filtration $(\mathcal{F}_t)_{t\geq 0}$. Prove that T_a is not almost surely bounded.

Exercise 1.36. Let $(\mathcal{F}_t)_{t\geq 0}$ be a filtration defined on a probability space $(\Omega, \mathcal{F}, \mathbb{P})$ and let $(M_t)_{t\geq 0}$ be a continuous submartingale with respect to the filtration $(\mathcal{F}_t)_{t\geq 0}$. By mimicking the proof of Doob's stopping theorem, show that if T_1 and T_2 are two almost surely bounded stopping times of the filtration $(\mathcal{F}_t)_{t\geq 0}$ such that $T_1 \leq T_2$ and $\mathbb{E}(|M_{T_1}|) < +\infty$, $\mathbb{E}(|M_{T_2}|) < +\infty$, then,

$$\mathbb{E}(M_{T_1}) \leq \mathbb{E}(M_{T_2}).$$

By using a similar proof, the Doob stopping theorem is easily extended as follows:

Proposition 1.37. *Let $(\mathcal{F}_t)_{t\geq 0}$ be a filtration defined on a probability space $(\Omega, \mathcal{F}, \mathbb{P})$ and let $(M_t)_{t\geq 0}$ be a continuous martingale with respect to the filtration $(\mathcal{F}_t)_{t\geq 0}$. If T_1 and T_2 are two almost surely bounded stopping times of the filtration $(\mathcal{F}_t)_{t\geq 0}$ such that $T_1 \leq T_2$ and $\mathbb{E}(|M_{T_1}|) < +\infty$, $\mathbb{E}(|M_{T_2}|) < +\infty$, then*

$$\mathbb{E}(M_{T_2} \mid \mathcal{F}_{T_1}) = M_{T_1}.$$

Finally, as a direct consequence of Doob's stopping theorem, we finally have the following result that shall repeatedly be used in the sequel.

Proposition 1.38. *Let $(\mathcal{F}_t)_{t\geq 0}$ be a filtration defined on a probability space $(\Omega, \mathcal{F}, \mathbb{P})$ and let $(M_t)_{t\geq 0}$ be a continuous martingale with respect to the filtration $(\mathcal{F}_t)_{t\geq 0}$. If T is a bounded stopping time of the filtration $(\mathcal{F}_t)_{t\geq 0}$ then the stopped process $(M_{t\wedge T})_{t\geq 0}$ is a martingale with respect to the filtration $(\mathcal{F}_t)_{t\geq 0}$.*

Proof. Let S be a bounded stopping time of the filtration $(\mathcal{F}_t)_{t\geq 0}$. From Doob's stopping theorem, we have $\mathbb{E}(M_{S\wedge T}) = \mathbb{E}(M_0)$. Since S is arbitrary, we conclude also from Doob's theorem that $(M_{t\wedge T})_{t\geq 0}$ is a martingale with respect to the filtration $(\mathcal{F}_t)_{t\geq 0}$. □

We now turn to convergence theorems for martingales. These convergence results rely on the notion of uniform integrability that we now recall.

Definition 1.39. Let $(X_i)_{i\in\mathcal{I}}$ be a family of random variables. We say that the family $(X_i)_{i\in\mathcal{I}}$ is *uniformly integrable* if for every $\varepsilon > 0$ there exists $K \geq 0$ such that

$$\forall i \in \mathcal{I}, \quad \mathbb{E}(|X_i| 1_{|X_i|>K}) < \varepsilon.$$

We have the following properties:

- A finite family of integrable random variables is uniformly integrable.
- If the family $(X_i)_{i\in\mathcal{I}}$ is uniformly integrable then it is bounded in L^1, that is, $\sup_{\mathcal{I}} \mathbb{E}(|X_i|) < +\infty$.
- If the family $(X_i)_{i\in\mathcal{I}}$ is bounded in L^p with $p > 1$, that is, $\sup_{\mathcal{I}} \mathbb{E}(|X_i|^p) < +\infty$, then it is uniformly integrable.

Thanks to the following result the notion of uniform integrability is often used to prove a convergence in L^1:

Proposition 1.40. *Let $(X_n)_{n\in\mathbb{N}}$ be a sequence of integrable random variables. Let X be an integrable random variable. The sequence $(X_n)_{n\in\mathbb{N}}$ converges toward X in L^1, that is, $\lim_{n\to+\infty} \mathbb{E}(|X_n - X|) = 0$, if and only if the following holds:*

(1) *In probability, $X_n \xrightarrow[n\to+\infty]{} X$, that is, for every $\varepsilon > 0$,*

$$\lim_{n\to+\infty} \mathbb{P}(|X_n - X| \geq \varepsilon) = 0.$$

(2) *The family $(X_n)_{n\in\mathbb{N}}$ is uniformly integrable.*

We have seen in Exercise 1.31 that if X is an integrable random variable defined on a filtered probability space $(\Omega, (\mathcal{F}_t)_{t\geq 0}, \mathcal{F}, \mathbb{P})$ then the process $(\mathbb{E}(X \mid \mathcal{F}_t))_{t\geq 0}$ is a martingale with respect to the filtration $(\mathcal{F}_t)_{t\geq 0}$. The following theorem characterizes the martingales that are of this form.

Theorem 1.41 (Doob convergence theorem). *Let $(\mathcal{F}_t)_{t\geq 0}$ be a filtration defined on a probability space $(\Omega, \mathcal{F}, \mathbb{P})$ and let $(M_t)_{t\geq 0}$ be a martingale with respect to the filtration $(\mathcal{F}_t)_{t\geq 0}$ whose paths are left limited and right continuous. The following properties are equivalent:*

(1) *When $t \to +\infty$, $(M_t)_{t\geq 0}$ converges in L^1.*

(2) *When $t \to +\infty$, $(M_t)_{t\geq 0}$ almost surely converges toward an integrable and $\mathcal{F}$-measurable random variable X that satisfies*

$$M_t = \mathbb{E}(X \mid \mathcal{F}_t), t \geq 0.$$

(3) *The family $(M_t)_{t\geq 0}$ is uniformly integrable.*

Proof. As a first step, we show that if the martingale $(M_t)_{t\geq 0}$ is bounded in L^1, that is,

$$\sup_{t\geq 0} \mathbb{E}(|M_t|) < +\infty,$$

then $(M_t)_{t\geq 0}$ almost surely converges toward an integrable and $\mathcal{F}$-measurable random variable X.

Let us first observe that

$$\{\omega \in \Omega, M_t(\omega) \text{ converges}\} = \{\omega \in \Omega, \lim \sup_{t\to+\infty} M_t(\omega) = \lim \inf_{t\to+\infty} M_t(\omega)\}.$$

Therefore, in order to show that $(M_t)_{t\geq 0}$ almost surely converges when $t \to +\infty$, we may prove that

$$\mathbb{P}(\{\omega \in \Omega, \lim \sup_{t\to+\infty} M_t(\omega) > \lim \inf_{t\to+\infty} M_t(\omega)\}) = 0.$$

Let us assume that

$$\mathbb{P}(\{\omega \in \Omega, \lim \sup_{t\to+\infty} M_t(\omega) > \lim \inf_{t\to+\infty} M_t(\omega)\}) > 0.$$

In this case we may find $a < b$ such that

$$\mathbb{P}(\{\omega \in \Omega, \lim \sup_{t\to+\infty} M_t(\omega) > a > b > \lim \inf_{t\to+\infty} M_t(\omega)\}) > 0.$$

The idea now is to study the oscillations of $(M_t)_{t\geq 0}$ between a and b. For $N \in \mathbb{N}$, $N > 0$ and $n \in \mathbb{N}$, we write

$$\mathcal{D}_{n,N} = \left\{\frac{kN}{2^n},\ 0 \leq k \leq 2^n\right\}$$

and

$$\mathcal{D} = \bigcup_{n,N} \mathcal{D}_{n,N}.$$

Let $\mathcal{N}(a,b,n,N)$ be the greatest integer k for which we may find elements of $\mathcal{D}_{n,N}$,

$$0 \le q_1 < r_1 < q_2 < r_2 < \cdots < q_k < r_k \le N$$

that satisfy

$$M_{q_i} < a, \quad M_{r_i} > b.$$

Let now

$$Y_{n,N} = \sum_{k=1}^{2^n} C_{\frac{kN}{2^n}} (M_{\frac{kN}{2^n}} - M_{\frac{(k-1)N}{2^n}}),$$

where $C_k \in \{0,1\}$ is recursively defined by

$$C_1 = 1_{M_0 < a},$$

$$C_k = 1_{C_{k-1}=1} 1_{M_{\frac{(k-1)N}{2^n}} \le b} + 1_{C_{k-1}=0} 1_{M_{\frac{(k-1)N}{2^n}} < a}.$$

Since $(M_t)_{t\ge 0}$ is martingale, it is easily checked that

$$\mathbb{E}(Y_{n,N}) = 0.$$

Furthermore, thanks to the very definition of $\mathcal{N}(a,b,n,N)$, we have

$$Y_{n,N} \ge (b-a)\mathcal{N}(a,b,n,N) - \max(a - M_N, 0).$$

Therefore, we have

$$\begin{aligned}(b-a)\mathbb{E}(\mathcal{N}(a,b,n,N)) &\le \mathbb{E}(\max(a - M_N, 0)) \\ &\le |a| + \mathbb{E}(|M_N|) \\ &\le |a| + \sup_{t>0} \mathbb{E}(|M_t|),\end{aligned}$$

and thus

$$(b-a)\mathbb{E}(\sup_{n,N} \mathcal{N}(a,b,n,N)) \le |a| + \sup_{t>0} \mathbb{E}(|M_t|).$$

This implies that almost surely $\sup_{n,N} \mathcal{N}(a,b,n,N) < +\infty$, from which it is deduced

$$\mathbb{P}(\{\omega \in \Omega, \lim \sup_{t\to+\infty, t\in\mathcal{D}} M_t(\omega) > a > b > \lim \inf_{t\to+\infty, t\in\mathcal{D}} M_t(\omega)\}) = 0.$$

Since the paths of $(M_t)_{t\ge 0}$ are right continuous, we have

$$\begin{aligned}&\mathbb{P}(\{\omega \in \Omega, \lim \sup_{t\to+\infty, t\in\mathcal{D}} M_t(\omega) > a > b > \lim \inf_{t\to+\infty, t\in\mathcal{D}} M_t(\omega)\}) \\ &= \mathbb{P}(\{\omega \in \Omega, \lim \sup_{t\to+\infty} M_t(\omega) > a > b > \lim \inf_{t\to+\infty} M_t(\omega)\}).\end{aligned}$$

This is absurd. Thus, if $(M_t)_{t\geq 0}$ is bounded in L^1, it almost surely converges toward an $\mathcal{F}$-measurable random variable X. Fatou's lemma provides the integrability of X.

With this preliminary result in hands, we can now turn to the proof of the theorem.

Let us assume that $(M_t)_{t\geq 0}$ converges in L^1. In this case, it is of course bounded in L^1, and thus almost surely converges toward an $\mathcal{F}$-measurable and integrable random variable X. Let $t \geq 0$ and $A \in \mathcal{F}_t$. For $s \geq t$ we have

$$\mathbb{E}(M_s 1_A) = \mathbb{E}(M_t 1_A).$$

By letting $s \to +\infty$, the dominated convergence theorem yields

$$\mathbb{E}(X 1_A) = \mathbb{E}(M_t 1_A).$$

Therefore, as expected, we obtain

$$\mathbb{E}(X \mid \mathcal{F}_t) = M_t.$$

Let us now assume that $(M_t)_{t\geq 0}$ almost surely converges toward an $\mathcal{F}$-measurable and integrable random variable X that satisfies

$$M_t = \mathbb{E}(X \mid \mathcal{F}_t), t \geq 0.$$

We almost surely have $\sup_{t\geq 0} |M_t| < +\infty$ and thus for $A \geq 0$,

$$\begin{aligned}\mathbb{E}(|M_t| 1_{|M_t|\geq A}) &= \mathbb{E}(|\mathbb{E}(X \mid \mathcal{F}_t)| 1_{|M_t|\geq A})\\ &\leq \mathbb{E}(|X| 1_{|M_t|\geq A})\\ &\leq \mathbb{E}(|X| 1_{\sup_{t\geq 0}|M_t|\geq A}).\end{aligned}$$

This implies the uniform integrability for the family $(M_t)_{t\geq 0}$.

Finally, if the family $(M_t)_{t\geq 0}$ is uniformly integrable, then it is bounded in L^1 and therefore almost surely converges. The almost sure convergence, together with the uniform integrability, provides the convergence in L^1. □

Exercise 1.42. By using the same reasoning as in the previous proof, show that a right continuous and left limited positive supermartingale needs to converge almost surely when $t \to \infty$.

When dealing with stochastic processes, it is often important to work with versions of these processes whose paths are as regular as possible. In that direction, the Kolmogorov's continuity theorem (see Theorem 1.24) provided a sufficient condition allowing to work with continuous versions of stochastic processes. For martingales, the possibility of working with regular versions is related to the regularity properties of the filtration with respect to which the martingale property is satisfied.

Definition 1.43. Let $(\mathcal{F}_t)_{t\geq 0}$ be a filtration on a probability space $(\Omega, \mathcal{F}, \mathbb{P})$. Let following assumptions be fulfilled:

(1) If $A \in \mathcal{F}$ satisfies $\mathbb{P}(A) = 0$, then every subset of A is in $\mathcal{F}_0$.

(2) The filtration $(\mathcal{F}_t)_{t\geq 0}$ is right continuous, that is, for every $t \geq 0$,

$$\mathcal{F}_t = \bigcap_{\varepsilon>0} \mathcal{F}_{t+\varepsilon}.$$

Then the filtered probability space

$$(\Omega, (\mathcal{F}_t)_{t\geq 0}, \mathcal{F}, \mathbb{P})$$

is said to satisfy the *usual conditions*.

Remark 1.44. The above set of assumptions are called the usual conditions because, as we will see in Chapter 5, these are the conditions under which it is convenient to work in order to properly define the stochastic integral.

Remark 1.45. Let $(\mathcal{F}_t)_{t\geq 0}$ be a filtration on a probability space $(\Omega, \mathcal{F}, \mathbb{P})$ and let $(M_t)_{t\geq 0}$ be a (sub, super) martingale with respect to the filtration $(\mathcal{F}_t)_{t\geq 0}$ whose paths are right continuous and left limited. The filtered probability space

$$(\Omega, (\mathcal{F}_t)_{t\geq 0}, \mathcal{F}, \mathbb{P})$$

may canonically be enlarged into a filtered probability space

$$(\Omega, (\mathcal{G}_t)_{t\geq 0}, \mathcal{G}, \mathbb{P})$$

that satisfies the usual conditions. Indeed, $\mathcal{G}$ can be taken to be the $\mathbb{P}$-completion of $\mathcal{F}$ and

$$\mathcal{G}_t = \bigcap_{u>t} \sigma(\mathcal{F}_u, \mathcal{N})$$

where $\mathcal{N}$ is the set of events whose probability is zero. Moreover $(M_t)_{t\geq 0}$ is a (sub, super) martingale with respect to the filtration $(\mathcal{G}_t)_{t\geq 0}$ (this is not straightforward and left to the reader as an exercise). The filtered probability space

$$(\Omega, (\mathcal{G}_t)_{t\geq 0}, \mathcal{G}, \mathbb{P})$$

is called the *usual completion* of

$$(\Omega, (\mathcal{F}_t)_{t\geq 0}, \mathcal{F}, \mathbb{P}).$$

Exercise 1.46. Let $(\Omega, (\mathcal{F}_t)_{t\geq 0}, \mathcal{F}, \mathbb{P})$ be a filtered probability space that satisfies the usual conditions and let $(X_t)_{t\geq 0}$ be stochastic process adapted to the filtration $(\mathcal{F}_t)_{t\geq 0}$ whose paths are left limited and right continuous. Let K be compact subset of $\mathbb{R}$. Show that the random time

$$T = \inf\{t \geq 0, X_t \in K\}$$

is a stopping time of the filtration $(\mathcal{F}_t)_{t\geq 0}$.

Theorem 1.47 (Doob regularization theorem). *Let $(\Omega, (\mathcal{F}_t)_{t\geq 0}, \mathcal{F}, \mathbb{P})$ be a filtered probability space that satisfies the usual conditions and let $(M_t)_{t\geq 0}$ be a supermartingale with respect to the filtration $(\mathcal{F}_t)_{t\geq 0}$. Let us assume that the function $t \to \mathbb{E}(M_t)$ is right continuous.*

There exists a modification $(\widetilde{M}_t)_{t\geq 0}$ of $(M_t)_{t\geq 0}$ with the following properties:

(1) *$(\widetilde{M}_t)_{t\geq 0}$ is adapted to the filtration $(\mathcal{F}_t)_{t\geq 0}$.*

(2) *The paths of $(\widetilde{M}_t)_{t\geq 0}$ are locally bounded, right continuous and left limited.*

(3) *$(\widetilde{M}_t)_{t\geq 0}$ is a supermartingale with respect to the filtration $(\mathcal{F}_t)_{t\geq 0}$.*

Proof. As for the proof of Doob's convergence theorem, the idea is to study the oscillations of $(M_t)_{t\geq 0}$. In what follows, we will use the notations introduced in the proof of this theorem that we recall below.

For $N \in \mathbb{N}$, $N > 0$ and $n \in \mathbb{N}$, we write

$$\mathcal{D}_{n,N} = \left\{ \frac{kN}{2^n}, 0 \leq k \leq 2^n \right\},$$

$$\mathcal{D}_N = \bigcup_n \mathcal{D}_{n,N} \quad \text{and} \quad \mathcal{D} = \bigcup_{n,N} \mathcal{D}_{n,N}.$$

For $a < b$, let $\mathcal{N}(a,b,n,N)$ be the greatest integer k for which we can find elements of $\mathcal{D}_{n,N}$,

$$0 \leq q_1 < r_1 < q_2 < r_2 < \cdots < q_k < r_k \leq N$$

such that

$$M_{q_i} < a, \quad M_{r_i} > b.$$

Let now Ω^* be the set of $\omega \in \Omega$ such that for all $t \geq 0$, $\lim_{s\to t, s>t, s\in\mathcal{D}} M_s(\omega)$ exists and is finite.

It is easily seen that

$$\Omega^* = \bigcap_{a,b\in\mathbb{Q}} \bigcap_{N\in\mathbb{N}^*} \{\omega \in \Omega,\ \sup_{t\in\mathcal{D}_N} |M_t(\omega)| < +\infty$$

and

$$\sup_{n\in\mathbb{N}} \mathcal{N}(a,b,n,N) < +\infty\}.$$

Therefore, $\Omega^* \in \mathcal{F}$. We may prove, as we proved for the Doob convergence theorem, that $\mathbb{P}(\Omega^*) = 1$.

For $t \geq 0$, we define $(\widetilde{M}_t)_{t\geq 0}$ in the following way:

- If $\omega \in \Omega^*$, then
$$\widetilde{M}_t(\omega) = \lim_{s\to t, s>t, s\in\mathcal{D}} M_s(\omega).$$

- If $\omega \notin \Omega^*$, then
$$\widetilde{M}_t(\omega) = 0.$$

It is clear that the paths of $(\widetilde{M}_t)_{t\geq 0}$ are locally bounded, right continuous and left limited. Let us now show that this process is the expected modification of $(M_t)_{t\geq 0}$.

We first observe that for $t \geq 0$, the random variable
$$\lim_{s\to t, s>t, s\in\mathcal{D}} M_s$$
is measurable with respect to $\bigcap_{s>t} \mathcal{F}_s = \mathcal{F}_t$. Furthermore, $\Omega\backslash\Omega^*$ has a zero probability and is therefore in $\mathcal{F}_0$, according to the usual conditions. This shows that the process $(\widetilde{M}_t)_{t\geq 0}$ is adapted to the filtration $(\mathcal{F}_t)_{t\geq 0}$.

We now show that $(\widetilde{M}_t)_{t\geq 0}$ is a modification of $(M_t)_{t\geq 0}$. Let $t \geq 0$. We have almost surely
$$\lim_{s\to t, s>t, s\in\mathcal{D}} M_s = \widetilde{M}_t.$$

Let us prove that this convergence also holds in L^1. To prove this, it is enough to check that for every decreasing family $(s_n)_{n\in\mathbb{N}}$ such that $s_n \in \mathcal{D}$ and that converges toward t, the family $(M_{s_n})_{n\in\mathbb{N}}$ is uniformly integrable.

Let $\varepsilon > 0$. Since $u \to \mathbb{E}(M_u)$ is assumed to be right continuous, we can find $s \in \mathbb{R}$ such that $t < s$ and such that for every $s > u > t$,
$$0 \leq \mathbb{E}(M_u) - \mathbb{E}(M_s) \leq \frac{\varepsilon}{2}.$$

For $s > u > t$ and $\lambda > 0$, we have
$$\begin{aligned}
\mathbb{E}(|M_u|1_{|M_u|>\lambda}) &= -\mathbb{E}(M_u 1_{M_u<-\lambda}) + \mathbb{E}(M_u) - \mathbb{E}(M_u 1_{M_u\leq-\lambda}) \\
&\leq -\mathbb{E}(M_s 1_{M_u<-\lambda}) + \mathbb{E}(M_u) - \mathbb{E}(M_s 1_{M_u\leq-\lambda}) \\
&\leq \mathbb{E}(|M_s|1_{|M_u|>\lambda}) + \frac{\varepsilon}{2}.
\end{aligned}$$

Now, since $M_s \in L^1$, we can find $\delta > 0$ such that for every $F \in \mathcal{F}$ that satisfies $\mathbb{P}(F) < \delta$, we have $\mathbb{E}(|M_s|1_F) < \frac{\varepsilon}{2}$. But for $t < u < s$,
$$\mathbb{P}(|M_u| > \lambda) \leq \frac{\mathbb{E}(|M_u|)}{\lambda} = \frac{\mathbb{E}(M_u) + 2\mathbb{E}(\max(-M_u, 0))}{\lambda}.$$

From Jensen's inequality, it is seen that the process $(\max(-M_u, 0))_{t<u<s}$ is a submartingale, therefore
$$\mathbb{E}(\max(-M_u, 0)) \leq \mathbb{E}(\max(-M_s, 0)).$$

We deduce that for $t < u < s$,

$$\mathbb{P}(|M_u| > \lambda) \le \frac{\mathbb{E}(M_t) + 2\mathbb{E}(\max(-M_s, 0))}{\lambda}.$$

It is thus possible to find $A > 0$ such that for every $t < u < s$,

$$\mathbb{P}(|M_u| > A) < \delta.$$

For $t < u < s$, we then have

$$\mathbb{E}(|M_u| 1_{|M_u| > \lambda}) < \varepsilon.$$

This implies that for every decreasing family $(s_n)_{n \in \mathbb{N}}$ such that $s_n \in \mathcal{D}$ and that converges toward t, the family $(M_{s_n})_{n \in \mathbb{N}}$ is uniformly integrable. The convergence

$$\lim_{s \to t, s > t, s \in \mathcal{D}} M_s = \widetilde{M}_t$$

thus also holds in L^1. Now, since $(M_t)_{t \ge 0}$ is a supermartingale, for $s \ge t$ we have

$$\mathbb{E}(M_s \mid \mathcal{F}_t) \le M_t.$$

This implies

$$\lim_{s \to t, s > t, s \in \mathcal{D}} \mathbb{E}(M_s \mid \mathcal{F}_t) \le M_t,$$

and

$$\mathbb{E}(\widetilde{M}_t \mid \mathcal{F}_t) \le M_t.$$

Hence, since $\widetilde{M}_t$ is adapted to the filtration $\mathcal{F}_t$,

$$\widetilde{M}_t \le M_t.$$

Due to the fact that the function $u \to \mathbb{E}(M_u)$ is right continuous, we have

$$\lim_{s \to t, s > t, s \in \mathcal{D}} \mathbb{E}(M_s) = \mathbb{E}(M_t).$$

But from the L^1 convergence, we also have

$$\lim_{s \to t, s > t, s \in \mathcal{D}} \mathbb{E}(M_s) = \mathbb{E}\Big(\lim_{s \to t, s > t, s \in \mathcal{D}} M_s\Big) = \mathbb{E}(\widetilde{M}_t).$$

This gives

$$\mathbb{E}(\widetilde{M}_t) = \mathbb{E}(M_t).$$

The random variable $M_t - \widetilde{M}_t$ is therefore non-negative and has a zero expectation. This implies that almost surely $M_t = \widetilde{M}_t$. The stochastic process $(\widetilde{M}_t)_{t \ge 0}$ is therefore a modification of $(M_t)_{t \ge 0}$. Finally, since a modification of a supermartingale is still a supermartingale, this concludes the proof of the theorem. □

The following exercise shows that martingales naturally appear when studying equivalent measures on a filtered probability space. We first recall a basic theorem from measure theory which is known as the Radon–Nikodym theorem. Let Ω be a set endowed with a σ-field $\mathcal{F}$.

Definition 1.48. Let $\mathbb{P}$ and $\mathbb{Q}$ be probability measures on $(\Omega, \mathcal{F})$. It is said that $\mathbb{P}$ is *absolutely continuous* with respect to $\mathbb{Q}$ if $\mathbb{Q}(A) = 0$ implies $\mathbb{P}(A) = 0$ for every $A \in \mathcal{F}$. We then write $\mathbb{P} \ll \mathbb{Q}$. If $\mathbb{P} \ll \mathbb{Q}$ and $\mathbb{Q} \ll \mathbb{P}$, it is said that $\mathbb{P}$ and $\mathbb{Q}$ are *equivalent*: In this case we write $\mathbb{P} \asymp \mathbb{Q}$.

Theorem 1.49 (Radon–Nikodym). *Let $\mathbb{P}$ and $\mathbb{Q}$ be two probability measures on $(\Omega, \mathcal{F})$. We have $\mathbb{P} \ll \mathbb{Q}$ if and only if there is a random variable D, which is $\mathcal{F}$-measurable and such that for every $A \in \mathcal{F}$,*

$$\mathbb{P}(A) = \int_A D \, d\mathbb{Q}.$$

D is called the density of $\mathbb{P}$ with respect to $\mathbb{Q}$ and we write

$$D = \frac{d\mathbb{P}}{d\mathbb{Q}}.$$

Moreover, under the same assumptions $\mathbb{P} \asymp \mathbb{Q}$ if and only if D is positive $\mathbb{P}$-a.s. In this case

$$\frac{d\mathbb{Q}}{d\mathbb{P}} = \frac{1}{D}.$$

Exercise 1.50. Let $(\Omega, (\mathcal{F}_t)_{t \geq 0}, \mathcal{F}, \mathbb{P})$ be a filtered probability space that satisfies the usual conditions. We write

$$\mathcal{F}_\infty = \sigma(\mathcal{F}_t, t \geq 0)$$

and for $t \geq 0$, $\mathbb{P}_{/\mathcal{F}_t}$ is the restriction of $\mathbb{P}$ to $\mathcal{F}_t$. Let $\mathbb{Q}$ be a probability measure on $\mathcal{F}_\infty$ such that for every $t \geq 0$,

$$\mathbb{Q}_{/\mathcal{F}_t} \ll \mathbb{P}_{/\mathcal{F}_t}.$$

(1) Show that there exists a right continuous and left limited martingale $(D_t)_{t \geq 0}$ such that for every $t \geq 0$,

$$D_t = \frac{d\mathbb{Q}_{/\mathcal{F}_t}}{d\mathbb{P}_{/\mathcal{F}_t}}, \quad \mathbb{P}\text{-a.s.}$$

(2) Show that the following properties are equivalent:

(a) $\mathbb{Q}_{/\mathcal{F}_\infty} \ll \mathbb{P}_{/\mathcal{F}_\infty}$.

(b) The martingale $(D_t)_{t \geq 0}$ is uniformly integrable.

(c) $(D_t)_{t\geq 0}$ converges in L^1.

(d) $(D_t)_{t\geq 0}$ almost surely converges to an integrable and $\mathcal{F}_\infty$-measurable random variable D such that

$$D_t = \mathbb{E}(D \mid \mathcal{F}_t), \quad t \geq 0.$$

1.8 Martingale inequalities

In this section, we prove some fundamental martingale inequalities that, once again, are due to Doob

Theorem 1.51 (Doob maximal inequalities). *Let $(\mathcal{F}_t)_{t\geq 0}$ be a filtration on a probability space $(\Omega, \mathcal{F}, \mathbb{P})$ and let $(M_t)_{t\geq 0}$ be a continuous martingale with respect to the filtration $(\mathcal{F}_t)_{t\geq 0}$.*

(1) *Let $p \geq 1$ and $T > 0$. If $\mathbb{E}(|M_T|^p) < +\infty$, then for every $\lambda > 0$,*

$$\mathbb{P}(\sup_{0\leq t\leq T} |M_t| \geq \lambda) \leq \frac{\mathbb{E}(|M_T|^p)}{\lambda^p}.$$

(2) *Let $p > 1$ and $T > 0$. If $\mathbb{E}(|M_T|^p) < +\infty$, then*

$$\mathbb{E}((\sup_{0\leq t\leq T} |M_t|)^p) \leq \left(\frac{p}{p-1}\right)^p \mathbb{E}(|M_T|^p).$$

Proof. (1) Let $p \geq 1$ and $T > 0$. If $\mathbb{E}(|M_T|^p) < +\infty)$ then, from Jensen's inequality the process $(|M_t|^p)_{0\leq t\leq T}$ is a submartingale. Let $\lambda > 0$ and

$$\tau = \inf\{s \geq 0 \text{ such that } |M_s| \geq \lambda\} \wedge T,$$

with the convention that $\inf \emptyset = +\infty$. It is seen that τ is an almost surely bounded stopping time. Therefore

$$\mathbb{E}(|M_\tau|^p) \leq \mathbb{E}(|M_T|^p).$$

But from the very definition of τ,

$$|M_\tau|^p \geq 1_{\sup_{0\leq t\leq T} |M_t|\geq\lambda}\lambda^p + 1_{\sup_{0\leq t\leq T} |M_t|<\lambda}|M_T|^p,$$

which implies

$$\mathbb{P}(\sup_{0\leq t\leq T} |M_t| \geq \lambda) \leq \frac{\mathbb{E}(|M_T|^p 1_{\sup_{0\leq t\leq T} |M_t|\geq\lambda})}{\lambda^p} \leq \frac{\mathbb{E}(|M_T|^p)}{\lambda^p}.$$

(2) Let $p \geq 1$ and $T > 0$. Let us first assume that

$$\mathbb{E}((\sup_{0 \leq t \leq T} |M_t|)^p) < +\infty.$$

The previous proof shows that for $\lambda > 0$,

$$\mathbb{P}(\sup_{0 \leq t \leq T} |M_t| \geq \lambda) \leq \frac{\mathbb{E}(|M_T| 1_{\sup_{0 \leq t \leq T} |M_t| \geq \lambda})}{\lambda}.$$

We deduce

$$\int_0^{+\infty} \lambda^{p-1} \mathbb{P}(\sup_{0 \leq t \leq T} |M_t| \geq \lambda) d\lambda \leq \int_0^{+\infty} \lambda^{p-2} \mathbb{E}(|M_T| 1_{\sup_{0 \leq t \leq T} |M_t| \geq \lambda}) d\lambda.$$

From the Fubini theorem we now have

$$\begin{aligned} \int_0^{+\infty} \lambda^{p-1} \mathbb{P}(\sup_{0 \leq t \leq T} |M_t| \geq \lambda) d\lambda &= \int_\Omega \Big(\int_0^{\sup_{0 \leq t \leq T} |M_t|(\omega)} \lambda^{p-1} d\lambda \Big) d\mathbb{P}(\omega) \\ &= \frac{1}{p} \mathbb{E}((\sup_{0 \leq t \leq T} |M_t|)^p). \end{aligned}$$

Similarly, we obtain

$$\int_0^{+\infty} \lambda^{p-2} \mathbb{E}(|M_T| 1_{\sup_{0 \leq t \leq T} |M_t| \geq \lambda}) d\lambda = \frac{1}{p-1} \mathbb{E}((\sup_{0 \leq t \leq T} |M_t|)^{p-1} |M_T|).$$

Hence,

$$\mathbb{E}((\sup_{0 \leq t \leq T} |M_t|)^p) \leq \frac{p}{p-1} \mathbb{E}((\sup_{0 \leq t \leq T} |M_t|)^{p-1} |M_T|).$$

By using now Hölder's inequality we obtain

$$\mathbb{E}((\sup_{0 \leq t \leq T} |M_t|)^{p-1} |M_T|) \leq \mathbb{E}(|M_T|^p)^{\frac{1}{p}} \mathbb{E}((\sup_{0 \leq t \leq T} |M_t|)^p)^{\frac{p-1}{p}},$$

which implies

$$\mathbb{E}((\sup_{0 \leq t \leq T} |M_t|)^p) \leq \frac{p}{p-1} \mathbb{E}(|M_T|^p)^{\frac{1}{p}} \mathbb{E}((\sup_{0 \leq t \leq T} |M_t|)^p)^{\frac{p-1}{p}}.$$

As a conclusion if $\mathbb{E}((\sup_{0 \le t \le T} |M_t|)^p) < +\infty$, we have

$$\mathbb{E}((\sup_{0 \le t \le T} |M_t|)^p) \le \Big(\frac{p}{p-1}\Big)^p \mathbb{E}(|M_T|^p).$$

Now, if $\mathbb{E}((\sup_{0 \le t \le T} |M_t|)^p) = +\infty$, we consider for $N \in \mathbb{N}$ the stopping time $\tau_N = \inf\{t \ge 0, |M_t| \ge N\} \wedge T$. By applying the above result to the martingale $(M_{t \wedge \tau_N})_{t \ge 0}$, we obtain

$$\mathbb{E}((\sup_{0 \le t \le T} |M_{t \wedge \tau_N}|)^p) \le \Big(\frac{p}{p-1}\Big)^p \mathbb{E}(|M_T|^p),$$

from which we may conclude by using the monotone convergence theorem. □

Notes and comments

We mostly restricted our attention to continuous stochastic processes but a general theory of left limited right continuous processes can be developed similarly, see the books by Jacod–Shiryaev [44] and by Protter [61] for a detailed account on jump processes. As pointed out in the text, most of the material concerning filtrations, martingales and stopping times is due to Doob and was exposed in his extremely influential book [19]. Further readings about the general theory of stochastic processes and martingales include the classical references by Dellacherie–Meyer [17], Ikeda–Watanabe [38], Protter [61] and Revuz–Yor [64].

Chapter 2
Brownian motion

The chapter is devoted to the study of the Brownian motion. This is without doubt the most important continuous stochastic process and will serve as a canonical example in most of the developments of the next chapters. It is at the same time a Gaussian process, a martingale and a Markov process. In the first part of the chapter we prove the existence of the Brownian motion as a consequence of the Daniell–Kolmogorov and of the Kolmogorov continuity theorem. We apply the martingale techniques developed in Chapter 1 to study basic properties of the Brownian motion paths like the law of iterated logarithm. The second part of the chapter focusses on the point of view that the Brownian motion can be seen as a continuous random walk in continuous time. More precisely, we will prove that the Brownian motion is the limit of suitably rescaled symmetric random walks. The study of random walks will then allow us to obtain several properties of the Brownian motion paths by a limiting procedure.

2.1 Definition and basic properties

Definition 2.1. Let $(\Omega, \mathcal{F}, \mathbb{P})$ be a probability space. A continuous real-valued process $(B_t)_{t\geq 0}$ is called a *standard Brownian motion* if it is a Gaussian process with mean function

$$\mathbb{E}(B_t) = 0$$

and covariance function

$$\mathbb{E}(B_s B_t) = \min(s,t).$$

Remark 2.2. It is seen that $R(s,t) = \min(s,t)$ is a covariance function, because it is obviously symmetric and for $a_1, \dots, a_n \in \mathbb{R}$ and $t_1, \dots, t_n \in \mathbb{R}_{\geq 0}$,

$$\begin{aligned}\sum_{1\leq i,j\leq n} a_i a_j \min(t_i,t_j) &= \sum_{1\leq i,j\leq n} a_i a_j \int_0^{+\infty} \mathbf{1}_{[0,t_i]}(s)\mathbf{1}_{[0,t_j]}(s)ds \\ &= \int_0^{+\infty} \Big(\sum_{i=1}^n a_i \mathbf{1}_{[0,t_i]}(s)\Big)^2 ds \geq 0.\end{aligned}$$

Definition 2.3. The distribution of a standard Brownian motion, which is thus a probability measure on the space of continuous functions $\mathcal{C}(\mathbb{R}_{\geq 0}, \mathbb{R})$, is called the *Wiener measure*.

Remark 2.4. An n-dimensional stochastic process $(B_t)_{t\geq 0}$ is called a standard Brownian motion if

$$(B_t)_{t\geq 0} = (B_t^1, \ldots, B_t^n)_{t\geq 0},$$

where the processes $(B_t^i)_{t\geq 0}$ are independent standard Brownian motions.

Of course, the definition of Brownian motion is worth only because such an object exists.

Theorem 2.5. *There exist a probability space $(\Omega, \mathcal{F}, \mathbb{P})$ and a stochastic process on it which is a standard Brownian motion.*

Proof. From Proposition 1.19, there exists a probability space $(\Omega, \mathcal{F}, \mathbb{P})$ and a Gaussian process $(X_t)_{t\geq 0}$ on it whose mean function is 0 and covariance function is

$$\mathbb{E}(X_s X_t) = \min(s,t).$$

We have for $n \geq 0$ and $0 \leq s \leq t$:

$$\mathbb{E}((X_t - X_s)^{2n}) = \frac{(2n)!}{2^n n!}(t-s)^n.$$

Therefore, by using the Kolmogorov continuity theorem, there exists a modification $(B_t)_{t\geq 0}$ of $(X_t)_{t\geq 0}$ whose paths are locally γ-Hölder if $\gamma \in [0, \frac{n-1}{2n})$. □

Remark 2.6. From the previous proof, we also deduce that the paths of a standard Brownian motion are locally γ-Hölder for every $\gamma < \frac{1}{2}$. It can be shown that they are not $\frac{1}{2}$-Hölder (see Exercise 2.12 and Theorem 2.23).

The following exercises give some first basic properties of Brownian motion and study some related processes. In all the exercises, $(B_t)_{t\geq 0}$ is a standard one-dimensional Brownian motion.

Exercise 2.7. Show the following properties:

(1) $B_0 = 0$ a.s.;

(2) for any $h \geq 0$, the process $(B_{t+h} - B_h)_{t\geq 0}$ is a standard Brownian motion;

(3) for any $t > s \geq 0$, the random variable $B_t - B_s$ is independent of the σ-algebra $\sigma(B_u, u \leq s)$.

Exercise 2.8 (Symmetry property of the Brownian motion).

(1) Show that the process $(-B_t)_{t\geq 0}$ is a standard Brownian motion.

(2) More generally, show that if

$$(B_t)_{t\geq 0} = (B_t^1, \ldots, B_t^d)_{t\geq 0}$$

is a d-dimensional Brownian motion and if M is an orthogonal $d \times d$ matrix, then $(MB_t)_{t\geq 0}$ is a standard Brownian motion.

Exercise 2.9 (Scaling property of the Brownian motion). Show that for every $c > 0$, the process $(B_{ct})_{t\geq 0}$ has the same law as the process $(\sqrt{c}B_t)_{t\geq 0}$.

Exercise 2.10 (Time inversion property of Brownian motion).

(1) Show that almost surely, $\lim_{t\to+\infty} \frac{B_t}{t} = 0$.

(2) Deduce that the process $(tB_{\frac{1}{t}})_{t\geq 0}$ has the same law as the process $(B_t)_{t\geq 0}$.

Exercise 2.11 (Non-canonical representation of Brownian motion).

(1) Show that for $t > 0$, the Riemann integral $\int_0^t \frac{B_s}{s} ds$ almost surely exists.

(2) Show that the process $(B_t - \int_0^t \frac{B_s}{s} ds)_{t\geq 0}$ is a standard Brownian motion.

Exercise 2.12. Show that

$$\mathbb{P}\Big(\sup_{s,t\in[0,1]} \frac{|B_t - B_s|}{\sqrt{t-s}} = +\infty\Big) = 1.$$

Hint. Divide the interval $[0, 1]$ in subintervals $[k/n, (k+1)/n]$ to bound from below $\sup_{s,t\in[0,1]} \frac{|B_t-B_s|}{\sqrt{t-s}}$ by the supremum of the absolute value of independent Gaussian random variables.

Exercise 2.13 (Non-differentiability of the Brownian paths).

(1) Show that if $f : (0, 1) \to \mathbb{R}$ is differentiable at $t \in (0, 1)$, then there exist an interval $(t - \delta, t + \delta)$ and a constant $C > 0$ such that for $s \in (t - \delta, t + \delta)$,

$$|f(t) - f(s)| \leq C|t - s|.$$

(2) For $n \geq 1$, let

$$M_n = \min_{1\leq k\leq n} \{\max\{|B_{k/n} - B_{(k-1)/n}|, |B_{(k+1)/n} - B_{k/n}|, |B_{(k+2)/n} - B_{k+1/n}|\}\}.$$

Show that $\lim_{n\to+\infty} \mathbb{P}(M_n) = 0$.

(3) Deduce that

$$\mathbb{P}(\exists t \in (0, 1), B \text{ is differentiable at } t) = 0.$$

Exercise 2.14 (Fractional Brownian motion). Let $0 < H < 1$.

(1) Show that for $s \in \mathbb{R}$, the function

$$f_s(t) = (|t - s|^{H-\frac{1}{2}} - \mathbf{1}_{(-\infty,0]}(t)|t|^{H-\frac{1}{2}})\mathbf{1}_{(-\infty,s]}(t)$$

is square integrable on $\mathbb{R}$.

(2) Deduce that

$$R(s,t) = \frac{1}{2}(s^{2H} + t^{2H} - |t-s|^{2H}), \quad s,t \geq 0,$$

is a covariance function.

(3) A continuous and centered Gaussian process with covariance function R is called a fractional Brownian motion with parameter H. Show that such process exists and study its Hölder sample path regularity.

(4) Let $(B_t)_{t\geq 0}$ be a fractional Brownian motion with parameter H. Show that for any $h \geq 0$, the process $(B_{t+h} - B_h)_{t\geq 0}$ is a fractional Brownian motion.

(5) Show that for every $c > 0$, the process $(B_{ct})_{t\geq 0}$ has the same law as the process $(c^H B_t)_{t\geq 0}$

Exercise 2.15 (Brownian bridge). Let $T > 0$ and $x \in \mathbb{R}$.

(1) Show that the process

$$X_t = \frac{t}{T}x + B_t - \frac{t}{T}B_T, \quad 0 \leq t \leq T,$$

is a Gaussian process. Compute its mean function and its covariance function.

(2) Show that $(X_t)_{0\leq t\leq T}$ is a Brownian motion conditioned to be x at time T, that is, for every $0 \leq t_1 \leq \cdots \leq t_n < T$, and $A_1, \dots, A_n$ Borel sets of $\mathbb{R}$,

$$\mathbb{P}(X_{t_1} \in A_1, \dots, X_{t_n} \in A_n) = \mathbb{P}(B_{t_1} \in A_1, \dots, B_{t_n} \in A_n | B_T = x).$$

(3) Let $(\alpha_n)_{n\geq 0}, (\beta_n)_{n\geq 1}$ be two independent sequences of i.i.d. Gaussian random variables with mean 0 and variance 1. By using the Fourier series decomposition of the process $B_t - tB_1$, show that the random series

$$X_t = t\alpha_0 + \sqrt{2}\sum_{n=1}^{+\infty}\left(\frac{\alpha_n}{2\pi n}(\cos(2\pi nt) - 1) + \frac{\beta_n}{2\pi n}\sin(2\pi nt)\right)$$

is a Brownian motion on [0, 1].

Exercise 2.16 (Ornstein–Uhlenbeck process). We write

$$X_t = e^{\frac{t}{2}} B_{1-e^{-t}}, \quad t \geq 0.$$

(1) Show that $(X_t)_{t\geq 0}$ is a Gaussian process. Compute its mean function and its covariance function.

(2) Show that the process

$$X_t - \frac{1}{2}\int_0^t X_u du$$

is a Brownian motion.

2.2 Basic properties

In this section we study some basic properties of the Brownian motion paths.

Proposition 2.17. *Let* $(B_t)_{t\geq 0}$ *be a standard Brownian motion. Then*

$$\mathbb{P}(\inf_{t\geq 0} B_t = -\infty, \sup_{t\geq 0} B_t = +\infty) = 1.$$

Proof. Since the process $(-B_t)_{t\geq 0}$ is also a Brownian motion, in order to prove that

$$\mathbb{P}(\inf_{t\geq 0} B_t = -\infty, \sup_{t\geq 0} B_t = +\infty) = 1,$$

we just have to check that

$$\mathbb{P}(\sup_{t\geq 0} B_t = +\infty) = 1.$$

Let $N \in \mathbb{N}$. From the scaling property of Brownian motion we have

$$\mathbb{P}(c \sup_{t\geq 0} B_t \leq N) = \mathbb{P}(\sup_{t\geq 0} B_t \leq N), \quad c > 0.$$

Therefore we have

$$\mathbb{P}(\sup_{t\geq 0} B_t \leq N) = \mathbb{P}(\sup_{t\geq 0} B_t = 0).$$

Now we may observe that

$$\mathbb{P}(\sup_{t\geq 0} B_t = 0) \leq \mathbb{P}(B_1 \leq 0, \sup_{t\geq 1} B_t \leq 0) = \mathbb{P}(B_1 \leq 0, \sup_{t\geq 0}(B_{t+1} - B_1) \leq -B_1).$$

Since the process $(B_{t+1} - B_1)_{t\geq 0}$ is a Brownian motion independent of B_1 (see Exercise 2.7), we have, as before, for $c > 0$,

$$\mathbb{P}(B_1 \leq 0, \sup_{t\geq 0}(B_{t+1} - B_1) \leq -B_1) = \mathbb{P}(B_1 \leq 0, c \sup_{t\geq 0}(B_{t+1} - B_1) \leq -B_1).$$

Therefore we get

$$\begin{aligned}
\mathbb{P}(B_1 \leq 0, \sup_{t\geq 0}(B_{t+1} - B_1) \leq -B_1) &= \mathbb{P}(B_1 \leq 0, \sup_{t\geq 0}(B_{t+1} - B_1) = 0) \\
&= \mathbb{P}(B_1 \leq 0)\mathbb{P}(\sup_{t\geq 0}(B_{t+1} - B_1) = 0) \\
&= \frac{1}{2}\mathbb{P}(\sup_{t\geq 0} B_t = 0).
\end{aligned}$$

Thus,

$$\mathbb{P}(\sup_{t\geq 0} B_t = 0) \leq \frac{1}{2}\mathbb{P}(\sup_{t\geq 0} B_t = 0),$$

and we can deduce that

$$\mathbb{P}(\sup_{t\geq 0} B_t = 0) = 0$$

and

$$\mathbb{P}(\sup_{t\geq 0} B_t \leq N) = 0.$$

Since this holds for every N, it implies that

$$\mathbb{P}(\sup_{t\geq 0} B_t = +\infty) = 1. \qquad \square$$

By using this proposition in combination with Exercise 2.7 we deduce the following proposition whose proof is left as an exercise to the reader.

Proposition 2.18 (Recurrence property of Brownian motion). *Let $(B_t)_{t\geq 0}$ be a Brownian motion. For every $t \geq 0$ and $x \in \mathbb{R}$,*

$$\mathbb{P}(\exists s \geq t, B_s = x) = 1.$$

Martingale theory provides powerful tools to study Brownian motion. We list in the proposition below some martingales naturally associated with the Brownian motion.

Proposition 2.19. *Let $(B_t)_{t\geq 0}$ be a standard Brownian motion. The following processes are martingales (with respect to their natural filtration):*

(1) $(B_t)_{t\geq 0}$;

(2) $(B_t^2 - t)_{t\geq 0}$;

(3) $(e^{\lambda B_t - \frac{\lambda^2}{2}t})_{t\geq 0}$, $\lambda \in \mathbb{C}$.

Proof. (1) First, we note that $\mathbb{E}(\mid B_t \mid) < +\infty$ for $t \geq 0$ because B_t is a Gaussian random variable. Now for $t \geq s$,

$$\mathbb{E}(B_t - B_s \mid \mathcal{F}_s) = \mathbb{E}(B_t - B_s) = 0,$$

therefore we get

$$\mathbb{E}(B_t \mid \mathcal{F}_s) = B_s.$$

(2) For $t \geq 0$, $\mathbb{E}(B_t^2) = t < +\infty$ and for $t \geq s$,

$$\mathbb{E}((B_t - B_s)^2 \mid \mathcal{F}_s) = \mathbb{E}((B_t - B_s)^2) = t - s,$$

therefore we obtain

$$\mathbb{E}(B_t^2 - t \mid \mathcal{F}_s) = B_s^2 - s.$$

(3) For $t \geq 0$, $\mathbb{E}(|e^{\lambda B_t - \frac{\lambda^2}{2}t}|) < +\infty$, because B_t is a Gaussian random variable. Then we have for $t \geq s$,

$$\mathbb{E}(e^{\lambda(B_t - B_s)} \mid \mathcal{F}_s) = \mathbb{E}(e^{\lambda(B_t - B_s)}) = e^{\frac{\lambda^2}{2}(t-s)},$$

and therefore

$$\mathbb{E}(e^{\lambda B_t - \frac{\lambda^2}{2}t} \mid \mathcal{F}_s) = e^{\lambda B_s - \frac{\lambda^2}{2}s}.$$

□

Remark 2.20. The curious reader may wonder how the previous martingales have been constructed. A first hint is that the functions $(t, x) \to x$, $(t, x) \to x^2 - t$, and $(t, x) \to \exp(\lambda x - \frac{\lambda^2}{2}t)$ have in common that they satisfy the following partial differential equation

$$\frac{\partial f}{\partial t} + \frac{1}{2}\frac{\partial^2 f}{\partial x^2} = 0.$$

The full explanation will be given in Theorem 3.9 in Chapter 3.

Exercise 2.21. Let $(B_t)_{t \geq 0}$ be a standard Brownian motion. Let $a < 0 < b$. We write

$$T = \inf\{t \geq 0, B_t \notin (a, b)\}.$$

(1) Show that T is a stopping time.

(2) Show that

$$\mathbb{P}(B_T = a) = \frac{b}{b - a}, \quad \mathbb{P}(B_T = b) = \frac{-a}{b - a}.$$

(3) Deduce another proof of the recurrence property of the standard Brownian motion, that is, for every $t \geq 0$ and $x \in \mathbb{R}$,

$$\mathbb{P}(\exists s \geq t, B_s = x) = 1.$$

The previous martingales may be used to explicitly compute the distribution of some functionals associated to the Brownian motion.

Proposition 2.22. *Let $(B_t)_{t \geq 0}$ be a standard Brownian motion. We write, for $a > 0$,*

$$T_a = \inf\{t \geq 0, \ B_t = a\}.$$

For every $\lambda > 0$, we have

$$\mathbb{E}(e^{-\lambda T_a}) = e^{-a\sqrt{2\lambda}}.$$

Therefore, the distribution of T_a is given by the density function

$$\mathbb{P}(T_a \in dt) = \frac{a}{(2\pi t)^{3/2}} e^{-\frac{a^2}{2t}} dt, \quad t > 0.$$

Proof. Let $\alpha > 0$. For $N \geq 1$, we denote by T_N the almost surely bounded stopping time:

$$T_N = T_a \wedge N.$$

Applying the Doob stopping theorem to the martingale $(e^{\alpha B_t - \frac{\alpha^2}{2}t})_{t \geq 0}$ yields

$$\mathbb{E}(e^{\alpha B_{T_a \wedge N} - \frac{\alpha^2}{2}(T_a \wedge N)}) = 1.$$

But for $N \geq 1$, we have

$$e^{\alpha B_{T_a \wedge N} - \frac{\alpha^2}{2}(T_a \wedge N)} \leq e^{\alpha a}.$$

Therefore from Lebesgue's dominated convergence theorem, $n \to +\infty$, we obtain

$$\mathbb{E}(e^{\alpha B_{T_a} - \frac{\alpha^2}{2}T_a}) = 1.$$

Because of continuity of the Brownian paths we have

$$B_{T_a} = a,$$

and we conclude that

$$\mathbb{E}(e^{-\frac{\alpha^2}{2}T_a}) = e^{-\alpha a}.$$

The formula for the density function of T_a is then obtained by inverting the previous Laplace transform. □

2.3 The law of iterated logarithm

We already observed that as a consequence of Kolmogorov's continuity theorem, the Brownian paths are γ-Hölder continuous for every $\gamma \in \left(0, \frac{1}{2}\right)$. The next proposition, which is known as the law of iterated logarithm shows in particular that Brownian paths are not $\frac{1}{2}$-Hölder continuous.

Theorem 2.23 (Law of iterated logarithm). *Let $(B_t)_{t \geq 0}$ be a Brownian motion. For $s \geq 0$,*

$$\mathbb{P}\left(\liminf_{t \to 0} \frac{B_{t+s} - B_s}{\sqrt{2t \ln \ln \frac{1}{t}}} = -1, \ \limsup_{t \to 0} \frac{B_{t+s} - B_s}{\sqrt{2t \ln \ln \frac{1}{t}}} = 1 \right) = 1.$$

Proof. Due to the symmetry and invariance by translation of the Brownian paths, it suffices to show that

$$\mathbb{P}\left(\limsup_{t \to 0} \frac{B_t}{\sqrt{2t \ln \ln \frac{1}{t}}} = 1 \right) = 1.$$

Let us first prove that

$$\mathbb{P}\Big(\limsup_{t\to 0}\frac{B_t}{\sqrt{2t\ln\ln\frac{1}{t}}}\le 1\Big)=1.$$

We write

$$h(t)=\sqrt{2t\ln\ln\frac{1}{t}}.$$

Let $\alpha,\beta>0$. From Doob's maximal inequality applied to the exponential martingale $(e^{\alpha B_t-\frac{\alpha^2}{2}t})_{t\ge 0}$, we have for $t\ge 0$:

$$\mathbb{P}(\sup_{0\le s\le t}(B_s-\frac{\alpha}{2}s)>\beta)=\mathbb{P}(\sup_{0\le s\le t}e^{\alpha B_s-\frac{\alpha^2}{2}s}>e^{\alpha\beta})\le e^{-\alpha\beta}.$$

Let now $\theta,\delta\in(0,1)$. Using the previous inequality for every $n\in\mathbb{N}$ with

$$t=\theta^n,\quad \alpha=\frac{(1+\delta)h(\theta^n)}{\theta^n},\quad \beta=\frac{1}{2}h(\theta^n),$$

yields when $n\to+\infty$,

$$\mathbb{P}\Big(\sup_{0\le s\le\theta^n}\Big(B_s-\frac{(1+\delta)h(\theta^n)}{2\theta^n}s\Big)>\frac{1}{2}h(\theta^n)\Big)=O\Big(\frac{1}{n^{1+\delta}}\Big).$$

Therefore from the Borel–Cantelli lemma, for almost every $\omega\in\Omega$, we may find $N(\omega)\in\mathbb{N}$ such that for $n\ge N(\omega)$,

$$\sup_{0\le s\le\theta^n}\Big(B_s(\omega)-\frac{(1+\delta)h(\theta^n)}{2\theta^n}s\Big)\le\frac{1}{2}h(\theta^n).$$

But

$$\sup_{0\le s\le\theta^n}\Big(B_s(\omega)-\frac{(1+\delta)h(\theta^n)}{2\theta^n}s\Big)\le\frac{1}{2}h(\theta^n)$$

implies that for $\theta^{n+1}<t\le\theta^n$,

$$B_t(\omega)\le\sup_{0\le s\le\theta^n}B_s(\omega)\le\frac{1}{2}(2+\delta)h(\theta^n)\le\frac{(2+\delta)h(t)}{2\sqrt{\theta}}.$$

We conclude that

$$\mathbb{P}\Big(\limsup_{t\to 0}\frac{B_t}{\sqrt{2t\ln\ln\frac{1}{t}}}\le\frac{2+\delta}{2\sqrt{\theta}}\Big)=1.$$

Letting now $\theta \to 1$ and $\delta \to 0$ yields

$$\mathbb{P}\Big(\limsup_{t\to 0} \frac{B_t}{\sqrt{2t \ln\ln \frac{1}{t}}} \leq 1\Big) = 1.$$

Let us now prove that

$$\mathbb{P}\Big(\limsup_{t\to 0} \frac{B_t}{\sqrt{2t \ln\ln \frac{1}{t}}} \geq 1\Big) = 1.$$

Let $\theta \in (0,1)$. For $n \in \mathbb{N}$, we write

$$A_n = \{\omega, B_{\theta^n}(\omega) - B_{\theta^{n+1}}(\omega) \geq (1-\sqrt{\theta})h(\theta^n)\}.$$

Let us prove that

$$\sum_{n\in\mathbb{N}} \mathbb{P}(A_n) = +\infty.$$

The basic inequality

$$\int_a^{+\infty} e^{-\frac{u^2}{2}}\,du \geq \frac{a}{1+a^2} e^{-\frac{a^2}{2}}, \quad a > 0,$$

which is obtained by integrating by parts the left-hand side of

$$\int_a^{+\infty} \frac{1}{u^2} e^{-\frac{u^2}{2}}\,du \leq \frac{1}{a^2}\int_a^{+\infty} e^{-\frac{u^2}{2}}\,du,$$

implies

$$\mathbb{P}(A_n) = \frac{1}{\sqrt{2\pi}}\int_{a_n}^{+\infty} e^{-\frac{u^2}{2}}\,du \geq \frac{a_n}{1+a_n^2} e^{-\frac{a_n^2}{2}},$$

with

$$a_n = \frac{(1-\sqrt{\theta})h(\theta^n)}{\theta^{n/2}\sqrt{1-\theta}}.$$

When $n \to +\infty$,

$$\frac{a_n}{1+a_n^2} e^{-\frac{a_n^2}{2}} = O\Big(\frac{1}{n^{\frac{1+\theta-2\sqrt{\theta}}{1-\theta}}}\Big),$$

therefore we proved

$$\sum \mathbb{P}(A_n) = +\infty.$$

As a consequence of the independence of the increments of the Brownian motion and of the Borel–Cantelli lemma, the event

$$B_{\theta^n} - B_{\theta^{n+1}} \geq (1-\sqrt{\theta})h(\theta^n)$$

will occur almost surely for infinitely many n's. But, thanks to the first part of the proof, for almost every ω, we may find $N(\omega)$ such that for $n \geq N(\omega)$,

$$B_{\theta^{n+1}} > -2h(\theta^{n+1}) \geq -2\sqrt{\theta}h(\theta^n).$$

Thus, almost surely, the event

$$B_{\theta^n} > h(\theta^n)(1 - 3\sqrt{\theta})$$

will occur for infinitely many n's. This implies

$$\mathbb{P}\Big(\limsup_{t \to 0} \frac{B_t}{\sqrt{2t \ln \ln \frac{1}{t}}} \geq 1 - 3\sqrt{\theta}\Big) = 1.$$

We finally get

$$\mathbb{P}\Big(\limsup_{t \to 0} \frac{B_t}{\sqrt{2t \ln \ln \frac{1}{t}}} \geq 1\Big) = 1$$

by letting $\theta \to 0$. □

As a straightforward consequence, we may observe that the time inversion invariance property of Brownian motion implies:

Corollary 2.24. *Let $(B_t)_{t \geq 0}$ be a standard Brownian motion. Then*

$$\mathbb{P}\Big(\liminf_{t \to +\infty} \frac{B_t}{\sqrt{2t \ln \ln t}} = -1, \limsup_{t \to +\infty} \frac{B_t}{\sqrt{2t \ln \ln t}} = 1\Big) = 1.$$

2.4 Symmetric random walks

In the previous section, the existence of the Brownian motion was proven as a consequence of the Daniell–Kolmogorov theorem. However, as it has been stressed, the proof of the Daniell–Kolmogorov theorem relies on the axiom of choice. As a consequence it does not provide any insight of how Brownian motion may explicitly be constructed or simulated by computers. In this section, we provide an *explicit construction* of Brownian motion as a limit of a sequence of suitably rescaled symmetric random walks.

Definition 2.25. A *random walk* on $\mathbb{Z}$ is a sequence of $\mathbb{Z}$-valued random variables $(S_n)_{\geq 0}$ that are defined on a probability space $(\Omega, \mathcal{F}, \mathbb{P})$ for which the following two properties hold:

(1) $(S_n)_{\geq 0}$ has stationary increments, that is, for $m, n \in \mathbb{N}$, $S_{m+n} - S_m$ has the same distribution as S_n.

(2) $(S_n)_{\geq 0}$ has independent increments, that is, for $m, n \in \mathbb{N}$, $S_{m+n} - S_m$ is independent from the σ-field $\sigma(S_0, \ldots, S_m)$.

For instance, let us consider a game in which we toss a fair coin. If this is heads, the player earns 1\$, if this is tails, the player loses 1\$. Then the algebraic wealth of the player after n tosses is a random walk on $\mathbb{Z}$.

More precisely, let us consider a sequence of independent random variables $(X_i)_{i \geq 1}$ that are defined on a probability space $(\Omega, \mathcal{F}, \mathbb{P})$ and that satisfy

$$\mathbb{P}(X_i = -1) = \mathbb{P}(X_i = 1) = \frac{1}{2}.$$

Consider now the sequence $(S_n)_{n \geq 0}$ such that $S_0 = 0$ and for $n \geq 1$,

$$S_n = \sum_{i=1}^{n} X_i.$$

For $n \geq 0$, we will denote by $\mathcal{F}_n$ the σ-field generated by the random variables $S_0, \ldots, S_n$.

Proposition 2.26. *The sequences $(S_n)_{n \geq 0}$ and $(-S_n)_{n \geq 0}$ are random walks on $\mathbb{Z}$.*

Proof. As a consequence of the independence of the X_i's, the following equality holds in distribution

$$(X_1, \ldots, X_n) = (X_{m+1}, \ldots, X_{m+n}).$$

Therefore, in distribution we have

$$S_{n+m} - S_m = \sum_{k=m+1}^{n+m} X_k = \sum_{k=1}^{n} X_k = S_n.$$

The independence of the increments is shown in the very same way and the proof that $(-S_n)_{n \geq 0}$ is also a random walk is identical. □

The random walk $(S_n)_{n \geq 0}$ is called the *symmetric random walk* on $\mathbb{Z}$ and is the one that shall be the most important for us. Its distribution may be computed by standard combinatorial arguments.

Proposition 2.27. *For $n \in \mathbb{N}$, $k \in \mathbb{Z}$:*

(1) *If the integers n and k have the same parity:*

$$\mathbb{P}(S_n = k) = \frac{1}{2^n} \binom{n}{\frac{n+k}{2}},$$

where $\binom{n}{p} = \frac{n!}{p!(n-p)!}$ is the binomial coefficient.

(2) *If the integers n and k do not have the same parity*

$$\mathbb{P}(S_n = k) = 0.$$

Proof. We observe that the random variable $\frac{n+S_n}{2}$ is the sum of n independent Bernoulli random variables, therefore the distribution of $\frac{n+S_n}{2}$ is binomial with parameters $(0, n)$. The result follows directly. □

Exercise 2.28. For $x \in \mathbb{R}$, by using Stirling's formula ($n! \sim e^{-n} n^n \sqrt{2\pi n}$) give an equivalent when $n \to \infty$ of $\mathbb{P}(S_{2n} = 2[x])$, where $[x]$ denotes the integer part of x.

Several interesting martingales are naturally associated to the symmetric random walk. They play the same role in the study of $(S_n)_{n \geq o}$ as the martingales in Theorem 2.19 for the Brownian motion.

Proposition 2.29. *The following sequences of random variables are martingales with respect to the filtration* $(\mathcal{F}_n)_{n\geq 0}$:

(1) $(S_n)_{n \geq 0}$,

(2) $(S_n^2 - n)_{n \geq 0}$,

(3) $\exp(-\lambda S_n - n \ln(\cosh(\lambda)))$, $n \geq 1, \lambda > 0$.

Proof. (1) For $n \geq m$,

$$\mathbb{E}(S_n \mid \mathcal{F}_m) = \mathbb{E}(S_n - S_m \mid \mathcal{F}_m) + \mathbb{E}(S_m \mid \mathcal{F}_m) = \mathbb{E}(S_n - S_m) + S_m = S_m.$$

(2) First, let us observe that

$$\begin{aligned}\mathbb{E}(S_n^2) &= \mathbb{E}\Big(\Big(\sum_{i=1}^{n} X_i\Big)^2\Big) \\ &= \mathbb{E}\Big(\sum_{i,j=1}^{n} X_i X_j\Big) \\ &= \sum_{i=1}^{n} \mathbb{E}(X_i^2) + \sum_{i,j=1,i\neq j}^{n} \mathbb{E}(X_i)\mathbb{E}(X_j) = n.\end{aligned}$$

Now, for $n \geq m$, we have

$$\mathbb{E}((S_n - S_m)^2 \mid \mathcal{F}_m) = \mathbb{E}((S_n - S_m)^2) = \mathbb{E}(S_{n-m}^2) = n - m.$$

But on the other hand we have

$$\begin{aligned}\mathbb{E}((S_n - S_m)^2 \mid \mathcal{F}_m) &= \mathbb{E}(S_n^2 \mid \mathcal{F}_m) - 2\mathbb{E}(S_n S_m \mid \mathcal{F}_m) + \mathbb{E}(S_m^2 \mid \mathcal{F}_m) \\ &= \mathbb{E}(S_n^2 \mid \mathcal{F}_m) - 2S_m^2 + S_m^2 \\ &= \mathbb{E}(S_n^2 \mid \mathcal{F}_m) - S_m^2.\end{aligned}$$

We therefore conclude that

$$\mathbb{E}(S_n^2 - n \mid \mathcal{F}_m) = S_m^2 - m.$$

(3 Similarly, we have for $n \geq m$,

$$\begin{aligned}\mathbb{E}(e^{-\lambda(S_n - S_m)} \mid \mathcal{F}_m) &= \mathbb{E}(e^{-\lambda(S_n - S_m)}) \\ &= \mathbb{E}(e^{-\lambda S_{n-m}}) \\ &= \mathbb{E}(e^{-\lambda X_1})^{n-m} \\ &= (\cosh \lambda)^{n-m},\end{aligned}$$

which yields the expected result. □

The following proposition shows that the symmetric random walk is a Markov process.

Proposition 2.30 (Markov property of the random walk). *For $m \leq n$, $k \in \mathbb{Z}$,*

$$\mathbb{P}(S_n = k \mid \mathcal{F}_m) = \mathbb{P}(S_n = k \mid S_m).$$

Proof. We have for $\lambda > 0$,

$$\mathbb{E}(e^{-\lambda S_n} \mid \mathcal{F}_m) = (\cosh \lambda)^{n-m} e^{-\lambda S_m}.$$

Thus, we have

$$\mathbb{E}(e^{-\lambda S_n} \mid \mathcal{F}_m) = \mathbb{E}(e^{-\lambda S_n} \mid S_m).$$

Since the conditional distribution of S_n given $\mathcal{F}_m$ is completely characterized by its conditional Laplace transform, we obtain the claimed result. □

A random variable T, valued in $\mathbb{N} \cup \{+\infty\}$, is called a stopping time for the random walk $(S_n)_{n \geq 0}$ if for any m the event $\{T \leq m\}$ belongs to the σ-field $\mathcal{F}_m$. The set

$$\mathcal{F}_T = \{A \in \mathcal{F},\ \forall\, m \in \mathbb{N}, A \cap \{T \leq m\} \in \mathcal{F}_m\}$$

is then a sub σ-field of $\mathcal{F}$.

Proposition 2.31. *Let T be a stopping time for the random walk $(S_n)_{n \geq 0}$ such that $\mathbb{P}(T < +\infty) = 1$. The sequence $(S_{n+T} - S_T)_{n \geq 0}$ is a symmetric random walk on $\mathbb{Z}$ that is independent from $\mathcal{F}_T$.*

Proof. Let us write

$$\widetilde{S}_n = S_{n+T} - S_T$$

and consider the stopping time

$$T_m = T + m.$$

From the Doob stopping theorem applied to the martingale $((\cos\lambda)^{-n}e^{i\lambda S_n})_{n\geq 0}$ and the bounded stopping time $T_m \wedge N$ where $N \in \mathbb{N}$, we get, for all $n \in \mathbb{N}$,

$$\mathbb{E}((\cos\lambda)^{-n}e^{i\lambda(S_{n+T_m\wedge N}-S_{T_m\wedge N})} \mid \mathcal{F}_{T_m\wedge N}) = 1.$$

Letting $N \to +\infty$ yields $\forall\, n \in \mathbb{N}$

$$\mathbb{E}((\cos\lambda)^{-n}e^{i\lambda(\widetilde{S}_{n+m}-\widetilde{S}_m)} \mid \mathcal{F}_{T+m}) = 1.$$

This implies that the increments of $\widetilde{S}_n$ are stationary and independent. Hence $(\widetilde{S}_n)_{n\geq 0}$ is a random walk on $\mathbb{Z}$ which is independent from $\mathcal{F}_T$.

Let us finally prove that this random walk is symmetric.The random variable $\widetilde{S}_{n+1} - \widetilde{S}_n$ is valued in the set $\{-1, 1\}$ and satisfies,

$$\mathbb{E}(e^{i\lambda(\widetilde{S}_{n+1}-\widetilde{S}_n)}) = \cos\lambda.$$

This implies

$$\mathbb{P}(\widetilde{S}_{n+1} - \widetilde{S}_n = -1) = \mathbb{P}(\widetilde{S}_{n+1} - \widetilde{S}_n = 1) = \frac{1}{2}. \qquad \square$$

As a corollary of the previous result, we obtain the so-called strong Markov property that reinforces the Markov property.

Corollary 2.32 (Strong Markov property). *Let T be a stopping time for the random walk $(S_n)_{n\geq 0}$ such that $\mathbb{P}(T < +\infty) = 1$. For every $k \in \mathbb{Z}$,*

$$\mathbb{P}(S_{T+1} = k \mid \mathcal{F}_T) = \mathbb{P}(S_{T+1} = k \mid S_T).$$

Proof. Since the sequence of random variables $(S_{n+T} - S_T)_{n\geq 0}$ is a symmetric random walk on $\mathbb{Z}$ that is independent from $\mathcal{F}_T$, for $\lambda > 0$,

$$\mathbb{E}(e^{-\lambda S_{T+1}} \mid \mathcal{F}_T) = (\cosh\lambda)e^{-\lambda S_T}.$$

Therefore

$$\mathbb{E}(e^{-\lambda S_{T+1}} \mid \mathcal{F}_T) = \mathbb{E}(e^{-\lambda S_{T+1}} \mid S_T),$$

and we conclude as in the proof of Proposition 2.30. $\square$

The next proposition shows that with probability 1 the symmetric random walk visits each integer of $\mathbb{Z}$. This property is called the recurrence property of the symmetric random walk. A similar result for the Brownian motion was proved in Proposition 2.18.

Proposition 2.33 (Recurrence property of the symmetric random walk).

$$\forall\, k \in \mathbb{Z},\ \forall\, m \in \mathbb{N},\quad \mathbb{P}(\exists\, n \geq m,\ S_n = k) = 1.$$

Proof. We do not give the simplest or the most direct proof (see for instance Exercise 2.34 for a simpler argument and Exercise 2.35 for still another proof), but the following proof has the advantage to provide several useful further information on the random walk.

We first show that

$$\mathbb{P}(\forall\, k \in [1, 2n],\ S_k \neq 0) = \mathbb{P}(S_{2n} = 0).$$

We have

$$\begin{aligned}
&\mathbb{P}(\forall\, k \in [1, 2n],\ S_k \neq 0) \\
&\quad = 2\mathbb{P}(\forall\, k \in [1, 2n],\ S_k > 0) \\
&\quad = 2\sum_{j=1}^{n} \mathbb{P}(\forall\, k \in [1, 2n],\ S_k > 0 \mid S_{2n} = 2j)\mathbb{P}(S_{2n} = 2j) \\
&\quad = 2\sum_{j=1}^{n} \frac{j}{n}\mathbb{P}(S_{2n} = 2j) \\
&\quad = 2\sum_{j=1}^{n} \frac{j}{n}\frac{1}{2^{2n}}\binom{2n}{n+j}.
\end{aligned}$$

Now we have the following binomial identity which may, for instance, be proved inductively:

$$\sum_{j=1}^{n} j\binom{2n}{n+j} = \frac{n}{2}\binom{2n}{n}.$$

This yields

$$\mathbb{P}(\forall\, k \in [1, 2n],\ S_k \neq 0) = \mathbb{P}(S_{2n} = 0).$$

Since

$$\mathbb{P}(S_{2n} = 0) = \frac{(2n)!}{(n!)^2}\frac{1}{2^{2n}} \xrightarrow[n \to +\infty]{} 0,$$

we deduce that

$$\mathbb{P}(\exists\, n > 1,\ S_n = 0) = 1.$$

The random variable

$$Z_1 = \inf\{n > 1,\ S_n = 0\}$$

is therefore finite. Let us write

$$T_1 = \inf\{n \geq 0,\ S_n = 1\}$$

and

$$T_{-1} = \inf\{n \geq 0,\ S_n = -1\}.$$

It is clear that T_1 and T_{-1} are stopping times for the random walk $(S_n)_{n\geq 0}$. Our goal will be to prove that

$$\mathbb{P}(T_1 < +\infty) = 1.$$

Thanks to the Markov property, for $n \geq 1$, we obtain

$$\begin{aligned}
&\mathbb{P}(Z_1 \geq n) \\
&\quad = \mathbb{P}(Z_1 \geq n \mid S_1 = 1)\mathbb{P}(S_1 = 1) + \mathbb{P}(Z_1 \geq n \mid S_1 = -1)\mathbb{P}(S_1 = -1) \\
&\quad = \frac{1}{2}\mathbb{P}(Z_1 \geq n \mid S_1 = 1) + \frac{1}{2}\mathbb{P}(Z_1 \geq n \mid S_1 = -1) \\
&\quad = \frac{1}{2}\mathbb{P}(T_1 \geq n-1) + \frac{1}{2}\mathbb{P}(T_{-1} \geq n-1).
\end{aligned}$$

Now, since we have the equality in distribution

$$(S_n)_{n\geq 0} = (-S_n)_{n\geq 0},$$

it is clear that $\mathbb{P}(T_1 < +\infty) = 1$ if and only if $\mathbb{P}(T_{-1} < +\infty) = 1$, in which case the following equality in distribution holds:

$$T_1 = T_{-1}.$$

Thus, from

$$\mathbb{P}(Z_1 < +\infty) = 1$$

we deduce that

$$\mathbb{P}(T_1 < +\infty) = 1.$$

Now we may apply the previous result to the random walk $(S_{n+T_1} - 1)_{n\geq 0}$. This gives

$$\mathbb{P}(\exists\, n > 1,\ S_n = 2) = 1.$$

By a straightforward induction, we obtain

$$\forall k \in \mathbb{N},\ \ \mathbb{P}(\exists\, n > 1,\ S_n = k) = 1.$$

Then by symmetry we get

$$\forall k \in \mathbb{N},\ \ \mathbb{P}(\exists\, n > 1,\ S_n = -k) = 1.$$

By using this with the random walk $(S_{n+m} - S_m)_{n\geq 0}$, we finally get the expected

$$\forall\, k \in \mathbb{Z},\ \forall\, m \in \mathbb{N},\ \ \mathbb{P}(\exists\, n \geq m,\ S_n = k) = 1. \qquad \square$$

Exercise 2.34. By using a method similar to the method of Exercise 2.21, give another proof of the recurrence property of the symmetric random walk.

Exercise 2.35 (Law of iterated logarithm for the random walk). (1) Let $(B_t)_{t\geq 0}$ be a one-dimensional Brownian motion. Consider the sequence of stopping times inductively defined by $T_0 = 0$ and

$$T_{n+1} = \inf\{t \geq T_n, |B_t - B_{T_n}| = 1\}.$$

Consider then the process $(S_n)_{n\geq 0}$ defined by $S_n = B_{T_n}$. Show that $(S_n)_{n\geq 0}$ is a symmetric random walk.

(2) Deduce that if $(S_n)_{n\geq 0}$ is a symmetric random walk,

$$\mathbb{P}\Big(\liminf_{n\to+\infty} \frac{S_n}{\sqrt{2n \ln \ln n}} = -1, \limsup_{n\to+\infty} \frac{S_n}{\sqrt{2n \ln \ln n}} = 1 \Big) = 1.$$

Let us consider for $k \in \mathbb{Z}$, the stopping hitting time

$$T_k = \inf\{n \geq 1, \ S_n = k\}.$$

A straightforward consequence of the previous proposition is that

$$\mathbb{P}(T_k < +\infty) = 1.$$

This sequence of hitting times $(T_k)_{k\geq 0}$ enjoys several nice properties, the first being that it is itself a random walk.

Proposition 2.36. *The sequence $(T_k)_{k\geq 0}$ is a random walk on $\mathbb{Z}$.*

Proof. For $0 \leq a < b$, we have

$$T_b = T_a + \inf\{n \geq 1, S_{n+T_a} = b\}.$$

Now, since $(S_{n+T_a} - a)_{n\geq 0}$ is a random walk on $\mathbb{Z}$ which is independent from T_a,we deduce that

(1) $\inf\{n \geq 1, S_{n+T_a} = b\}$ is independent from $\mathcal{F}_{T_a}$,

(2) the following equality holds in distribution:

$$\inf\{n \geq 1, S_{n+T_a} = b\} = T_{b-a}. \qquad \square$$

The distribution of T_k may explicitly be computed by using the method of generating series.

Proposition 2.37. *For $0 < x < 1$, and $k \in \mathbb{Z}$,*

$$\sum_{n=0} x^n \, \mathbb{P}(T_k = n) = \Big(\frac{1 - \sqrt{1 - x^2}}{x} \Big)^k.$$

Proof. Let $\lambda > 0$. Applying the Doob stopping theorem to the martingale

$$((\cosh \lambda)^{-n} e^{-\lambda S_n})_{n \geq 0}$$

with the stopping time $T_k \wedge N$, $N \in \mathbb{N}$, yields

$$\mathbb{E}(e^{-\lambda S_{N \wedge T_k} - (T_k \wedge N) \ln \cosh \lambda}) = 1.$$

The dominated convergence theorem implies then:

$$\mathbb{E}(e^{-\lambda k - T_k \ln \cosh \lambda}) = 1.$$

This yields

$$\mathbb{E}(e^{-(\ln \cosh \lambda) T_k}) = e^{\lambda k}.$$

By denoting

$$x = e^{-\ln \cosh \lambda}$$

we therefore get

$$\mathbb{E}(x^{T_k}) = \left(\frac{1 - \sqrt{1 - x^2}}{x}\right)^k,$$

which is the result we want to prove. □

From the previous generating series, we deduce the following explicit distribution:

Corollary 2.38. *For $k \in \mathbb{N}$, $n \in \mathbb{N}$,*

$$\mathbb{P}(T_k = n) = \frac{1}{2^n} \sum_I \frac{1}{2i_1 - 1} \cdots \frac{1}{2i_k - 1} \binom{2i_1 - 1}{i_1} \binom{2i_k - 1}{i_k},$$

where

$$I = \left\{(i_1, \ldots, i_k) \in \mathbb{N}^k,\ i_1 + \cdots + i_k = \frac{n-k}{2}\right\}.$$

In particular, for $n \in \mathbb{N}$,

$$\mathbb{P}(T_1 = 2n) = 0,$$

$$\mathbb{P}(T_1 = 2n - 1) = \frac{1}{2n-1} \binom{2n-1}{n} \frac{1}{2^{2n-1}}.$$

Proof. The proof reduces to the computation of the coefficients in the Taylor expansion of the function

$$x \to \left(\frac{1 - \sqrt{1 - x^2}}{x}\right)^k.$$

□

Exercise 2.39. Show that

$$\mathbb{P}(T_1 = 2n - 1) \sim_{n \to +\infty} \frac{C}{n^{3/2}},$$

where $C > 0$ is a constant to be computed.

Exercise 2.40. Let

$$Z_0 = 0, Z_{k+1} = \inf\{n > Z_k,\ S_n = 0\}.$$

Show that:

(1) The sequence $(Z_k)_{k \geq 0}$ is a random walk on $\mathbb{Z}$.

(2) For $k \geq 1$, the following equality in distribution holds:

$$Z_k = k + T_k.$$

The next proposition is known as the reflection principle for the random walk.

Proposition 2.41 (Reflection principle). *For $n \geq 0$, let*

$$M_n = \max_{0 \leq k \leq n} S_k.$$

We have for $k, n \in \mathbb{N}$,

$$\mathbb{P}(M_n \geq k,\ S_n < k) = \frac{1}{2}(\mathbb{P}(T_k \leq n) - \mathbb{P}(S_n = k)).$$

Proof. Let $k, n \in \mathbb{N}$. We first observe the following equality between events,

$$\{T_k \leq n\} = \{M_n \geq k\}.$$

Let us now consider the random walk $(\widetilde{S}_m)_{m \geq 0}$ defined by

$$\widetilde{S}_m = S_{m+T_k} - k.$$

As we have already seen it, this sequence is a symmetric random walk for which $n - T_k \wedge n$ is a stopping time. Now the following equality holds in distribution:

$$\widetilde{S}_{n-T_k \wedge n} = -\widetilde{S}_{n-T_k \wedge n}.$$

Therefore

$$\mathbb{P}(T_k \leq n, \widetilde{S}_{n-T_k} < 0) = \mathbb{P}(T_k \leq n, \widetilde{S}_{n-T_k} > 0),$$

which may be rewritten

$$\mathbb{P}(T_k \leq n, S_n < k) = \mathbb{P}(T_k \leq n, S_n > k).$$

We finally conclude that

$$2\mathbb{P}(T_k \leq n, S_n < k) + \mathbb{P}(S_n = k) = \mathbb{P}(T_k \leq n),$$

which is our claim. □

As a corollary of the reflection principle, we immediately obtain the following expression for the distribution of M_n.

Corollary 2.42. *For $k, n \in \mathbb{N}$,*

$$\mathbb{P}(M_n \geq k) = 2\mathbb{P}(S_n \geq k) - \mathbb{P}(S_n = k).$$

Exercise 2.43 (Local time of the symmetric random walk). For $n \in \mathbb{N}$ we write

$$l_n^0 = \mathrm{Card}\{0 \leq i \leq n,\ S_i = 0\}.$$

(1) Show that for $n \in \mathbb{N}$ and $k \in \mathbb{N}^*$

$$\mathbb{P}(l_{2n}^0 = k) = \mathbb{P}(l_{2n+1}^0 = k) = \mathbb{P}(S_{2n+1-k} = k - 1)$$

(2) Deduce that when $n \to +\infty$ the following convergence takes place in distribution

$$\frac{l_n^0}{\sqrt{n}} \to |\mathcal{N}(0, 1)|.$$

2.5 Donsker theorem

After this study of the symmetric random walk on $\mathbb{Z}$, we now turn to the Donsker invariance principle which says that the Brownian motion may constructed as a limit of conveniently rescaled random walks. Before stating this theorem, we need some general results about weak convergence of probability measures in infinite-dimensional spaces. Let us first recall the definition of convergence in distribution for a sequence of random variables that are valued in a Polish space[1]:

Definition 2.44. A sequence $(X_n)_{n \in \mathbb{N}}$ of random variables, valued in a Polish space E, is said to *converge in distribution* toward a random variable X if for every continuous and bounded function $f : E \to \mathbb{R}$ we have

$$\mathbb{E}(f(X_n)) \xrightarrow[n \to +\infty]{} \mathbb{E}(f(X)).$$

Equivalently, the sequence of the distributions of the X_n's is said to weakly converge to the distribution of X.

You may observe that the random variables X_n do not need to be defined on the same probability space. A usual strategy to show that a sequence $(X_n)_{n \in \mathbb{N}}$ converges in distribution is to prove the following:

(1) The family $(X_n)_{n \in \mathbb{N}}$ is relatively compact in the weak convergence topology.

[1] A Polish space is a complete and separable metric space.

(2) The sequence $(X_n)_{n\in\mathbb{N}}$ has a unique cluster point.

Since a stochastic process is nothing else but a random variable that is valued in $\mathcal{C}(\mathbb{R}_{\geq 0}, \mathbb{R})$, we have a notion of convergence in distribution for a sequence of stochastic processes and, due to the fact that the distribution of stochastic process is fully described by its finite-dimensional distributions, we can prove the following useful fact:

Proposition 2.45. *Let $(X^n)_{n\in\mathbb{N}}$ be a sequence of continuous processes and let X be a continuous process. Let us assume that*

(1) *the sequence of the distributions of the X^n's is relatively compact in the weak convergence topology; and*

(2) *for every $t_1, \dots, t_k \in \mathbb{R}^k$, the following convergence holds in distribution*

$$(X^n_{t_1}, \dots, X^n_{t_k}) \xrightarrow[n\to+\infty]{} (X_{t_1}, \dots, X_{t_k}).$$

Then the sequence $(X^n)_{n\in\mathbb{N}}$ converges in distribution toward X.

In order to efficiently use the previous proposition, we need to characterize the relatively compact sequences in the weak convergence topology. A first ingredient is the Prokhorov theorem .

Theorem 2.46 (Prokhorov theorem). *Let $\mathcal{P}$ be a family of probability measures on a Polish space E endowed with its Borel σ-field $\mathcal{E}$. Then $\mathcal{P}$ is a relatively compact set for the weak convergence of probability measures topology if and only if the family $\mathcal{P}$ is tight, that is, for every $\varepsilon \in (0,1)$, we can find a compact set $K_\varepsilon \subset E$ such that for every $\mathbb{P} \in \mathcal{P}$,*

$$\mathbb{P}(K_\varepsilon) \geq 1 - \varepsilon.$$

The second ingredient is Ascoli's theorem that describes the relatively compact sets in $\mathcal{C}(\mathbb{R}_{\geq 0}, \mathbb{R})$ for the topology of uniform convergence on compact sets.

Theorem 2.47 (Ascoli theorem). *For $N \in \mathbb{N}$, $f \in \mathcal{C}(\mathbb{R}_{\geq 0}, \mathbb{R})$ and $\delta > 0$, we write*

$$V^N(f, \delta) = \sup\{|f(t) - f(s)|, |t-s| \leq \delta,\ s, t \leq N\}.$$

A set $K \subset \mathcal{C}(\mathbb{R}_{\geq 0}, \mathbb{R})$ is relatively compact if and only if the following holds:

(1) *The set $\{f(0), f \in K\}$ is bounded.*

(2) *For every $N \in \mathbb{N}$,*

$$\lim_{\delta\to 0} \sup_{f\in K} V^N(f, \delta) = 0.$$

As usual, we write $\mathcal{C}(\mathbb{R}_{\geq 0}, \mathbb{R})$ for the space of continuous functions and the associated Borel σ-field is denoted by $\mathcal{B}(\mathbb{R}_{\geq 0}, \mathbb{R})$. Also, $(\pi_t)_{t\geq 0}$ denotes the coordinate process. Combining Prokhorov and Ascoli's theorems we obtain:

Proposition 2.48. *On the space $\mathcal{C}(\mathbb{R}_{\geq 0}, \mathbb{R})$, a sequence of probability measures $(\mathbb{P}_n)_{n\in\mathbb{N}}$ is relatively compact in the weak convergence topology if and only if the following holds:*

(1) *For every $\varepsilon > 0$, there exist $A > 0$ and $n_0 \in \mathbb{N}$ such that for every $n \geq n_0$,*

$$\mathbb{P}_n(|\pi_0| > A) \leq \varepsilon.$$

(2) *For every $\eta, \varepsilon > 0$ and $N \in \mathbb{N}$, we may find $\delta > 0$ and $n_0 \in \mathbb{N}$ such that for $n \geq n_0$,*

$$\mathbb{P}_n(V^n(\pi, \delta) > \eta) \leq \varepsilon.$$

Proof. Assume that the sequence $(\mathbb{P}_n)_{n\in\mathbb{N}}$ is relatively compact in the topology of weak convergence. From Prokhorov's theorem, this sequence is tight, that is, for $\varepsilon \in (0, 1)$, we can find a compact set $K_\varepsilon \subset \mathcal{C}(\mathbb{R}_+, \mathbb{R})$ such that for $n \in \mathbb{N}$:

$$\mathbb{P}_n(K_\varepsilon) \geq 1 - \varepsilon.$$

By writing K_ε in the form given by Ascoli's theorem, it is easily checked that properties (1) and (2) are satisfied with $n_0 = 0$.

Let us now assume that (1) and (2) are satisfied. First, as a finite sequence is relatively compact, we may assume that the properties (1) and (2) are fulfilled with $n_0 = 0$. Also, thanks to Prokhorov's theorem, we only need to show that the sequence is tight. Let $\varepsilon > 0$ and $N \in \mathbb{N}$. For every $k \geq 1$, we can find $A_N > 0$ and $\delta_{N,k}$ such that

$$\sup_{n\in\mathbb{N}} \mathbb{P}_n(|\pi_0| > A_N) \leq \frac{\varepsilon}{2^{N+1}},$$

$$\sup_{n\in\mathbb{N}} \mathbb{P}_n(V^N(\pi, \delta_{N,k}) > \tfrac{1}{k}) \leq \frac{\varepsilon}{2^{N+k+1}}.$$

We consider then

$$K_\varepsilon = \bigcap_{N\in\mathbb{N}} \{f \in \mathcal{C}(\mathbb{R}_+, \mathbb{R}),\ |f(0)| \leq A_N,\ V^N(\pi, \delta_{N,k}) \leq \tfrac{1}{k}, \forall k \geq 1\}.$$

Ascoli's theorem implies that K_ε is relatively compact, and it is easy to see that for $n \geq 0$,

$$\mathbb{P}_n(K_\varepsilon) \geq 1 - \varepsilon.$$

□

Exercise 2.49. Let $(X^n)_{n\in\mathbb{N}}$ be a sequence of continuous stochastic processes such that the following holds:

(1) The family formed by the distributions of the X_0^n's, $n \in \mathbb{N}$ is tight.
(2) There exist $\alpha, \beta, \gamma > 0$ such that for $s, t \geq 0$ and $n \geq 0$,

$$\mathbb{E}(|X_t^n - X_s^n|^\alpha) \leq \beta |t - s|^{1+\gamma}.$$

Show that the family of the distributions of the X^n's $n \in \mathbb{N}$ is relatively compact in the topology of the weak convergence.

We are now in a position to prove the Donsker approximation theorem, that constructs Brownian motion as a limit of symmetric random walks on $\mathbb{Z}$.

Let $(S_n)_{n\in\mathbb{N}}$ be a symmetric random walk on $\mathbb{Z}$. We define the sequence $(S_t^n)_{t\in[0,1]}$, $n \in \mathbb{N}$, as follows:

$$S_t^n = \sqrt{n}\Big(\Big(t - \frac{k}{n}\Big)S_{k+1} + \Big(\frac{k+1}{n} - t\Big)S_k\Big), \quad \frac{k}{n} \le t \le \frac{k+1}{n}.$$

The process $(S_t^n)_{t\ge 0}$ is therefore the piecewise affine continuous interpolation of the rescaled discrete sequence $(S_n)_{n\ge 0}$.

Theorem 2.50 (Donsker theorem). *The sequence of processes $(S_t^n)_{t\in[0,1]}$, $n \in \mathbb{N}$, converges in distribution to a standard Brownian motion $(B_t)_{t\in[0,1]}$.*

Proof. We need to check two things:

(1) For every $t_1, \dots, t_k \in [0,1]$, the following convergence in distribution takes place
$$(S_{t_1}^n, \dots, S_{t_k}^n) \xrightarrow[n\to+\infty]{} (B_{t_1}, \dots, B_{t_k}).$$

(2) The family of the distributions of the $(S_t^n)_{t\in[0,1]}$, $n \in \mathbb{N}$, is relatively compact in the weak convergence topology.

The first point (1) is left as exercise to the reader: It is basically a consequence of the multi-dimensional central limit theorem. Let us however point out that in order to simplify the computations, it will be easier to prove that for every $t_1 \le \dots \le t_k \in [0,1]$, the following convergence in distribution takes place:

$$(S_{t_1}^n, S_{t_2}^n - S_{t_1}^n, \dots, S_{t_k}^n - S_{t_{k-1}}^n) \xrightarrow[n\to+\infty]{} (B_{t_1}, B_{t_2} - B_{t_1} \dots, B_{t_k} - B_{t_{k-1}}).$$

So we turn to the second point (2). Let $\lambda > 0$. The process $(S_n^4)_{n\in\mathbb{N}}$ is a submartingale, therefore from Doob's maximal inequality, for $n \ge 1$,

$$\mathbb{P}(\max_{k\le n} |S_k| > \lambda\sqrt{n}) \le \frac{\mathbb{E}(S_n^4)}{\lambda^4 n^2} = \frac{3n^2 - 2n}{\lambda^4 n^2} \le \frac{3}{\lambda^4}.$$

Thanks to the stationarity of the increments of $(S_n)_{n\in\mathbb{N}}$, we deduce that for $k, n \ge 1$, $\lambda > 0$,

$$\mathbb{P}(\max_{i\le n} |S_{i+k} - S_k| > \lambda\sqrt{n}) \le \frac{3}{\lambda^4}.$$

Let now $\varepsilon, \eta \in (0,1)$. From the previous inequality we can find $\delta > 0$ such that for every $k, n \ge 1$

$$\mathbb{P}(\max_{i\le[n\delta]} |S_{i+k} - S_k| \ge \varepsilon\sqrt{n}) \le \eta\delta.$$

Let $N, \varepsilon, \eta > 0$. From the definition of S^n we deduce that we can find $\delta \in (0,1)$ such that for every $n \geq 1$ and $t \leq N$,

$$\mathbb{P}\big(\sup_{t \leq s \leq t+\delta} |S^n_s - S^n_t| \geq \eta\big) \leq \varepsilon\delta.$$

For $0 \leq i < \frac{N}{\delta}$ et $n \geq 1$, let

$$A^n_i = \{\omega \in \Omega, \sup_{i\delta \leq s \leq (i+1)\delta \wedge N} | S^n_{i\delta} - S^n_s | \geq \eta\}.$$

We may check that

$$\bigcap_i {}^c A^n_i \subset \{\sup\{|S^n_t - S^n_s|, |t-s| \leq \delta, \ s,t \leq N\} < 3\eta\}.$$

Therefore, for every $n \geq 1$,

$$\begin{aligned}\mathbb{P}(\sup\{|S^n_t - S^n_s|, |t-s| \leq \delta, \ s,t \leq N\} \geq 3\eta) &\leq \mathbb{P}\big(\bigcup_i A^n_i\big) \\ &\leq (1 + [N\delta^{-1}])\delta\varepsilon \\ &< (N+1)\varepsilon.\end{aligned}$$

This implies the expected relative compactness property thanks to Proposition 2.48. □

Besides its importance to simulate Brownian motion on a computer for instance, it is interesting that Donsker's theorem may also be used to explicitly compute some distributions of Brownian functionals.

Theorem 2.51 (Arcsine law). *Let $(B_t)_{t \geq 0}$ be a standard Brownian motion. For $t \geq 0$, we write*

$$A_t = \int_0^t 1_{[0,+\infty)}(B_s)ds.$$

We have for $x \leq t$,

$$\mathbb{P}(A_t \leq x) = \frac{2}{\pi} \arcsin \sqrt{\frac{x}{t}}.$$

Proof. Let $(S_n)_{n \in \mathbb{N}}$ be a symmetric random walk on $\mathbb{Z}$. We denote by T_1 the hitting time of 1 by $(S_n)_{n \in \mathbb{N}}$, and

$$Z_1 = \inf\{n \geq 1, S_n = 0\}.$$

Exercise 2.40 proved that the following equality in distribution takes place:

$$Z_1 = T_1 + 1.$$

Let us write

$$A_n = \text{Card}\{1 \le i \le n, \max(S_{i-1}, S_i) > 0\}.$$

We will prove by induction that for $0 \le k \le n$,

$$\mathbb{P}(A_{2n} = 2k) = \mathbb{P}(S_{2k} = 0)\mathbb{P}(S_{2n-2k} = 0).$$

First,

$$\mathbb{P}(A_{2n} = 0) = \mathbb{P}(T_1 \ge 2n+1) = \mathbb{P}(Z_1 \ge 2n+2) = \mathbb{P}(S_1 \ne 0, \dots, S_{2n} \ne 0).$$

As seen before,

$$\mathbb{P}(S_1 \ne 0, \dots, S_{2n} \ne 0) = \mathbb{P}(S_{2n} = 0).$$

Thus,

$$\mathbb{P}(A_{2n} = 0) = \mathbb{P}(S_{2n} = 0),$$

and in the same way

$$\mathbb{P}(A_{2n} = 2n) = \mathbb{P}(S_{2n} = 0).$$

For $1 \le k \le n-1$,

$$\{A_{2n} = 2k\} \subset \{Z_1 \le 2k\},$$

thus

$$\begin{aligned}
\mathbb{P}(A_{2n} = 2k) &= \sum_{i=1}^{k} \mathbb{P}(A_{2n} = 2k \mid Z_1 = 2i)\mathbb{P}(Z_1 = 2i) \\
&= \frac{1}{2}\sum_{i=1}^{k} \mathbb{P}(Z_1 = 2i)(\mathbb{P}(A_{2n-2i} = 2k) + \mathbb{P}(A_{2n-2i} = 2k-2i)).
\end{aligned}$$

Thanks to the induction assumption at the step n, we have

$$\begin{aligned}
\mathbb{P}(A_{2n} = 2k) = \frac{1}{2}\sum_{i=1}^{k} \mathbb{P}(Z_1 = 2i)(&\mathbb{P}(S_{2k} = 0)\mathbb{P}(S_{2n-2k-2i} = 0) \\
&+ \mathbb{P}(S_{2k-2i} = 0)\mathbb{P}(S_{2n-2k} = 0)).
\end{aligned}$$

The Markov property then implies

$$\begin{aligned}
\sum_{i=1}^{k} \mathbb{P}(Z_1 = 2i)\mathbb{P}(S_{2k-2i} = 0) &= \sum_{i=1}^{k} \mathbb{P}(Z_1 = 2i)\mathbb{P}(S_{2k} = 0 \mid Z_1 = 2i) \\
&= \mathbb{P}(S_{2k} = 0).
\end{aligned}$$

In the same way, we can prove

$$\sum_{i=1}^{k} \mathbb{P}(Z_1 = 2i)\mathbb{P}(S_{2n-2k-2i} = 0) = \mathbb{P}(S_{2n-2k} = 0).$$

Thus

$$\mathbb{P}(A_{2n} = 2k) = \mathbb{P}(S_{2k} = 0)\mathbb{P}(S_{2n-2k} = 0)$$

which completes the induction. As a conclusion, we have for $0 \le k \le n$,

$$\mathbb{P}(A_{2n} = 2k) = \mathbb{P}(S_{2k} = 0)\mathbb{P}(S_{2n-2k} = 0) = \frac{1}{2^n}\frac{(2n-2k)!}{(n-k)!^2}\frac{(2k)!}{(k)!^2}.$$

The Stirling formula

$$n! \sim_{n\to\infty} n^n e^{-n}\sqrt{2\pi n},$$

implies that for $x \in [0,1]$,

$$\mathbb{P}\Big(\frac{A_{2n}}{2n} \le x\Big) \sim_{n\to+\infty} \frac{1}{\pi}\sum_{k=0}^{[nx]} \frac{1}{\sqrt{k(n-k)}}.$$

Since

$$\int_0^x \frac{du}{\pi\sqrt{u(1-u)}} = \frac{2}{\pi}\arcsin\sqrt{x}$$

we deduce that for $x \in [0,1]$,

$$\lim_{n\to+\infty} \mathbb{P}\Big(\frac{A_{2n}}{2n} \le x\Big) = \frac{2}{\pi}\arcsin\sqrt{x}.$$

It is now time to use Donsker's theorem. We consider the sequence

$$S_t^n = \sqrt{n}\Big(\Big(t - \frac{k}{n}\Big)S_{k+1} + \Big(\frac{k+1}{n} - t\Big)S_k\Big), \quad \frac{k}{n} \le t \le \frac{k+1}{n}.$$

It is easy to see that

$$\int_0^1 1_{[0,+\infty)}(S_t^n)dt = \frac{A_{2n}}{2n}.$$

Donsker's theorem implies therefore that

$$\mathbb{P}(A_1 \le x) = \frac{2}{\pi}\arcsin\sqrt{x}.$$

The distribution of A_t is then finally deduced from the distribution of A_1 by using the scaling property of Brownian motion. □

Exercise 2.52. (1) Let $(S_n)_{n\in\mathbb{N}}$ be a symmetric random walk on $\mathbb{Z}$. We write

$$A_n = \mathrm{Card}\{1 \le i \le n, \max(S_{i-1}, S_i) > 0\}.$$

(a) Show that for $0 < x, y < 1$

$$\sum_{n=0}^{+\infty}\sum_{k=0}^{n} \mathbb{P}(A_{2n} = 2k, S_{2n} = 0)x^{2k}y^{2n} = \frac{2}{\sqrt{1-y^2}+\sqrt{1-x^2y^2}}.$$

(b) Deduce that, conditionally to $S_{2n} = 0$, the random variable A_{2n} is uniformly distributed on the set $\{0, 2, 4, \dots, 2n\}$.

(2) Let $(B_t)_{t\ge 0}$ be a standard Brownian motion. By using Donsker's theorem show that conditionally to $B_1 = 0$, the random variable

$$\int_0^1 1_{[0,+\infty)}(B_s)ds,$$

is uniformly distributed on $[0, 1]$.

Exercise 2.53. Let $(B_t)_{t\ge 0}$ be a standard Brownian motion. The goal of the exercise is to compute the distribution of the random variable

$$g_1 = \sup\{t \in [0, 1], B_t = 0\}.$$

(1) Let $(S_n)_{n\in\mathbb{N}}$ be a symmetric random walk on $\mathbb{Z}$. For $n \in \mathbb{N}$, we write

$$d_n = \max\{1 \le k \le n, S_k = 0\}.$$

Show that for $0 \le k \le n$

$$\mathbb{P}(d_{2n} = 2k) = \mathbb{P}(S_{2k} = 0)\mathbb{P}(S_{2n-2k} = 0).$$

(2) Deduce that for $x \in [0, 1]$

$$\lim_{n\to+\infty} \mathbb{P}\left(\frac{d_n}{n} \le x\right) = \frac{2}{\pi}\arcsin\sqrt{x}.$$

(3) Deduce the distribution of g_1.

Notes and comments

There are many rigorous constructions of the Brownian motion. The first one is due to Wiener [79]. A direct construction of the Wiener measure is given by Itô in [41]. Many properties of Brownian motions were known from Paul Lévy [54]. For more details about random walks, we refer to Durrett [22], Lawler–Limic [52] and the classical reference by Spitzer [73]. For further reading on Donsker's theorem and convergence in distribution in path spaces we refer to Section 6, Chapter II, of [66] or to the book by Jacod and Shiryaev [44].

Chapter 3
Markov processes

In this chapter we study continuous time Markov processes and emphasize the role played by the transition semigroup to investigate sample path properties of Markov processes. The class of processes we shall particularly be interested in are the Feller–Dynkin processes, which are Markov processes admitting regular versions and enjoying the strong Markov property. We will show that Feller–Dynkin processes admit generators and that these generators are a second order differential operator if the process is continuous. We finish the chapter with the study of the Lévy processes which are Feller–Dynkin processes associated to convolution semigroups.

3.1 Markov processes

Intuitively, a stochastic process defined on a probability space $(\Omega, \mathcal{F}, \mathbb{P})$ is a Markov process if it is memoryless, that is, if for every bounded Borel function $f : \mathbb{R}^n \to \mathbb{R}$,

$$\mathbb{E}(f(X_{t+s}) \mid \mathcal{F}_s^X) = \mathbb{E}(f(X_{t+s}) \mid X_s), \quad s, t \geq 0.$$

Let us turn to the more precise definition that relies on the notion of transition function. A transition function for a Markov process is the analogue in continuous time of the transition matrix associated to a Markov chain.

Definition 3.1. A *transition function* $\{P_t, t \geq 0\}$ on $\mathbb{R}^n$ is a family of kernels

$$P_t : \mathbb{R} \times \mathcal{B}(\mathbb{R}^n) \to [0, 1]$$

such that

(1) for $t \geq 0$ and $x \in \mathbb{R}^n$, $P_t(x, \cdot)$ is a probability measure on $\mathbb{R}^n$;

(2) for $t \geq 0$ and A Borel set in $\mathbb{R}^n$ the mapping $x \to P_t(x, A)$ is measurable;

(3) For $s, t \geq 0$, $x \in \mathbb{R}^n$ and A Borel set in $\mathbb{R}^n$,

$$P_{t+s}(x, A) = \int_{\mathbb{R}^n} P_t(y, A) P_s(x, dy). \tag{3.1}$$

The relation (3.1) is called the Chapman–Kolmogorov relation.

Given a transition function, we can define a one parameter family of linear operators $(\boldsymbol{P}_t)_{t \geq 0}$ from the space of bounded Borel functions into itself as follows:

$$(\boldsymbol{P}_t f)(x) = \int_{\mathbb{R}} f(y) P_t(x, dy).$$

If $\{P_t, t \geq 0\}$ is a transition function, then the following properties are satisfied:

- $\boldsymbol{P}_t 1 = 1$.
- For every $t \geq 0$, $\boldsymbol{P}_t$ is a positivity-preserving operator, in the sense that if f is non-negative, so is $\boldsymbol{P}_t f$.
- For every $t \geq 0$, $\boldsymbol{P}_t$ is a contraction from the space of bounded Borel functions into itself (that is, it is a continuous operator with a norm smaller than 1).
- The semigroup property holds: For every $s, t \geq 0$,

$$\boldsymbol{P}_{t+s} = \boldsymbol{P}_t \boldsymbol{P}_s.$$

Definition 3.2. A stochastic process $(X_t)_{t\geq 0}$ defined on a probability space $(\Omega, \mathcal{F}, \mathbb{P})$ is called a *Markov process* if there exists a transition function $\{P_t, t \geq 0\}$ on $\mathbb{R}^n$ such that for every bounded Borel function $f : \mathbb{R}^n \to \mathbb{R}$,

$$\mathbb{E}(f(X_{t+s}) \mid \mathcal{F}_s^X) = \boldsymbol{P}_t f(X_s), \quad s, t \geq 0,$$

where $\mathcal{F}^X$ denotes the natural filtration[1] of the process $(X_t)_{t\geq 0}$. The family of operators $(\boldsymbol{P}_t)_{t\geq 0}$ is called the semigroup of the Markov process.

Remark 3.3. We may also speak of the Markov property with respect to a given filtration. A stochastic process $(X_t)_{t\geq 0}$ defined on a filtered probability space $(\Omega, \mathcal{F}, (\mathcal{F}_t)_{t\geq 0}, \mathbb{P})$ is called a Markov process with respect to the filtration $(\mathcal{F}_t)_{t\geq 0}$ if there exists a transition function $\{P_t, t \geq 0\}$ on $\mathbb{R}^n$ such that for every bounded Borel function $f : \mathbb{R}^n \to \mathbb{R}$,

$$\mathbb{E}(f(X_{t+s}) \mid \mathcal{F}_s) = \boldsymbol{P}_t f(X_s), \quad s, t \geq 0.$$

Brownian motion is the primary example of a Markov process.

Proposition 3.4. *Let $(B_t)_{t\geq 0}$ be a (one-dimensional) stochastic process defined on a probability space $(\Omega, \mathcal{F}, \mathbb{P})$ such that $B_0 = 0$ a.s. Then $(B_t)_{t\geq 0}$ is a standard Brownian motion if and only if it is a Markov process with semigroup:*

$$\boldsymbol{P}_0 = \mathrm{Id}, \quad (\boldsymbol{P}_t f)(x) = \int_{\mathbb{R}} f(y) \frac{e^{-\frac{(x-y)^2}{2t}}}{\sqrt{2\pi t}} dy, \quad t > 0,\ x \in \mathbb{R}.$$

Proof. Let $(B_t)_{t\geq 0}$ be a Markov process defined on a probability space $(\Omega, \mathcal{F}, \mathbb{P})$ with semigroup:

$$\boldsymbol{P}_0 = \mathrm{Id}, \quad (\boldsymbol{P}_t f)(x) = \int_{\mathbb{R}} f(y) \frac{e^{-\frac{(x-y)^2}{2t}}}{\sqrt{2\pi t}} dy, \quad t > 0,\ x \in \mathbb{R}.$$

[1] Recall that $\mathcal{F}_s^X$ is the smallest σ-algebra that makes measurable all the random variables $(X_{t_1}, \dots, X_{t_m})$, $0 \leq t_1 \leq \dots \leq t_m \leq s$.

Thanks to the Markov property, by writing $\mathcal{F}_t$ for the natural filtration of $(B_t)_{t\geq 0}$, we have for $s, t \geq 0$, $\lambda \in \mathbb{R}$,

$$\mathbb{E}(e^{i\lambda B_{t+s}} \mid \mathcal{F}_s) = \int_{\mathbb{R}} e^{i\lambda(B_s+y)} \frac{e^{-\frac{y^2}{2t}}}{\sqrt{2\pi t}} dy.$$

This implies

$$\mathbb{E}(e^{i\lambda(B_{t+s}-B_s)} \mid \mathcal{F}_s) = \int_{\mathbb{R}} e^{i\lambda y} \frac{e^{-\frac{y^2}{2t}}}{\sqrt{2\pi t}} dy = e^{-\frac{1}{2}\lambda^2 t}.$$

In particular, the increments of $(B_t)_{t\geq 0}$ are stationary and independent. Now for $\lambda_1, \dots, \lambda_n \in \mathbb{R}$, $0 < t_1 < \dots < t_n$,

$$\begin{aligned} \mathbb{E}(e^{i\sum_{k=1}^n \lambda_k (B_{t_{k+1}} - B_{t_k})}) &= \prod_{k=1}^n \mathbb{E}(e^{i\lambda_k (B_{t_{k+1}} - B_{t_k})}) \\ &= \prod_{k=1}^n \mathbb{E}(e^{i\lambda_k B_{t_{k+1}-t_k}}) \\ &= e^{-\frac{1}{2}\sum_{k=1}^n (t_{k+1}-t_k)\lambda_k^2}. \end{aligned}$$

Hence $(B_t)_{t\geq 0}$ is a standard Brownian motion.

Conversely, let $(B_t)_{t\geq 0}$ be a standard Brownian motion with natural filtration $\mathcal{F}_t$. If f is a bounded Borel function and $s, t \geq 0$, we have

$$\mathbb{E}(f(B_{t+s}) \mid \mathcal{F}_s) = \mathbb{E}(f(B_{t+s} - B_s + B_s) \mid \mathcal{F}_s).$$

Since $B_{t+s} - B_s$ is independent from $\mathcal{F}_s$, we deduce that

$$\mathbb{E}(f(B_{t+s}) \mid \mathcal{F}_s) = \mathbb{E}(f(B_{t+s}) \mid B_s).$$

For $x \in \mathbb{R}$, we have

$$\mathbb{E}(f(B_{t+s}) \mid B_s = x) = \mathbb{E}(f(B_{t+s} - B_s + B_s) \mid B_s = x) = \mathbb{E}(f(X_t + x)),$$

where X_t is a random variable independent from B_s and normally distributed with mean 0 and variance t. Therefore we have

$$\mathbb{E}(f(B_{t+s}) \mid B_s = x) = \int_{\mathbb{R}} f(x+y) \frac{e^{-\frac{y^2}{2t}}}{\sqrt{2\pi t}} dy$$

and

$$\mathbb{E}(f(B_{t+s}) \mid \mathcal{F}_s) = \int_{\mathbb{R}} f(B_s+y) \frac{e^{-\frac{y^2}{2t}}}{\sqrt{2\pi t}} dy. \qquad \square$$

Exercise 3.5 (Multi-dimensional Brownian motion). Let $(B_t)_{t\geq 0}$ be an n-dimensional stochastic process defined on a probability space $(\Omega, \mathcal{F}, \mathbb{P})$ such that $B_0 = 0$ a.s. Show that $(B_t)_{t\geq 0}$ is a standard Brownian motion on $\mathbb{R}^n$ if and only if it is a Markov process with semigroup:

$$\boldsymbol{P}_0 = \mathrm{Id}, \quad (\boldsymbol{P}_t f)(x) = \int_{\mathbb{R}^n} f(y) \frac{e^{-\frac{\|x-y\|^2}{2t}}}{(2\pi t)^{n/2}} dy, \quad t > 0,\ x \in \mathbb{R}^n.$$

Exercise 3.6 (Ornstein–Uhlenbeck process). Let $(B_t)_{t\geq 0}$ be a one-dimensional Brownian motion and let $\theta \in \mathbb{R}\backslash\{0\}$. We consider the process

$$X_t = e^{\theta t} B_{\frac{1-e^{-2\theta t}}{2\theta}}.$$

Show that $(X_t)_{t\geq 0}$ is a Markov process with semigroup

$$(\boldsymbol{P}_t f)(x) = \int_{\mathbb{R}} f\left(e^{\theta t} x + \sqrt{\frac{e^{2\theta t} - 1}{2\theta}}\, y\right) \frac{e^{-\frac{y^2}{2}}}{\sqrt{2\pi}} dy.$$

Exercise 3.7 (Black–Scholes process). Let $(B_t)_{t\geq 0}$ be a one-dimensional Brownian motion and let $\mu \in \mathbb{R}$, $\sigma > 0$. We consider the process

$$X_t = e^{(\mu - \frac{\sigma^2}{2})t + \sigma B_t}.$$

Show that $(X_t)_{t\geq 0}$ is a Markov process with semigroup

$$(\boldsymbol{P}_t f)(x) = \int_{\mathbb{R}} f(x e^{(\mu - \frac{\sigma^2}{2})t + \sigma y}) \frac{e^{-\frac{y^2}{2t}}}{\sqrt{2\pi t}} dy.$$

Exercise 3.8 (Bessel process). Let $(B_t)_{t\geq 0}$ be an n-dimensional Brownian motion. We consider

$$X_t = \|B_t\| = \sqrt{\sum_{i=1}^{n} (B_t^i)^2}.$$

Show that $(X_t)_{t\geq 0}$ is a Markov process with semigroup given by

$$(\boldsymbol{P}_t f)(x) = \frac{1}{t} \int_{\mathbb{R}_{\geq 0}} f(y) \left(\frac{y}{x}\right)^{\frac{n}{2}-1} I_{\frac{n}{2}-1}\left(\frac{xy}{t}\right) y e^{-\frac{x^2+y^2}{2}} dy$$

for $x > 0$, and

$$(\boldsymbol{P}_t f)(0) = \frac{2^{1-\frac{n}{2}}}{\Gamma(n/2) t^{n/2}} \int_{\mathbb{R}_{\geq 0}} f(y)\, y^{n-1} e^{-\frac{y^2}{2t}} dy,$$

where Γ is the Euler Gamma function and $I_{\frac{n}{2}-1}$ is the modified Bessel function of the first kind with index $\frac{n}{2}-1$, that is, the solution of Bessel's equation

$$y'' + \frac{1}{x}y - \left(1 + \left(\frac{n}{2} - 1\right)^2 \frac{1}{x^2}\right)y = 0,$$

such that

$$y(x) \sim_{x \to 0} \frac{x^{\frac{n}{2}-1}}{2^{n/2}\Gamma(n/2)}.$$

The semigroup of the Brownian motion is intimately related to solutions of the heat equation. The next proposition shows how to construct martingales from solutions of the backward heat equation.

Theorem 3.9. *Let $(B_t)_{t\geq 0}$ be a standard Brownian motion. Let $f : \mathbb{R}_{\geq 0} \times \mathbb{R} \to \mathbb{C}$ be such that*

(1) *f is once continuously differentiable with respect to its first variable and twice continuously differentiable with respect to its second variable* (*we write $f \in \mathcal{C}^{1,2}(\mathbb{R}_{\geq 0} \times \mathbb{R}, \mathbb{C})$*);
(2) *for $t \geq 0$, there exist constants $K > 0$ and $\alpha > 0$ such that for every $x \in \mathbb{R}$,*

$$\sup_{0 \leq s \leq t} |f(s,x)| \leq K e^{\alpha |x|}.$$

The process $(f(t, B_t))_{t\geq 0}$ is a martingale if and only if

$$\frac{\partial f}{\partial t} + \frac{1}{2}\frac{\partial^2 f}{\partial x^2} = 0.$$

Proof. Let $t > 0$. In what follows, we denote by $\mathcal{F}$ the natural filtration of the Brownian motion. Thanks to the Markov property, we have for $s < t$,

$$\mathbb{E}(f(t, B_t) \mid \mathcal{F}_s) = \int_{\mathbb{R}} f(t,y) \frac{e^{-\frac{(y-B_s)^2}{2(t-s)}}}{\sqrt{2\pi(t-s)}} dy.$$

Therefore the process $(f(t, B_t))_{t\geq 0}$ is a martingale if and only if for $0 < s < t$ and $x \in \mathbb{R}$,

$$\int_{\mathbb{R}} f(t,y) \frac{e^{-\frac{(y-x)^2}{2(t-s)}}}{\sqrt{2\pi(t-s)}} dy = f(s,x).$$

We are thus led to characterize the functions f that satisfy the above functional equation and the given growth conditions.

So, let f be a function that satisfies the required regularity and growth conditions and such that for $0 < s < t$ and $x \in \mathbb{R}$,

$$\int_{\mathbb{R}} f(t,y) \frac{e^{-\frac{(y-x)^2}{2(t-s)}}}{\sqrt{2\pi(t-s)}} dy = f(s,x).$$

For a fixed $t > 0$, the function

$$g \colon (s,x) \to \int_{\mathbb{R}} f(t,y) \frac{e^{-\frac{(y-x)^2}{2(t-s)}}}{\sqrt{2\pi(t-s)}} dy$$

which is defined on $[0,t) \times \mathbb{R}$ is easily seen to satisfy the equation

$$\frac{\partial g}{\partial s} + \frac{1}{2}\frac{\partial^2 g}{\partial x^2} = 0,$$

so that f, of course, satisfies the same equation.

Now assume that f is a function that satisfies the required growth conditions and the equation

$$\frac{\partial f}{\partial t} + \frac{1}{2}\frac{\partial^2 f}{\partial x^2} = 0.$$

Let $t > 0$ be fixed. If we still write

$$g \colon (s,x) \to \int_{\mathbb{R}} f(t,y) \frac{e^{-\frac{(y-x)^2}{2(t-s)}}}{\sqrt{2\pi(t-s)}} dy,$$

we quickly realize that $h = f - g$ satisfies

$$\frac{\partial h}{\partial s} + \frac{1}{2}\frac{\partial^2 h}{\partial x^2} = 0$$

on $[0,t) \times \mathbb{R}$ and, in addition, the boundary condition

$$\forall x \in \mathbb{R}, \quad \lim_{s \to t} h(s,x) = 0.$$

From classical uniqueness of solution results for the heat equation, we deduce that $h = 0$. □

Exercise 3.10. Let $(B_t)_{t \geq 0}$ be an n-dimensional standard Brownian motion. Let $f \colon \mathbb{R}_{\geq 0} \times \mathbb{R}^n \to \mathbb{C}$ be such that

(1) $f \in \mathcal{C}^{1,2}(\mathbb{R}_{\geq 0} \times \mathbb{R}^n, \mathbb{C})$, and

(2) for $t \geq 0$, there exist constants $K > 0$ and $\alpha > 0$ such that

$$\sup_{0 \leq s \leq t} |f(s,x)| \leq K e^{\alpha \|x\|}.$$

Show that the process $(f(t, B_t))_{t \geq 0}$ is a martingale if and only if

$$\frac{\partial f}{\partial t} + \frac{1}{2} \Delta f = 0.$$

In particular, if f is a sub-exponential harmonic function (that is, $\Delta f = 0$), then the process $(f(B_t))_{t \geq 0}$ is a martingale.

Exercise 3.11. Let $(X_t)_{t \geq 0}$ be a Markov process with semigroup $\boldsymbol{P}_t$. Show that if $T > 0$, the process $((\boldsymbol{P}_{T-t} f)(X_t))_{0 \leq t \leq T}$ is a martingale. By using Doob's stopping theorem, deduce that if S is a stopping time such that $S \leq T$ almost surely, then

$$\mathbb{E}(f(X_T) \mid \mathcal{F}_S) = \boldsymbol{P}_{T-S} f(X_S).$$

Exercise 3.12 (Self-similar semigroups). A transition function $\{P_t, t \geq 0\}$ on $\mathbb{R}^n$ is said to be self-similar with index $H > 0$ if for every $t \geq 0$, $c > 0$, $x \in \mathbb{R}^n$ and every Borel set $A \subset \mathbb{R}^n$,

$$P_t(x, \Delta_c A) = P_{t/c}\left(\frac{x}{c^H}, A\right),$$

where Δ_c is the dilation of $\mathbb{R}^n$ defined by $\Delta_c(y) = c^H y$.

(1) Show that the transition function of the Brownian motion is self-similar with index $H = \frac{1}{2}$.

(2) Show that if $(X_t)_{t \geq 0}$ is a Markov process whose transition function is self-similar with index $H > 0$ such that $X_0 = 0$, then the two processes $(X_{ct})_{t \geq 0}$ and $(c^H X_t)_{t \geq 0}$ have the same distribution.

(3) Show that a transition function $\{P_t, t \geq 0\}$ is self-similar with index $H > 0$ if and only if for every $t \geq 0$, $c > 0$ and every bounded Borel function f,

$$(\boldsymbol{P}_{ct} f) \circ \Delta_c = \boldsymbol{P}_t (f \circ \Delta_c).$$

Exercise 3.13 (Rotationally invariant semigroups). A transition function $\{P_t, t \geq 0\}$ on $\mathbb{R}^n$ is said to be rotationally invariant if for every $t \geq 0$, $M \in \mathrm{SO}(\mathbb{R}^n)$ and every Borel set $A \subset \mathbb{R}^n$,

$$P_t(x, M \cdot A) = P_t(M^{-1} x, A),$$

where $\mathrm{SO}(\mathbb{R}^n)$ is the set of $n \times n$ matrices such that ${}^{\top}MM = \mathrm{I}_n$ and $\det(M) = 1$, and where $M \cdot A$ is the set $\{Mx, x \in A\}$.

(1) Show that the transition function of the Brownian motion is rotationally invariant.

(2) Show that if $(X_t)_{t \geq 0}$ is a Markov process whose transition function is rotationally invariant such that $X_0 = 0$, then for every $M \in \mathrm{SO}(\mathbb{R}^n)$ the two processes $(MX_t)_{t \geq 0}$ and $(X_t)_{t \geq 0}$ have the same distribution.

(3) Show that a transition function $\{P_t, t \geq 0\}$ is rotationally invariant if and only if for every $t \geq 0$, $M \in \mathrm{SO}(\mathbb{R}^n)$ and every bounded Borel function f,

$$(\boldsymbol{P}_t f) \circ M = \boldsymbol{P}_t(f \circ M).$$

It is remarkable that given any transition function, it is always possible to find a corresponding Markov process.

Theorem 3.14. *Let $\{P_t, t \geq 0\}$ be a transition function on $\mathbb{R}^n$. Let ν be a probability measure on $\mathbb{R}^n$. There exist a probability space $(\Omega, \mathcal{F}, \mathbb{P})$ and a stochastic process $(X_t)_{t \geq 0}$ such that the following holds.*

(1) *The distribution of X_0 is ν.*

(2) *If $f : \mathbb{R}^n \to \mathbb{R}$ is a bounded Borel function, then*

$$\mathbb{E}(f(X_{t+s}) \mid \mathcal{F}_s^X) = (\boldsymbol{P}_t f)(X_s), \quad s, t \geq 0,$$

where $\mathcal{F}^X$ is the natural filtration of X.

Proof. Let A be a Borel set in $\mathbb{R}^n$ and B a Borel set in $(\mathbb{R}^n)^{\otimes m}$. For $0 = t_0 < t_1 < \cdots < t_m$ we define

$$\mu_{t_0,t_1,\dots,t_m}(A \times B) = \int_A \int_B P_{t_1}(z, dx_1) P_{t_2-t_1}(x_1, dx_2) \dots P_{t_m-t_{m-1}}(x_{m-1}, dx_n) \nu(dz).$$

The measure $\mu_{t_0,t_1,\dots,t_n}$ is therefore a probability measure on $\mathbb{R}^n \times (\mathbb{R}^n)^{\otimes m}$. Since for a Borel set C in $\mathbb{R}^n$ and $x \in \mathbb{R}^n$ we have

$$P_{t+s}(x, C) = \int_{\mathbb{R}^n} P_t(y, C) P_s(x, dy),$$

we deduce that this family of probability satisfies the assumptions of the Daniell–Kolmogorov theorem. Therefore, we can find a process $(X_t)_{t \geq 0}$ defined on some probability space $(\Omega, \mathcal{F}, \mathbb{P})$ whose finite-dimensional distributions are given by the $\mu_{t_0,t_1,\dots,t_n}$'s. Let us now prove that this process satisfies the property stated in the theorem. First, the distribution of X_0 is ν because

$$\mu_0(A) = \int_A \nu(dz) = \nu(A), A \in \mathcal{B}(\mathbb{R}^n).$$

We now have to prove that if $f : \mathbb{R}^n \to \mathbb{R}$ is a bounded Borel function and if $0 < s, t$, then

$$\mathbb{E}(f(X_{s+t}) \mid \mathcal{F}_s^X) = (\boldsymbol{P}_t f)(X_s).$$

For this we have to prove that if $f : \mathbb{R}^n \to \mathbb{R}$, $F : (\mathbb{R}^n)^{\otimes m} \to \mathbb{R}$ are bounded Borel functions and if $0 = t_0 < t_1 < \cdots < t_m$, then

$$\mathbb{E}(f(X_{t_m}) F(X_{t_0}, \dots, X_{t_{m-1}})) = \mathbb{E}((\boldsymbol{P}_{t_m-t_{m-1}} f)(X_{t_{m-1}}) F(X_{t_0}, \dots, X_{t_{m-1}})).$$

But due to Fubini's theorem we have

$$\begin{aligned}
&\mathbb{E}(f(X_{t_m})F(X_{t_0},\dots,X_{t_{m-1}}))\\
&\quad= \int_{(\mathbb{R}^n)^{\otimes(m+1)}} f(x_m)F(z,x_1,\dots,x_{m-1})P_{t_1}(z,dx_1)P_{t_2-t_1}(x_1,dx_2)\dots\\
&\qquad\quad \dots P_{t_m-t_{m-1}}(x_{m-1},dx_m)\nu(dz)\\
&\quad= \int_{\mathbb{R}^n}\int_{(\mathbb{R}^n)^{\otimes m}} (\boldsymbol{P}_{t_m-t_{m-1}}f)(x_{m-1})F(z,x_1,\dots,x_{m-1})P_{t_1}(z,dx_1)\dots\\
&\qquad\quad \dots P_{t_{m-1}-t_{m-2}}(x_{m-2},dx_{m-1})\nu(dz)\\
&\quad= \mathbb{E}((\boldsymbol{P}_{t_m-t_{m-1}}f)(X_{t_{m-1}})F(X_{t_0},\dots,X_{t_{m-1}})).
\end{aligned}$$

This concludes the proof of the theorem. □

Remark 3.15. Observe the degree of freedom we have on the distribution of X_0: This reflects the fact that the transition function characterizes the distribution of a Markov process up to its initial distribution.

To finish the section, we introduce the notion of sub-Markov process, a notion which is very useful in many situations. A family of operators $(\boldsymbol{P}_t)_{t\geq 0}$ from the space of bounded Borel functions into itself is called a sub-Markov semigroup is the following conditions are satisfied:

- For every $t \geq 0$, $\boldsymbol{P}_t$ is a positivity-preserving operator.
- For every $t \geq 0$, $\boldsymbol{P}_t$ is a contraction from the space of bounded Borel functions into itself.
- The semigroup property holds: For every $s, t \geq 0$,

$$\boldsymbol{P}_{t+s} = \boldsymbol{P}_t\boldsymbol{P}_s.$$

Observe that unlike Markov semigroups which are associated with Markov processes, we do not necessarily have $\boldsymbol{P}_t 1 = 1$ but only $\boldsymbol{P}_t 1 \leq 1$. It turns out that a sub-Markov semigroup $(\boldsymbol{P}_t)_{t\geq 0}$ can always be embedded into a Markov semigroup which is defined on a larger space. Indeed let us consider a cemetery point $\star$. If $\tilde{f}$ is a bounded Borel function on $\mathbb{R}^n$, then we extend $\boldsymbol{P}_t f$ to $\mathbb{R}^n \cup \{\star\} \to \mathbb{R}$ by defining

$$\boldsymbol{P}_t f(\star) = 0.$$

Also any smooth and compactly supported function f on $\mathbb{R}^n$ is automatically extended to $\mathbb{R}^n \cup \{\star\} \to \mathbb{R}$ by setting $f(\star) = 0$. If f is a bounded Borel function on $\mathbb{R}^n \cup \{\star\} \to \mathbb{R}$, we then define for $x \in \mathbb{R}^n$,

$$\boldsymbol{Q}_t f(x) = \boldsymbol{P}_t\tilde{f}(x) + f(\star)(1 - \boldsymbol{P}_t 1(x)),$$

where $\tilde{f}$ is the restriction of f to $\mathbb{R}^n$. It is an easy exercise to check that $\boldsymbol{Q}_t$ is now a Markov semigroup on the space of bounded Borel functions $\mathbb{R}^n \cup \{\star\} \to \mathbb{R}$. By definition, a sub-Markov process $(X_t)_{t\geq 0}$ on $\mathbb{R}^n$ with transition semigroup $(\boldsymbol{P}_t)_{t\geq 0}$ is a Markov process on $\mathbb{R}^n \cup \{\star\} \to \mathbb{R}$ with transition semigroup $(\boldsymbol{Q}_t)_{t\geq 0}$. If $(X_t^x)_{t\geq 0}$ is a sub-Markov process with transition semigroup $(\boldsymbol{P}_t)_{t\geq 0}$ such that $X_0^x = x$ and if we define $\boldsymbol{e}(x) = \inf\{t \geq 0, X_t^x = \star\}$, then we have for any bounded Borel function f on $\mathbb{R}^n$,

$$\boldsymbol{P}_t f(x) = \mathbb{E}(f(X_t^x) 1_{t<\boldsymbol{e}(x)});$$

in particular, observe that we have

$$\boldsymbol{P}_t 1(x) = \mathbb{P}(t < \boldsymbol{e}(x)).$$

The random time $\boldsymbol{e}(x)$ is called the extinction time of $(X_t^x)_{t\geq 0}$. Killed Markov processes are canonical examples of sub-Markov processes.

Exercise 3.16 (Killed Markov process). Let $(X_t)_{t\geq 0}$ be a Markov process in $\mathbb{R}^n$ and $K \subset \mathbb{R}^n$ be a non-empty set. Let $T_K = \inf\{t \geq 0, X_t \in K\}$. Consider the process $(Y_t)_{t\geq 0}$ where

$$Y_t = \begin{cases} X_t, & t \leq T_K, \\ \star, & t > T_K. \end{cases}$$

Show that $(Y_t)_{t\geq 0}$ is a sub-Markov process.

Exercise 3.17 (Killing at an exponential time). Let $(X_t)_{t\geq 0}$ be a Markov process in $\mathbb{R}^n$. Let T be an exponential random variable with parameter λ which is independent from $(X_t)_{t\geq 0}$. Consider the process $(Y_t)_{t\geq 0}$ where

$$Y_t = \begin{cases} X_t, & t \leq T, \\ \star, & t > T. \end{cases}$$

Show that $(Y_t)_{t\geq 0}$ is a sub-Markov process and compute its semigroup in terms of the semigroup of $(X_t)_{t\geq 0}$.

3.2 Strong Markov processes

When studying martingales we already stressed and illustrated that in the study of stochastic processes it is often important to know how the process behaves with respect to the stopping times of the underlying filtration. For Markov processes this naturally leads to the notion of strong Markov process.

Definition 3.18. Let $(X_t)_{t\geq 0}$ be a Markov process with transition function $\{P_t, t \geq 0\}$. We say that $(X_t)_{t\geq 0}$ is a *strong Markov process* if for any bounded Borel function $f : \mathbb{R} \to \mathbb{R}$ and any finite stopping time S of the filtration $(\mathcal{F}_t^X)_{t\geq 0}$ we have

$$\mathbb{E}(f(X_{S+t}) \mid \mathcal{F}_S^X) = (\boldsymbol{P}_t f)(X_S), \quad t > 0.$$

Remark 3.19. As for the definition of the Markov property, we may of course define the strong Markov property of a process with respect to a filtration that does not need to be the natural filtration of the process.

In general, it is not straightforward to prove that a given Markov process satisfies the strong Markov property and not all Markov processes enjoy the strong Markov property.

Let us first focus on the Brownian motion case. In this case, as for the symmetric random walks, the strong Markov property is a consequence of the following result.

Proposition 3.20. *Let $(B_t)_{t\geq 0}$ be a standard Brownian motion and let T be a finite stopping time. The process*

$$(B_{T+t} - B_T)_{t\geq 0}$$

is a standard Brownian motion independent from $\mathcal{F}_T^B$.

Proof. Let T be a finite stopping time of the filtration $(\mathcal{F}_t^B)_{t\geq 0}$. We first assume T bounded. Let us consider the process

$$\widetilde{B}_t = B_{T+t} - B_T, \quad t \geq 0.$$

Let $\lambda \in \mathbb{R}$, $0 \leq s \leq t$. Applying Doob's stopping theorem to the martingale

$$(e^{i\lambda B_t + \frac{\lambda^2}{2}t})_{t\geq 0},$$

with the stopping times $t + T$ and $s + T$, yields

$$\mathbb{E}(e^{i\lambda B_{T+t} + \frac{\lambda^2}{2}(T+t)} \mid \mathcal{F}_{T+s}) = e^{i\lambda B_{T+s} + \frac{\lambda^2}{2}(T+s)}.$$

Therefore we have

$$\mathbb{E}(e^{i\lambda (B_{T+t} - B_{T+s})} \mid \mathcal{F}_{T+s}) = e^{-\frac{\lambda^2}{2}(t-s)}.$$

The increments of $(\widetilde{B}_t)_{t\geq 0}$ are therefore independent and stationary. The conclusion thus easily follows. If T is not bounded almost surely, then we can consider the stopping time $T \wedge N$ and from the previous result the finite-dimensional distributions $(B_{t_1+T\wedge N} - B_{T\wedge N}, \dots, B_{t_n+T\wedge N} - B_{T\wedge N})$ do not depend on N and are the same as a Brownian motion. We can then let $N \to +\infty$ to conclude. □

As a corollary, we obtain the strong Markov property of the Brownian motion:

Corollary 3.21. *Let $(B_t)_{t\geq 0}$ be a standard Brownian motion. Then $(B_t)_{t\geq 0}$ is a strong Markov process.*

Proof. Let $f : \mathbb{R} \to \mathbb{R}$ be a bounded Borel function, let $t \geq 0$ and let S be a finite stopping time. From the previous proposition, we have

$$\mathbb{E}(f(B_{t+S}) \mid \mathcal{F}_S) = \mathbb{E}(f(B_{t+S} - B_S + B_S) \mid \mathcal{F}_S).$$

Since $B_{t+S} - B_S$ is independent from $\mathcal{F}_S$, we first deduce that

$$\mathbb{E}(f(B_{t+S}) \mid \mathcal{F}_S) = \mathbb{E}(f(B_{t+S}) \mid B_S).$$

Now for $x \in \mathbb{R}$,

$$\mathbb{E}(f(B_{t+S}) \mid B_S = x) = \mathbb{E}(f(B_{t+S} - B_S + B_S) \mid B_S = x) = \mathbb{E}(f(X_t + x)),$$

where X_t is a Gaussian random variable independent from B_S which has mean 0 and variance t. Thus,

$$\mathbb{E}(f(B_{t+S}) \mid B_S = x) = \int_{\mathbb{R}} f(x+y) \frac{e^{-\frac{y^2}{2t}}}{\sqrt{2\pi t}} dy$$

and

$$\mathbb{E}(f(B_{t+S}) \mid \mathcal{F}_S) = \int_{\mathbb{R}} f(B_S + y) \frac{e^{-\frac{y^2}{2t}}}{\sqrt{2\pi t}} dy.$$

□

The following exercises show some applications of the strong Markov property.

In the exercises below, $(B_t)_{t \geq 0}$ is a standard Brownian motion and for $a \in \mathbb{R}$, we write

$$T_a = \inf\{t > 0, B_t = a\},$$

and for $t \geq 0$,

$$S_t = \sup\{B_s, s \leq t\}.$$

Exercise 3.22 (Reflection principle for the Brownian paths). Let $a \in \mathbb{R}$. Show that the process

$$\tilde{B}_t = \begin{cases} B_t, & t < T_a, \\ 2a - B_t, & t \geq T_a, \end{cases}$$

is a Brownian motion.

Exercise 3.23. Show that for $t \geq 0$, $a \geq 0$, $x \leq a$,

$$\mathbb{P}(S_t \in da, B_t \in dx) = \frac{2(2a-x)}{\sqrt{2\pi t^3}} e^{-\frac{(2a-x)^2}{2t}} da dx.$$

Exercise 3.24. Show that the two processes $(S_t - B_t)_{t \geq 0}$ and $(|B_t|)_{t \geq 0}$ have the same distribution and that $(|B_t|)_{t \geq 0}$ is a strong Markov process whose semigroup is given by

$$(\mathbf{P}_t f)(x) = \frac{2}{\sqrt{2\pi t}} \int_0^{+\infty} e^{-\frac{x^2+y^2}{2t}} \cosh\left(\frac{xy}{t}\right) f(y) dy, \quad t > 0.$$

Exercise 3.25 (Local time of the Brownian motion).

(1) Let $\varepsilon > 0$. We write $U = \inf\{t \geq 0, S_t - B_t > \varepsilon\}$. Show that the random variable S_U has an exponential distribution.

(2) We recursively define the following sequence of stopping times:

$$T_1'(\varepsilon) = 0, \quad T_n(\varepsilon) = \inf\{t > T_n'(\varepsilon), S_t - B_t > \varepsilon\},$$

$$T_{n+1}'(\varepsilon) = \inf\{t > T_n(\varepsilon), S_t - B_t = 0\}.$$

We write

$$U(t, \varepsilon) = \max\{n, T_n(\varepsilon) \leq t\}.$$

Show that, almost surely,

$$\lim_{n \to +\infty} 2^{-n} U(t, 2^{-n}) = S_t.$$

(3) Deduce that there exists a continuous and non-decreasing process $(L_t)_{t \geq 0}$ such that $(|B_t| - L_t)_{t \geq 0}$ is a Brownian motion. Show that the process $(L_t)_{t \geq 0}$ increases only when $B_t = 0$. The process $(L_t)_{t \geq 0}$ is called the local time of Brownian motion at 0.

(4) Compute the distribution of L_t.

3.3 Feller–Dynkin diffusions

From their very definition, Markov processes are associated to semigroups. More precisely, a multi-dimensional stochastic process $(X_t)_{t \geq 0}$ is a Markov process if and only if there exists a contraction semigroup of operators $(\mathbf{P}_t)_{t \geq 0}$ on the Banach space $L^\infty(\mathbb{R}^n, \mathbb{R})$ such that

$$0 \leq \mathbf{P}_t f \leq 1 \quad \text{for } 0 \leq f \leq 1,$$
$$\mathbf{P}_t 1 = 1,$$

and

$$\mathbb{E}(f(X_{t+s}) \mid \mathcal{F}_s^X) = \mathbf{P}_t f(X_s), \quad s, t \geq 0.$$

In general for an arbitrary $f \in L^\infty(\mathbb{R}^n, \mathbb{R})$, the map $t \to \mathbf{P}_t f$ fails to be continuous in the strong topology. However, for many interesting examples of Markov processes, this continuity issue is solved by restricting $(\mathbf{P}_t)_{t \geq 0}$ to a closed subspace $\mathbb{X}$ of $L^\infty(\mathbb{R}^n, \mathbb{R})$ that densely contains the set $\mathcal{C}_c(\mathbb{R}^n, \mathbb{R})$ of smooth and compactly supported functions $\mathbb{R}^n \to \mathbb{R}$. In what follows, we denote by $\mathcal{C}_0(\mathbb{R}^n, \mathbb{R})$ the Banach space of continuous functions $f : \mathbb{R}^n \to \mathbb{R}$ such that $\lim_{\|x\| \to +\infty} f(x) = 0$.

Exercise 3.26. Let

$$(\boldsymbol{P}_t f)(x) = \int_{\mathbb{R}} f(y) \frac{e^{-\frac{(x-y)^2}{2t}}}{\sqrt{2\pi t}} dy, \quad t > 0,\ x \in \mathbb{R},$$

be the semigroup of the Brownian motion.

(1) Give an example of $f \in L^\infty(\mathbb{R}, \mathbb{R})$ such that the map $t \to \boldsymbol{P}_t f$ fails to be continuous in the strong topology.

(2) Show that for all $f \in \mathcal{C}_0(\mathbb{R}, \mathbb{R})$,

$$\lim_{t \to 0} \|\boldsymbol{P}_t f - f\|_\infty = 0.$$

The previous exercise leads to the following definitions.

Definition 3.27 (Feller–Dynkin semigroups). Let $(\boldsymbol{P}_t)_{t \geq 0}$ be a contraction semigroup of operators on the Banach space $L^\infty(\mathbb{R}^n, \mathbb{R})$ such that

(1) for $0 \leq f \leq 1, 0 \leq \boldsymbol{P}_t f \leq 1$,

(2) $\boldsymbol{P}_t 1 = 1$.

We say that $(\boldsymbol{P}_t)_{t \geq 0}$ is a *Feller–Dynkin semigroup* if it satisfies the following additional properties:

(3) $\boldsymbol{P}_t : \mathcal{C}_0(\mathbb{R}^n, \mathbb{R}) \to \mathcal{C}_0(\mathbb{R}^n, \mathbb{R})$.

(4) $\forall f \in \mathcal{C}_0(\mathbb{R}^n, \mathbb{R})$,

$$\lim_{t \to 0} \|\boldsymbol{P}_t f - f\|_\infty = 0.$$

A Markov process $(X_t)_{t \geq 0}$ is said to be a *Feller–Dynkin process* if its semigroup is a Feller–Dynkin semigroup.

We know from Theorem 3.14 that it is always possible to associate a Markov process to a transition function. For Feller–Dynkin semigroups, we can moreover work with regular versions that enjoy the strong Markov property.

Theorem 3.28. *Let $\{P_t, t \geq 0\}$ be a Feller–Dynkin transition function. For every probability measure ν on $\mathbb{R}^n$, there exist a filtered probability space and a stochastic process $(X_t)_{t \geq 0}$ defined on that space such that*

(1) *the distribution of X_0 is ν;*

(2) *the paths of $(X_t)_{t \geq 0}$ are right continuous and left limited;*

(3) *with respect to the filtration $(\mathcal{F}_t)_{t \geq 0}$, $(X_t)_{t \geq 0}$ is a strong Markov process with transition function $\{P_t, t \geq 0\}$.*

Proof. Let $\{P_t, t \geq 0\}$ be a Feller–Dynkin transition function. We denote by $\boldsymbol{P}_t$ the corresponding semigroup. We already know from Theorem 3.14 that there exist a filtered probability space $(\Omega, (\mathcal{F}_t)_{t \geq 0}, \mathcal{F}, \mathbb{P})$ and a stochastic process $(X_t)_{t \geq 0}$ defined on that space such that

(1) the distribution of X_0 is ν;

(2) with respect to the filtration $(\mathcal{F}_t)_{t\geq 0}$, $(X_t)_{t\geq 0}$ is a Markov process with transition function $\{P_t, t \geq 0\}$.

So, we need to prove that $(X_t)_{t\geq 0}$ admits a right continuous and left limited modification. The idea is to use the so-called resolvent functions of the semigroup. For $\alpha > 0$ and $f \in \mathcal{C}_0(\mathbb{R}^n, \mathbb{R})$, consider

$$U_\alpha f(x) = \int_0^{+\infty} e^{-\alpha t} \boldsymbol{P}_t f(x) dt.$$

If $f \geq 0$, we have for $t \geq s$,

$$\begin{aligned}
\mathbb{E}(e^{-\alpha t} U_\alpha f(X_t) \mid \mathcal{F}_s) &= e^{-\alpha t} \boldsymbol{P}_{t-s} U_\alpha f(X_s) \\
&= e^{-\alpha t} \int_0^{+\infty} e^{-\alpha u} \boldsymbol{P}_{t-s+u} f(X_s) du \\
&= e^{-\alpha s} \int_{t-s}^{+\infty} e^{-\alpha u} \boldsymbol{P}_u f(X_s) du \\
&\leq e^{-\alpha s} U_\alpha f(X_s).
\end{aligned}$$

Therefore, if $f \geq 0$, then $e^{-\alpha t} U_\alpha f(X_t)$ is a supermartingale. From the Doob regularization theorem (Theorem 1.47), we deduce that for every $\alpha > 0$, and every $f \in \mathcal{C}_0(\mathbb{R}^n, \mathbb{R})$, $f \geq 0$, the process $U_\alpha f(X_t)$ has almost surely right limits along $\mathbb{Q}$. Let now $\mathcal{I}$ be a countable set of non-negative functions in $\mathcal{C}_0(\mathbb{R}^n, \mathbb{R})$ that separate the points in the sense that for $x, y \in \mathbb{R}^n$ with $x \neq y$, we always can find $f \in \mathcal{I}$ such that $f(x) \neq f(y)$. Since it is easily proved that for $f \in \mathcal{C}_0(\mathbb{R}^n, \mathbb{R})$, we always have

$$\lim_{\alpha \to \infty} \|\alpha U_\alpha f - f\|_\infty = 0,$$

we deduce that the countable set

$$\mathcal{A} = \{U_\alpha f, \alpha \in \mathbb{N}, f \in \mathcal{I}\}$$

also separate points. Since for every $a \in \mathcal{A}$, the process $a(X_t)$ has almost surely right limits along $\mathbb{Q}$, we conclude that the process X_t itself has almost surely right limits along $\mathbb{Q}$. This allows us to define

$$\tilde{X}_t = \lim_{s \searrow t, s \in \mathbb{Q}} X_s.$$

For every bounded functions g, f on $\mathbb{R}^n$ and every $0 \leq t \leq s$, we have

$$\mathbb{E}(g(X_t) f(X_s)) = (g(X_t) \boldsymbol{P}_{t-s} f(X_t)).$$

Therefore, by letting $s \searrow t, s \in \mathbb{Q}$, we deduce

$$\mathbb{E}(g(X_t)f(\widetilde{X}_t)) = (g(X_t)f(X_t)).$$

Since the previous equality should hold for every bounded functions f and g, we deduce easily by the monotone class theorem that for every $t \geq 0$, $\widetilde{X}_t = X_t$ almost surely. As a conclusion $(\widetilde{X}_t)_{t\geq 0}$ is a right continuous modification of $(X_t)_{t\geq 0}$. Finally, by using once again the Doob regularization theorem for the supermartingales $(a(\widetilde{X}_t))_{t\geq 0}$, $a \in \mathcal{A}$, we conclude that $(\widetilde{X}_t)_{t\geq 0}$ almost surely has left limits at any points.

Let us now prove that $(\widetilde{X}_t)_{t\geq 0}$ is a strong Markov process. We need to check that for any Borel function $f : \mathbb{R} \to \mathbb{R}$, and any finite stopping time S of the filtration $(\mathcal{F}_t)_{t\geq 0}$, we have

$$\mathbb{E}(f(\widetilde{X}_{S+t}) \mid \mathcal{F}_S) = (\boldsymbol{P}_t f)(\widetilde{X}_S), \quad t > 0.$$

Without loss of generality we may assume S to be bounded (otherwise, just consider $S_n = S \wedge n$ and let $n \to \infty$).

As a first step, let us assume that S takes a finite number of values $s_1, \dots, s_n$. Let Z be a bounded random variable measurable with respect to $\mathcal{F}_S$. We observe that the random variable $1_{\{S=s_k\}}Z$ is measurable with respect to $\mathcal{F}_{s_k}$ and deduce that

$$\begin{aligned}
\mathbb{E}(f(\widetilde{X}_{S+t})Z) &= \sum_{k=1}^{n} \mathbb{E}(f(\widetilde{X}_{s_k+t})Z1_{\{S=s_k\}}) \\
&= \sum_{k=1}^{n} \mathbb{E}((\boldsymbol{P}_t f)(\widetilde{X}_{s_k})Z1_{\{S=s_k\}}) \\
&= \mathbb{E}((\boldsymbol{P}_t f)(\widetilde{X}_S)Z).
\end{aligned}$$

This yields

$$\mathbb{E}(f(\widetilde{X}_{S+t}) \mid \mathcal{F}_S) = (\boldsymbol{P}_t f)(\widetilde{X}_S), \quad t > 0.$$

If S takes an infinite number of values, we approximate S by the following sequence of stopping times:

$$\tau_n = \sum_{k=1}^{2^n} \frac{kK}{2^n} 1_{\left\{\frac{(k-1)K}{2^n} \leq S < \frac{kK}{2^n}\right\}},$$

where K is such that $S \leq K$ almost surely. The stopping time τ_n takes its values in a finite set and when $n \to +\infty$, $\tau_n \searrow S$. As before, let Z be a bounded random variable measurable with respect to $\mathcal{F}_S$ and let $f \in \mathcal{C}_0(\mathbb{R}^n, \mathbb{R})$. Since $\tau_n \geq S$, from the above computation we have

$$\mathbb{E}(f(\widetilde{X}_{\tau_n+t})Z) = \mathbb{E}((\boldsymbol{P}_t f)(\widetilde{X}_{\tau_n})Z).$$

Since $\boldsymbol{P}_t$ is a Feller semigroup, $\boldsymbol{P}_t f$ is a bounded continuous function. By letting $n \to +\infty$ and using the right continuity of $(\widetilde{X}_t)_{t\geq 0}$ in combination with the dominated convergence we deduce that

$$\mathbb{E}(f(\widetilde{X}_{S+t})Z) = \mathbb{E}((\boldsymbol{P}_t f)(\widetilde{X}_S)Z).$$

By using the monotone class theorem, we see that the previous equality then also holds for every bounded function f. This concludes the proof of the theorem. □

The Markov process of the previous theorem may not be continuous. Actually, many interesting examples of Feller–Dynkin processes are discontinuous. This is for instance the case of the Lévy processes that will be studied later.

Exercise 3.29 (Quasi-left continuity). Let $(X_t)_{t\geq 0}$ be the Feller–Dynkin Markov process constructed in Theorem 3.28.

(1) Let T be a stopping time of the natural filtration of $(X_t)_{t\geq 0}$. Show that if T_n is an increasing sequence of stopping times almost surely converging to T, then almost surely on the set $\{T < +\infty\}$, we have

$$\lim_{n\to+\infty} X_{T_n} = X_T.$$

(2) Deduce that for any $T \geq 0$, the set $\{X_t(\omega), 0 \leq t \leq T\}$ is almost surely bounded.

The following proposition gives a useful criterion for the continuity of the Markov process which is associated to a Feller–Dynkin transition function.

Proposition 3.30. *For $x \in \mathbb{R}^n$ and $\varepsilon > 0$, we denote by $\mathrm{B}(x,\varepsilon)$ the open ball in $\mathbb{R}^n$ with center x and radius ε. Let $\{P_t, t \geq 0\}$ be a Feller–Dynkin transition function that satisfies the condition*

$$\lim_{t\to 0} \sup_{x\in K} \frac{1}{t} P_t(x, {}^c\mathrm{B}(x,\varepsilon)) = 0,$$

for every $\varepsilon > 0$ and every compact set K. The stochastic process $(X_t)_{t\geq 0}$ given by Theorem 3.28 is then continuous.

Proof. If K is a compact set, $n \in \mathbb{N}$, $n \geq 1$ and $\varepsilon > 0$, we define

$$A(n,K,\varepsilon) = \Big\{\omega,\ \max_{0\leq k\leq n-1} \|X_{\frac{k+1}{n}}(\omega) - X_{\frac{k}{n}}(\omega)\| > \varepsilon,\ \forall s \in [0,1],\ X_s(\omega) \in K\Big\}.$$

By using the Markov property, it is easily checked that

$$\mathbb{P}(A(n,K,\varepsilon)) \leq n \sup_{x\in K} P_{1/n}(x, {}^c\mathrm{B}(x,\varepsilon)).$$

As a consequence, if

$$\lim_{t \to 0} \sup_{x \in K} \frac{1}{t} P_t(x, {}^c \mathrm{B}(x, \varepsilon)) = 0,$$

then we have

$$\lim_{n \to \infty} \mathbb{P}(A(n, K, \varepsilon)) = 0.$$

Since from Exercise 3.29 the set $\{X_t(\omega), 0 \le t \le 1\}$ is almost surely bounded and thus contained in a compact set, we deduce from the Borel–Cantelli lemma that with probability 1, the set of n's such that

$$\max_{0 \le k \le n-1} \|X_{\frac{k+1}{n}}(\omega) - X_{\frac{k}{n}}(\omega)\| > \varepsilon$$

is finite. This implies that $(X_t)_{0 \le t \le 1}$ is continuous. □

A fundamental property of Feller–Dynkin semigroups is that they admit generators.

Proposition 3.31. *Let $(\mathbf{P}_t)_{t \ge 0}$ be a Feller–Dynkin semigroup. There exists a densely defined operator*

$$L \colon \mathcal{D}(L) \subset \mathcal{C}_0(\mathbb{R}^n, \mathbb{R}) \to \mathcal{C}_0(\mathbb{R}^n, \mathbb{R}),$$

where

$$\mathcal{D}(L) = \left\{ f \in \mathcal{C}_0(\mathbb{R}^n, \mathbb{R}), \quad \lim_{t \to 0} \frac{\mathbf{P}_t f - f}{t} \ \textit{exists} \right\}$$

such that for $f \in \mathcal{D}(L)$,

$$\lim_{t \to 0} \left\| \frac{\mathbf{P}_t f - f}{t} - Lf \right\|_\infty = 0.$$

The operator L is called the generator of the semigroup $(\mathbf{P}_t)_{t \ge 0}$. We also say that L generates $(\mathbf{P}_t)_{t \ge 0}$.

Proof. Let us consider the following bounded operators on $\mathcal{C}_0(\mathbb{R}^n, \mathbb{R})$:

$$A_t = \frac{1}{t} \int_0^t \mathbf{P}_s ds.$$

For $f \in \mathcal{C}_0(\mathbb{R}^n, \mathbb{R})$ and $h > 0$, we have

$$\frac{1}{t}(\mathbf{P}_t A_h f - A_h f) = \frac{1}{ht} \int_0^h (\mathbf{P}_{s+t} f - \mathbf{P}_s f) ds = \frac{1}{ht} \int_0^t (\mathbf{P}_{s+h} f - \mathbf{P}_s f) ds.$$

Therefore, we obtain

$$\lim_{t\to 0}\frac{1}{t}(\boldsymbol{P}_t A_h f - A_h f) = \frac{1}{h}(\boldsymbol{P}_h f - f).$$

This implies

$$\{A_h f, f \in \mathcal{C}_0(\mathbb{R}^n,\mathbb{R}), h>0\} \subset \left\{ f \in \mathcal{C}_0(\mathbb{R}^n,\mathbb{R}),\ \lim_{t\to 0}\frac{\boldsymbol{P}_t f - f}{t} \text{ exists} \right\}$$

Since $\lim_{h\to 0} A_h f = f$, we deduce that

$$\left\{ f \in \mathcal{C}_0(\mathbb{R}^n,\mathbb{R}),\ \lim_{t\to 0}\frac{\boldsymbol{P}_t f - f}{t} \text{ exists} \right\}$$

is dense in $\mathcal{C}_0(\mathbb{R}^n,\mathbb{R})$. We can then consider

$$Lf = \lim_{t\to 0}\frac{\boldsymbol{P}_t f - f}{t},$$

which is of course defined on the domain

$$\mathcal{D}(L) = \left\{ f \in \mathcal{C}_0(\mathbb{R}^n,\mathbb{R}),\ \lim_{t\to 0}\frac{\boldsymbol{P}_t f - f}{t} \text{ exists} \right\}. \qquad \square$$

If $(X_t)_{t\geq 0}$ is a Feller–Dynkin process, the generator of its semigroup is also called the generator of the Markov process $(X_t)_{t\geq 0}$.

Exercise 3.32. Let $(B_t)_{t\geq 0}$ be an n-dimensional standard Brownian motion. Show that $(B_t)_{t\geq 0}$ is a Feller–Dynkin process, that the domain of its generator L contains $\mathcal{C}_c(\mathbb{R}^n,\mathbb{R})$ and that for $f \in \mathcal{C}_c(\mathbb{R}^n,\mathbb{R})$,

$$Lf = \frac{1}{2}\Delta f.$$

An interesting sub-class of Feller–Dynkin processes is the class of Feller–Dynkin diffusion processes. Let us recall that we denote by $\mathcal{C}(\mathbb{R}_{\geq 0},\mathbb{R}^n)$ the space of continuous functions $\mathbb{R}_{\geq 0}\to\mathbb{R}^n$. As usual $(\pi_t)_{t\geq 0}$ will denote the coordinate process on this path space and

$$\mathcal{G}_t = \sigma(\pi_s, 0\leq s\leq t),\ t\geq 0, \quad \mathcal{G}_\infty = \sigma(\pi_s, s\geq 0).$$

Definition 3.33 (Diffusion process). Let $(\boldsymbol{P}_t)_{t\geq 0}$ be a Feller–Dynkin semigroup. We say that $(\boldsymbol{P}_t)_{t\geq 0}$ is a (Feller–Dynkin) *diffusion semigroup* if the domain of its generator contains $\mathcal{C}_c(\mathbb{R}^n,\mathbb{R})$ and if for every probability measure ν on $\mathbb{R}^n$ there exists a probability measure $\mathbb{P}^\nu$ on $\mathcal{G}_\infty$ such that

(1) the distribution of π_0 under $\mathbb{P}^\nu$ is ν;

(2) on the filtered probability space $(\mathcal{C}(\mathbb{R}^n_{\geq 0}, \mathbb{R}), (\mathcal{G}_t)_{t\geq 0}, \mathcal{G}_\infty, \mathbb{P}^\nu)$, $(\pi_t)_{t\geq 0}$ is a Markov process with semigroup $(\boldsymbol{P}_t)_{t\geq 0}$.

A continuous Markov process $(X_t)_{t\geq 0}$ is said to be a (Feller–Dynkin) diffusion process if its semigroup is a Feller–Dynkin diffusion semigroup.

As a consequence of Theorems 3.28 and 3.30, it is immediate that if $(\boldsymbol{P}_t)_{t\geq 0}$ is a Feller–Dynkin semigroup such that

$$\lim_{t\to 0} \sup_{x\in K} \frac{1}{t} P_t(x, \mathrm{B}(x,\varepsilon)^c) = 0,$$

for every $\varepsilon > 0$ and every compact set K, then $(\boldsymbol{P}_t)_{t\geq 0}$ is a diffusion semigroup.

Exercise 3.34 (Brownian motion with drift). Let $(B_t)_{t\geq 0}$ be a Brownian motion on $(\Omega, (\mathcal{F}_t)_{t\geq 0}, \mathcal{F}, \mathbb{P})$. For $\mu \in \mathbb{R}$, show that the process $(B_t + \mu t)_{t\geq 0}$ is diffusion process with infinitesimal generator L such that for $f \in \mathcal{C}_c(\mathbb{R}, \mathbb{R})$,

$$Lf = \mu \frac{df}{dx} + \frac{1}{2}\frac{d^2 f}{dx^2}.$$

Exercise 3.35 (Ornstein–Uhlenbeck process). Let $(B_t)_{t\geq 0}$ be a Brownian motion on $(\Omega, (\mathcal{F}_t)_{t\geq 0}, \mathcal{F}, \mathbb{P})$. Let $\theta \in \mathbb{R}\backslash\{0\}$ and consider the process

$$X_t = e^{\theta t} B_{\frac{1-e^{-2\theta t}}{2\theta}}.$$

Show that $(X_t)_{t\geq 0}$ is a diffusion with infinitesimal generator L such that for $f \in \mathcal{C}_c(\mathbb{R}, \mathbb{R})$,

$$Lf = \theta x \frac{df}{dx} + \frac{1}{2}\frac{d^2 f}{dx^2}.$$

Diffusion processes admits canonical martingales.

Proposition 3.36. *Let $(X_t)_{t\geq 0}$ be a diffusion process defined on the probability space $(\Omega, (\mathcal{F}_t)_{t\geq 0}, \mathcal{F}, \mathbb{P})$. Let us denote by $\{P_t,\ t \geq 0\}$ its transition function and by L its generator. For $f \in \mathcal{C}_c(\mathbb{R}^n, \mathbb{R})$, the process*

$$\Big(f(X_t) - \int_0^t (Lf)(X_s)ds\Big)_{t\geq 0}$$

is a martingale with respect to the filtration $(\mathcal{F}_t)_{t\geq 0}$.

Proof. For $f \in \mathcal{C}_c(\mathbb{R}^n, \mathbb{R})$ and $t \geq 0$, we have

$$\lim_{\varepsilon\to 0} \frac{\boldsymbol{P}_{t+\varepsilon} f - \boldsymbol{P}_t f}{\varepsilon} = \boldsymbol{P}_t \Big(\lim_{\varepsilon\to 0} \frac{\boldsymbol{P}_\varepsilon f - f}{\varepsilon} \Big) = \boldsymbol{P}_t L f.$$

Thus, we get

$$\boldsymbol{P}_t f = f + \int_0^t \boldsymbol{P}_u L f du.$$

This yields

$$\begin{aligned}\mathbb{E}(f(X_t) \mid \mathcal{F}_s) &= (\boldsymbol{P}_{t-s} f)(X_s) \\ &= f(X_s) + \int_0^{t-s} (\boldsymbol{P}_u L f)(X_s) du \\ &= f(X_s) + \int_s^t (\boldsymbol{P}_{u-s} L f)(X_s) du \\ &= f(X_s) + \int_s^t \mathbb{E}((Lf)(X_u) \mid \mathcal{F}_s) \\ &= f(X_s) + \mathbb{E}\Big(\int_s^t (Lf)(X_u) du \mid \mathcal{F}_s\Big). \end{aligned}$$ □

The following very nice theorem, which is due to Dynkin, states that infinitesimal generators of diffusion semigroups need to be second order differential operators.

Theorem 3.37 (Dynkin theorem). *Let $(X_t)_{t\geq 0}$ be an n-dimensional diffusion process with generator L. There exist continuous functions $b\colon \mathbb{R}^n \to \mathbb{R}$ and $a_{ij}\colon \mathbb{R}^n \to \mathbb{R}$ such that the matrix $(a_{ij}(x))_{1\leq i,j\leq n}$ is semidefinite non-negative and such that for $f \in \mathcal{C}_c(\mathbb{R}^n, \mathbb{R})$*

$$Lf = \sum_{i=1}^n b_i(x) \frac{\partial f}{\partial x_i} + \frac{1}{2} \sum_{i,j=1}^n a_{ij}(x) \frac{\partial^2 f}{\partial x_i \partial x_j}.$$

Proof. We make the proof in dimension $n = 1$ and let the reader extend it as an exercise in higher dimension. Let $(X_t)_{t\geq 0}$ be a one-dimensional diffusion process with generator L which is defined on a filtered probability space $(\Omega, (\mathcal{F}_t)_{t\geq 0}, \mathcal{F}, \mathbb{P})$. Our strategy will be to prove that L needs to satisfy the following three properties

(1) $L\colon \mathcal{C}_c(\mathbb{R}, \mathbb{R}) \to \mathcal{C}_0(\mathbb{R}, \mathbb{R})$ is a linear operator;

(2) L is a local operator, i.e. if $f, g \in \mathcal{C}_c(\mathbb{R}, \mathbb{R})$ agree on a neighborhood of $x \in \mathbb{R}$, then $(Lf)(x) = (Lg)(x)$;

(3) L satisfies the positive maximum principle: If $f \in \mathcal{C}_c(\mathbb{R}, \mathbb{R})$ attains a maximum at $x \in \mathbb{R}$ with $f(x) \geq 0$, then $(Lf)(x) \leq 0$.

Then we will show that the only operators that satisfy the three above properties are second order differential operators.

The linearity of L is obvious so let us focus on the local property of L.

Let $f, g \in \mathcal{C}_c(\mathbb{R}, \mathbb{R})$ agree on a neighborhood of $x \in \mathbb{R}$. We have

$$(\boldsymbol{P}_t f)(x) = \mathbb{E}^x(f(\pi_t)),$$

where $\mathbb{E}^x$ is the expectation under the probability measure $\mathbb{P}^x$ such that

- under $\mathbb{P}^x$, the distribution of π_0 is the Dirac mass at x;
- on the probability space $(\mathcal{C}(\mathbb{R}_{\geq 0}, \mathbb{R}), (\mathcal{G}_t)_{t\geq 0}, \mathcal{G}_\infty, \mathbb{P}^\nu)$, $(\pi_t)_{t\geq 0}$ is a Markov process with transition function $\{P_t, t \geq 0\}$.

We also have

$$(\boldsymbol{P}_t g)(x) = \mathbb{E}^x(g(\pi_t)).$$

Since $(\pi_t)_{t\geq 0}$ is a continuous process, we deduce that there is a positive and finite, $\mathbb{P}^x$ almost surely, stopping time T such that

$$f(\pi_t) = g(\pi_t), \quad t < T.$$

This implies

$$\lim_{t\to 0} \frac{\mathbb{E}^x(f(\pi_t)) - \mathbb{E}^x(g(\pi_t))}{t} = \lim_{t\to 0} \frac{\mathbb{E}^x(f(1_{t<T}\pi_t)) - \mathbb{E}^x(g(1_{t<T}\pi_t))}{t} = 0.$$

On the other hand, we have

$$\lim_{t\to 0} \frac{\mathbb{E}^x(f(\pi_t)) - \mathbb{E}^x(g(\pi_t))}{t} = (Lf)(x) - (Lg)(x).$$

We deduce that

$$(Lf)(x) = (Lg)(x),$$

so that L is indeed a local operator.

Let us now show that L satisfies the positive maximum principle. Let $f \in \mathcal{C}_c(\mathbb{R}, \mathbb{R})$ attain a maximum at $x \in \mathbb{R}$ with $f(x) \geq 0$. As before, let $\mathbb{P}^x$ be the probability measure such that

- under $\mathbb{P}^x$, the distribution of π_0 is the Dirac mass at x;
- on the probability space $(\mathcal{C}(\mathbb{R}_{\geq 0}, \mathbb{R}), (\mathcal{G}_t)_{t\geq 0}, \mathcal{G}_\infty, \mathbb{P}^\nu)$, $(\pi_t)_{t\geq 0}$ is a Markov process with transition function $\{P_t, t \geq 0\}$.

From the previous proposition, under $\mathbb{P}^x$, the process

$$\left(f(\pi_t) - \int_0^t (Lf)(\pi_s)ds\right)_{t\geq 0}$$

is a martingale with respect to the filtration $(\mathcal{G}_t)_{t\geq 0}$. Therefore, for every $t \geq 0$,

$$\mathbb{E}^x(f(\pi_t)) = f(x) + \int_0^t \mathbb{E}^x((Lf)(\pi_u))du.$$

Since for every $t \geq 0$,

$$\mathbb{E}^x(f(\pi_t)) \leq f(x),$$

we deduce that

$$\frac{1}{t}\int_0^t \mathbb{E}^x((Lf)(\pi_u))du \le 0.$$

Letting $t \to 0$ yields

$$(Lf)(x) \le 0.$$

As a conclusion, L is a linear and local operator that satisfies the positive maximum principle. Let us now show that such operators are second order differential operators. Let $x \in \mathbb{R}$. Let ψ_0 be a compactly supported C^∞ function such that in a neighborhood of x, $\psi_0 = 1$. Since $P_t 1 = 1$, it is easy to deduce from the local property of L that

$$(L\psi_0)(x) = 0.$$

Let now ψ_1 be a compactly supported C^∞ function such that in a neighborhood of x, $\psi_1(y) = y - x$. We write

$$b(x) = (L\psi_1)(x).$$

The function ψ_1^2 attains a local minimum in x, therefore

$$(L\psi_1^2)(x) \ge 0,$$

and we can define σ such that

$$\sigma^2(x) = (L\psi_1^2)(x).$$

These functions b and σ are well defined and obviously continuous. Let us now prove that for every $f \in \mathcal{C}_c(\mathbb{R},\mathbb{R})$,

$$(Lf)(x) = b(x)f'(x) + \frac{1}{2}\sigma(x)^2 f''(x).$$

Let $f \in \mathcal{C}_c(\mathbb{R},\mathbb{R})$. From the Taylor expansion formula, in a neighborhood of x we can write

$$f(y) = f(x)\psi_0(y) + f'(x)\psi_1(y) + \frac{1}{2}f''(x)\psi_1^2(y) + R(y)\psi_1^3(y),$$

where R is a continuous function.

We therefore have

$$(Lf)(x) = f(x)(L\psi_0)(x) + f'(x)L\psi_1)(x) + \frac{1}{2}f''(x)(L\psi_1^2)(x) + (LR\psi_1^3)(x).$$

Since we already know that $L\psi_0 = 0$, it remains to prove that

$$(LR\psi_1^3)(x) = 0.$$

For $\varepsilon > 0$ which is small enough,

$$y \to R(y)\psi_1^3(y) - \varepsilon(y-x)^2$$

has a local maximum in x, thus

$$(LR\psi_1^3)(x) \le \varepsilon\sigma^2(x).$$

By letting $\varepsilon \to 0$, we therefore get

$$(LR\psi_1^3)(x) \le 0.$$

In the very same way, considering the function

$$y \to R(y)\psi_1^3(y) + \varepsilon(y-x)^2,$$

we obtain that

$$(LR\psi_1^3)(x) \ge 0.$$

As a consequence we established that

$$(LR\psi_1^3)(x) = 0,$$

which concludes the proof of the theorem. □

Exercise 3.38. Let L be the generator of a Feller–Dynkin semigroup. We assume that $\mathcal{C}_c(\mathbb{R}^n, \mathbb{R}) \subset \mathcal{D}(L)$. Show that L satisfies the positive maximum principle: If $f \in \mathcal{C}_c(\mathbb{R}^n, \mathbb{R})$ attains a maximum at $x \in \mathbb{R}$ with $f(x) \ge 0$, then $(Lf)(x) \le 0$.

According to the previous exercise, if L is the domain of a Feller–Dynkin process and if $\mathcal{C}_c(\mathbb{R}^n, \mathbb{R}) \subset \mathcal{D}(L)$, then L satisfies the positive maximum principle (see the previous exercise for the definition). Operators satisfying the positive maximum principle have been classified by Courrège in a theorem generalizing Dynkin's and that we mention without proof. The interested reader will find the proof in the book by Jacob [42].

In the following statement $\mathcal{B}(\mathbb{R}^n)$ denotes the set of Borel sets on $\mathbb{R}^n$ and a kernel μ on $\mathbb{R}^n \times \mathcal{B}(\mathbb{R}^n)$ is a family $\{\mu(x,\cdot), x \in \mathbb{R}^n\}$ of Borel measures.

Theorem 3.39 (Courrège theorem). *Let $(X_t)_{t\ge 0}$ be a Feller–Dynkin process with generator L. If $\mathcal{D}(L)$ contains $\mathcal{C}_c(\mathbb{R}^n, \mathbb{R})$, then there exist a symmetric and non-negative matrix $(\sigma_{ij}(x))_{1\le i,j\le n}$, functions $b_i, c\colon \mathbb{R}^n \to \mathbb{R}$, with $c \ge 0$ and a kernel μ on $\mathbb{R}^n \times \mathcal{B}(\mathbb{R}^n)$ such that for every $f \in \mathcal{C}_c(\mathbb{R}^n, \mathbb{R})$ and $x \in \mathbb{R}^n$,*

$$\begin{aligned} Lf(x) &= \sum_{i,j=1}^n \sigma_{ij}(x)\frac{\partial^2 f}{\partial x_i \partial x_j} + \sum_{i=1}^n b_i(x)\frac{\partial f}{\partial x_i} - c(x)f(x) \\ &\quad + \int_{\mathbb{R}^n}\Big(f(y) - \chi(y-x)f(x) - \sum_{j=1}^n \frac{\partial f}{\partial x_j}(x)\chi(y-x)(y_j-x_j)\Big)\mu(x,dy), \end{aligned}$$

where $\chi \in \mathcal{C}_c(\mathbb{R}^n, \mathbb{R})$, $0 \le \chi \le 1$, *takes the constant value* 1 *on the ball* $\mathrm{B}(0,1)$. *In addition, the functions* b_j *and* c *are continuous and for every* $y \in \mathbb{R}^n$, *the function* $x \to \sum_{i,j} \sigma_{ij}(x) y_i y_j$ *is upper semicontinuous.*

3.4 Lévy processes

Lévy processes form a fundamental class of Feller–Dynkin processes. They are the Markov processes associated with the so-called convolution semigroups.

In what follows, we consider a filtered probability space $(\Omega, (\mathcal{F}_t)_{t\ge 0}, \mathcal{F}, \mathbb{P})$.

Definition 3.40 (Lévy process). Let $(X_t)_{t\ge 0}$ be a stochastic process. It is said that $(X_t)_{t\ge 0}$ is a *Lévy process* on $(\Omega, (\mathcal{F}_t)_{t\ge 0}, \mathcal{F}, \mathbb{P})$ if the following conditions are fulfilled:

(1) almost surely $X_0 = 0$;

(2) the paths of $(X_t)_{t\ge 0}$ are left limited and right continuous;

(3) $(X_t)_{t\ge 0}$ is adapted to the filtration $(\mathcal{F}_t)_{t\ge 0}$;

(4) for every $T \ge 0$, the process $(X_{t+T} - X_T)_{t\ge 0}$ is independent of the σ-algebra $\mathcal{F}_T$;

(5) for every $t, T \ge 0$, $X_{t+T} - X_T$ has the same distribution as X_t.

Remark 3.41. Of course, the notion of Lévy process in $\mathbb{R}^n$ is similarly defined.

Exercise 3.42. Show that if $(X_t)_{t\ge 0}$ is a Lévy process defined on

$$(\Omega, (\mathcal{F}_t)_{t\ge 0}, \mathcal{F}, \mathbb{P}),$$

then it is also a Lévy process on the space

$$(\Omega, (\mathcal{F}_t^X)_{t\ge 0}, \mathcal{F}, \mathbb{P}),$$

where $(\mathcal{F}_t^X)_{t\ge 0}$ is the natural filtration of $(X_t)_{t\ge 0}$

The following two exercises provide fundamental examples of Lévy processes.

Exercise 3.43 (Brownian motion). Let $(B_t)_{t\ge 0}$ be a standard Brownian motion. Show that it is a Lévy process with respect to its natural filtration.

Exercise 3.44 (Compound Poisson process). Let $(T_n)_{n\in\mathbb{N}}$ be an i.i.d. sequence of exponential random variables with parameter λ,

$$\mathbb{P}(T_n \in dt) = \lambda e^{-\lambda t} dt, \quad t \ge 0,\ n \in \mathbb{N},$$

defined on a probability space $(\Omega, \mathcal{F}, \mathbb{P})$. We write

$$S_n = \sum_{i=1}^{n} T_i, \quad n \ge 1,$$

and $S_0 = 0$. For $t \geq 0$, let

$$N_t = \max\{n \geq 0, S_n \leq t\}.$$

(1) Show that the process $(N_t)_{t \geq 0}$ is a Lévy process on the space

$$(\Omega, (\mathcal{F}_t^N)_{t \geq 0}, \mathcal{F}, \mathbb{P}),$$

where $(\mathcal{F}_t^N)_{t \geq 0}$ is the natural filtration of $(N_t)_{t \geq 0}$.

(2) Show that for $t \geq 0$, the random variable N_t is distributed as a Poisson random variable with parameter λt, that is,

$$\mathbb{P}(N_t = n) = e^{-\lambda t} \frac{(\lambda t)^n}{n!}, \quad n \in \mathbb{N}.$$

The process $(N_t)_{t \geq 0}$ is called a Poisson process with intensity $\lambda > 0$.

(3) Let now $(Y_n)_{n \geq 0}$ be an i.i.d. sequence of random variables with distribution μ and independent from $(N_t)_{t \geq 0}$. For $t \geq 0$ we define

$$X_t = 1_{N_t \geq 1} \Big(\sum_{i=1}^{N_t} Y_i \Big).$$

Show that the process $(X_t)_{t \geq 0}$ is a Lévy process on the space

$$(\Omega, (\mathcal{F}_t^N)_{t \geq 0}, \mathcal{F}, \mathbb{P}),$$

where $(\mathcal{F}_t^X)_{t \geq 0}$ is the natural filtration of $(X_t)_{t \geq 0}$.

(4) Show that for $\theta \in \mathbb{R}$ and $t \geq 0$:

$$\mathbb{E}(e^{i\theta X_t}) = \exp\Big(t\lambda \int_{\mathbb{R}} (e^{i\theta x} - 1)\mu(dx) \Big).$$

The process $(X_t)_{t \geq 0}$ is called a compound Poisson process.

In the study of Lévy processes, one of the most important tools is the Lévy–Khinchin theorem that we recall below. We first recall the following basic definition.

Definition 3.45. A random variable X is said to be *infinitely divisible* if for every $n \geq 1$ we may find independent and identically distributed random variables $X_1, \dots, X_n$ such that in distribution

$$X_1 + \dots + X_n =^{\text{law}} X.$$

The Lévy–Khinchin theorem, which is admitted here, completely characterizes infinitely divisible random variables in terms of their characteristic function. A proof using Lévy processes may be found in the book by Applebaum [1].

Theorem 3.46 (Lévy–Khinchin theorem). *Let X be an infinitely divisible random variable. There exist $\mu \in \mathbb{R}$, $\sigma > 0$ and a Borel measure ν on $\mathbb{R}\backslash\{0\}$ such that*

$$\int_{\mathbb{R}} (1 \wedge x^2)\nu(dx) < +\infty,$$

$$\mathbb{E}(e^{i\lambda X}) = \exp\left(i\lambda\mu - \frac{1}{2}\sigma^2\lambda^2 + \int_{\mathbb{R}} (e^{i\lambda x} - 1 - i\lambda x 1_{|x|\leq 1})\nu(dx)\right) \quad \text{for all } \lambda \in \mathbb{R}.$$

Conversely, let $\mu \in \mathbb{R}$, $\sigma > 0$ and ν be a Borel measure on $\mathbb{R}\backslash\{0\}$ such that

$$\int_{\mathbb{R}} (1 \wedge x^2)\nu(dx) < +\infty.$$

The function

$$\lambda \to \exp\left(i\lambda\mu - \frac{1}{2}\sigma^2\lambda^2 + \int_{\mathbb{R}} (e^{i\lambda x} - 1 - i\lambda x 1_{|x|\leq 1})\nu(dx)\right)$$

is the characteristic function of the distribution of an infinitely divisible random variable.

In what follows, we consider a Lévy process $(X_t)_{t\geq 0}$. Since for every $n \geq 1$,

$$X_1 = \sum_{k=1}^{n} (X_{\frac{k}{n}} - X_{\frac{k-1}{n}}),$$

and since the increments $X_{\frac{k}{n}} - X_{\frac{k-1}{n}}$ are independent and identically distributed, we deduce that the random variable X_1 is infinitely divisible. Therefore, from the Lévy–Khinchin theorem, there exist $\sigma > 0$, $\mu \in \mathbb{R}$ and a Borel measure ν on $\mathbb{R}\backslash\{0\}$ such that

$$\int_{\mathbb{R}} (1 \wedge x^2)\nu(dx) < +\infty$$

and

$$\mathbb{E}(e^{i\lambda X_1}) = e^{\psi(\lambda)}, \quad \lambda \in \mathbb{R},$$

where

$$\psi(\lambda) = i\mu\lambda - \frac{1}{2}\sigma^2\lambda^2 + \int_{\mathbb{R}} (e^{i\lambda x} - 1 - i\lambda x 1_{|x|\leq 1})\nu(dx).$$

The function ψ is called the characteristic exponent of the Lévy process, and the measure ν the Lévy measure. For instance, for a Brownian motion, we have

$$\mu = 0, \quad \sigma = 1, \quad \nu = 0,$$

whereas for a compound Poisson process as defined in Exercise 3.44,

$$\mu = -\int_{-1}^{1} x\mu(dx), \quad \sigma = 0, \quad \nu(dx) = \mu(dx).$$

The characteristic exponent of a Lévy process characterizes the law of such a process. Indeed, let $\lambda \in \mathbb{R}$ and let us consider the following function:

$$f(t) = \mathbb{E}(e^{i\lambda X_t}), \quad t \geq 0.$$

Since $(X_t)_{t\geq 0}$ is a Lévy process, for $s, t \geq 0$,

$$\begin{aligned} f(t+s) &= \mathbb{E}(e^{i\lambda X_{t+s}}) \\ &= \mathbb{E}(e^{i\lambda(X_{t+s}-X_t)+X_t}) \\ &= \mathbb{E}(e^{i\lambda(X_{t+s}-X_t)})\mathbb{E}(e^{i\lambda X_t}) \\ &= \mathbb{E}(e^{i\lambda X_s})\mathbb{E}(e^{i\lambda X_t}) \\ &= f(t)f(s). \end{aligned}$$

Moreover, since $(X_t)_{t\geq 0}$ has right continuous paths at 0, the function f is itself right continuous at 0. From $f(1) = e^{\psi(\lambda)}$, we may then deduce that

$$f(t) = e^{t\psi(\lambda)}, \quad t \geq 0.$$

We finally conclude that for $\lambda \in \mathbb{R}$ and $t \geq 0$,

$$\mathbb{E}(e^{i\lambda X_t}) = e^{t\psi(\lambda)}.$$

With this in hands we can now turn to the Feller–Dynkin property of Lévy processes and also characterize the Lévy processes as being the Markov processes associated to the convolution semigroups.

Proposition 3.47. *Let $(X_t)_{t\geq 0}$ be a Lévy process. For $t \geq 0$, let $p_t(dx)$ be the distribution of the random variable X_t. The family $(p_t(dx))_{t\geq 0}$ is a convolution semigroup of probability measures, that is,*

$$p_{t+s} = p_t * p_s, \quad s, t \geq 0.$$

Moreover the process $(X_t)_{t\geq 0}$ is a Feller–Dynkin process with semigroup

$$\boldsymbol{P}_t f(x) = \int_{\mathbb{R}} f(x+y)p_t(dy).$$

Conversely, let $(p_t(dx))_{t\geq 0}$ be a convolution semigroup of probability measures that is right continuous at 0 in the topology of convergence in distribution with $p_0(dx) = \delta_0$ (Dirac distribution at 0). Then there exists a filtered probability space $(\Omega, \mathcal{F}, (\mathcal{F}_t)_{t\geq 0}, \mathbb{P})$ and a Lévy process $(X_t)_{t\geq 0}$ on it such that the distribution of X_t is $p_t(dx)$.

Proof. The family $(p_t(dx))_{t\geq 0}$ is seen to be a convolution semigroup of probability measures from (3.47), by taking the inverse Fourier transform. From this last property, we deduce that the family of operators $(\boldsymbol{P}_t)_{t\geq 0}$ defined by

$$\boldsymbol{P}_t f(x) = \int_{\mathbb{R}} f(x+y)p_t(dy)$$

enjoys the semigroup property. Finally due to the fact that $p_t(dx)$ is a probability measure, it is easily checked that $(\boldsymbol{P}_t)_{t\geq 0}$ is a Feller–Dynkin semigroup.

Let us now prove that it is the semigroup corresponding to the process $(X_t)_{t\geq 0}$. Let f be a bounded Borel function, we have for $s,t \geq 0$,

$$\mathbb{E}\left(f(X_{t+s}) \mid \mathcal{F}_s\right) = \mathbb{E}(f((X_{t+s}-X_s)+X_s) \mid \mathcal{F}_s).$$

But the random variable $X_{t+s}-X_s$ is independent from $\mathcal{F}_s$ and distributed as $p_t(dx)$, therefore

$$\mathbb{E}(f(X_{t+s}) \mid \mathcal{F}_s) = \int_{\mathbb{R}} f(y+X_s)p_t(dy) = \boldsymbol{P}_t f(X_s).$$

We now turn to the proof of the second part of the proposition. As above, the family of operators $(\boldsymbol{P}_t)_{t\geq 0}$ given by

$$\boldsymbol{P}_t f(x) = \int_{\mathbb{R}} f(x+y)p_t(dy)$$

defines a Feller–Dynkin semigroup. From Theorem 3.28, there exists a probability space $(\Omega, \mathcal{F}, \mathbb{P})$ and a stochastic process $(X_t)_{t\geq 0}$ such that the following holds:

- The distribution of X_0 is δ_0.
- $(X_t)_{t\geq 0}$ is a Markov process with transition function $\{P_t,\, t\geq 0\}$.
- The paths of $(X_t)_{t\geq 0}$ are right continuous and left limited.

Let us check that this process $(X_t)_{t\geq 0}$ is a Lévy process with respect to its natural filtration. For $\lambda \in \mathbb{R}$ and $s,t \geq 0$ we have

$$\begin{aligned}
\mathbb{E}(e^{i\lambda(X_{t+s}-X_s)} \mid \mathcal{F}_s^X) &= e^{-i\lambda X_s}\mathbb{E}(e^{i\lambda X_{t+s}} \mid \mathcal{F}_s^X) \\
&= e^{-i\lambda X_s}\int_{\mathbb{R}} e^{i\lambda(X_s+y)}p_t(dy) \\
&= \int_{\mathbb{R}} e^{i\lambda y}p_t(dy).
\end{aligned}$$

We deduce therefore that if f is a bounded Borel function, then

$$\mathbb{E}(f(X_{t+s}-X_s) \mid \mathcal{F}_s^X) = \boldsymbol{P}_t f(0).$$

We conclude that, as expected, the process $(X_t)_{t\geq 0}$ is a Lévy process. □

Since Lévy processes are Feller–Dynkin processes they have, according to Proposition 3.31, an infinitesimal generator. The next proposition computes the generator in terms of the Lévy measure.

Proposition 3.48. *Let $(X_t)_{t\geq 0}$ be a Lévy process with characteristic exponent*

$$\psi(\lambda) = i\mu\lambda - \frac{1}{2}\sigma^2\lambda^2 + \int_{\mathbb{R}} (e^{i\lambda x} - 1 - i\lambda x 1_{|x|\leq 1})\nu(dx).$$

The domain $\mathcal{D}(L)$ of the infinitesimal generator L of $(X_t)_{t\geq 0}$ contains the space $\mathcal{S}$ of smooth and rapidly decreasing functions and for $f \in \mathcal{S}$,

$$Lf(x) = \mu f'(x) + \frac{1}{2}\sigma^2 f''(x) + \int_{\mathbb{R}} (f(x+y) - f(x) - yf'(x)1_{|y|\leq 1})\nu(dy).$$

Proof. Let $\lambda \in \mathbb{R}$ and write

$$e_\lambda(x) = e^{i\lambda x}.$$

We have

$$\boldsymbol{P}_t e_\lambda(x) = \int_{\mathbb{R}} e^{i\lambda(x+y)} p_t(dy) = e_\lambda(x) e^{t\psi(\lambda)}.$$

Therefore, we obtain

$$\lim_{t\to 0} \frac{\boldsymbol{P}_t e_\lambda - e_\lambda}{t} = \psi(\lambda) e_\lambda.$$

This last equality proves the proposition by using the inverse Fourier transform that maps $\mathcal{S}$ into itself. □

Exercise 3.49. Let $(X_t)_{t\geq 0}$ be a Lévy process in $\mathbb{R}^n$. Show that $(X_t)_{t\geq 0}$ is a Feller–Dynkin process whose generator can be written on the space of rapidly decreasing functions

$$Lf(x) = \langle \mu, \nabla f(x)\rangle + \frac{1}{2}\sum_{i,j=1}^{n} a_{ij} \frac{\partial^2 f}{\partial x_i \partial x_j}(x) + \int_{\mathbb{R}^n} (f(x+y) - f(x) - \langle y, \nabla f(x)\rangle 1_{\|y\|\leq 1})\nu(dy),$$

where $\mu \in \mathbb{R}^n$, $(a_{ij})_{0\leq i,j\leq n}$ is a non-negative symmetric matrix and ν is a Borel measure on $\mathbb{R}^n - \{0\}$.

An important class of Lévy processes is the class of subordinators.

Definition 3.50 (Subordinators). A non-negative Lévy process $(S_t)_{t\geq 0}$ is called a *subordinator*.

For instance, the Poisson process (see Exercise 3.44) is a subordinator. For Lévy processes the following identity in distribution holds: $S_t - S_s =^{\text{law}} S_{t-s}$, $t \geq s$. Hence we deduce that the paths $t \to S_t$ of a subordinator are almost surely non-decreasing. If $(S_t)_{t\geq 0}$ is a subordinator, then there exists a function $\Psi\colon \mathbb{R}_{\geq 0} \to \mathbb{R}_{\geq 0}$ such that for every $t, \lambda \geq 0$,

$$\mathbb{E}(e^{-\lambda S_t}) = e^{-t\Psi(\lambda)}. \tag{3.2}$$

Actually, we have $\Psi(\lambda) = \psi(i\lambda)$, where ψ is the analytic extension of the characteristic exponent of $(S_t)_{t\geq 0}$ on the upper half plane of the complex plane. From the representation (3.2), we can see that Ψ needs to be a Bernstein function, that is, for every $\lambda > 0$ and $k \geq 1$,

$$(-1)^k \Psi^{(k)}(\lambda) \leq 0.$$

Bernstein functions are characterized by the following well-known theorem of Bernstein (see [69] for a proof).

Theorem 3.51 (Bernstein theorem). *A smooth function $f\colon (0, +\infty) \to \mathbb{R}_{\geq 0}$ is completely monotone, that is, for every $\lambda > 0$ and $k \geq 0$,*

$$(-1)^k f^{(k)}(\lambda) \geq 0$$

if and only if there exists a non-negative Borel measure m on $[0, +\infty)$ such that

$$f(\lambda) = \int_0^{+\infty} e^{-\lambda z} m(dz), \quad \lambda > 0.$$

As a consequence of the Bernstein theorem, any Bernstein function Ψ such that $\Psi(0) = 0$ can be written as

$$\Psi(\lambda) = a\lambda + \int_0^{+\infty} (1 - e^{-\lambda z})\mu(dz),$$

where $a \geq 0$ and where μ is a Borel measure on $(0, +\infty)$ such that

$$\int_0^{+\infty} (1 \wedge z)\mu(dz) < +\infty.$$

Conversely, given any Bernstein function Ψ, we can find a corresponding subordinator $(S_t)_{t\geq 0}$. Indeed, let Ψ be a Bernstein function such that $\Psi(0) = 0$. Then for every $t > 0$, the function $\lambda \to e^{-t\Psi(\lambda)}$ is completely monotone and, as a consequence of the Bernstein theorem, we can find a probability measure m_t on $[0, +\infty)$ such that

$$\int_0^{+\infty} e^{-\lambda z} m_t(dz) = e^{-t\Psi(\lambda)}.$$

It is then seen that $(m_t)_{t\geq 0}$ is a convolution semigroup of probability measure. As a consequence of Proposition 3.47 we can thus find a Lévy process associated to m_t. Since m_t is supported on $[0, +\infty)$, this process necessarily is a subordinator. As a conclusion, there is a one to one correspondence between subordinators and Bernstein functions.

Exercise 3.52 (α-stable subordinators). Let $0 < \alpha < 1$.

(1) Show that $\Psi_\alpha(\lambda) = \lambda^\alpha$ is a Bernstein function.

(2) A subordinator $(S_t^\alpha)_{t\geq 0}$ associated to the Bernstein function Ψ_α is called an α-stable subordinator. Show that for $c > 0$, the processes $(S_{ct}^\alpha)_{t\geq 0}$ and $(c^\alpha S_t^\alpha)_{t\geq 0}$ have the same distribution.

(3) Show the subordination identity

$$e^{-y|\alpha|} = \frac{y}{2\sqrt{\pi}} \int_0^{+\infty} \frac{e^{-\frac{y^2}{4t} - t\alpha^2}}{t^{3/2}} dt, \quad y > 0,\ \alpha \in \mathbb{R},$$

and deduce the distribution of $S_t^{1/2}$, $t > 0$. We mention that there is no such simple expression for the distribution of S_t^α, $\alpha \neq 1/2$.

Exercise 3.53 (Subordinated Brownian motion). Let $(B_t)_{t\geq 0}$ be a standard Brownian motion in $\mathbb{R}^n$ and let $(S_t)_{t\geq 0}$ be an independent subordinator. The process

$$X_t = B_{S_t}, \quad t \geq 0,$$

is called a subordinated Brownian motion.

(1) Show that $(X_t)_{t\geq 0}$ is a Lévy process with characteristic exponent $\Psi(\frac{\|\lambda\|^2}{2})$, where Ψ is the Bernstein function of the subordinator.

(2) Let L be the generator of $(X_t)_{t\geq 0}$. Compute the Fourier transform of Lf in terms of the Fourier transform of f when f is a smooth and rapidly decreasing function.

(3) Let m_t be the probability distribution of S_t at time $t > 0$. Show that the distribution of X_t has a density given by

$$\int_0^{+\infty} \frac{e^{-\frac{\|x\|^2}{2s}}}{(2\pi s)^{n/2}} m_t(ds)$$

if the integral is convergent. Compute this density if $(S_t)_{t\geq 0}$ is a $1/2$-stable subordinator (see Exercise 3.52).

The following theorem which finishes the section characterizes the continuous Lévy processes. It is remarkable that the only such Lévy processes are, up to a renormalization factor, the Brownian motions with drift, see Exercise 3.34.

Theorem 3.54 (Lévy theorem). *Let $(X_t)_{t\geq 0}$ be a continuous Lévy process defined on $(\Omega, (\mathcal{F}_t)_{t\geq 0}, \mathcal{F}, \mathbb{P})$. Then the Lévy measure of $(X_t)_{t\geq 0}$ is 0, and therefore $(X_t)_{t\geq 0}$ may be written*

$$X_t = \sigma B_t + \mu t,$$

where $(B_t)_{t\geq 0}$ is a standard Brownian motion.

Proof. Let

$$\psi(\lambda) = i\mu\lambda - \frac{1}{2}\sigma^2\lambda^2 + \int_{\mathbb{R}} (e^{i\lambda x} - 1 - i\lambda x \mathbf{1}_{|x|\leq 1})\nu(dx)$$

be the characteristic exponent of $(X_t)_{t\geq 0}$. For $\varepsilon \in (0,1)$, we have

$$\psi = \psi_\varepsilon + \phi_\varepsilon,$$

where

$$\psi_\varepsilon(\lambda) = i\mu\lambda - \frac{1}{2}\sigma^2\lambda^2 + \int_{|x|\leq\varepsilon} (e^{i\lambda x} - 1 - i\lambda x)\nu(dx),$$

and

$$\phi_\varepsilon(\lambda) = \int_{|x|>\varepsilon} (e^{i\lambda x} - 1 - i\lambda x \mathbf{1}_{|x|\leq 1})\nu(dx).$$

This decomposition of ψ will actually correspond to a pathwise decomposition of $(X_t)_{t\geq 0}$.

For $t \geq 0$, let μ_t be the probability measure on $\mathbb{R}$ with characteristic function:

$$\int_{\mathbb{R}} e^{i\lambda x}\mu_t(dx) = e^{t\psi_\varepsilon(\lambda)}, \lambda \in \mathbb{R}.$$

For $s, t \geq 0$ we have

$$\mu_t * \mu_s = \mu_{t+s}.$$

From Proposition 3.47, we therefore can find a probability space $(\widetilde{\Omega}, (\widetilde{\mathcal{F}}_t)_{t\geq 0}, \widetilde{\mathcal{F}}, \widetilde{\mathbb{P}})$ and a process $(Y_t)_{t\geq 0}$ that is a Lévy process with characteristics exponent $\psi_\varepsilon(\lambda)$.

In exactly the same way, we may, by enlarging the filtered probability space $(\widetilde{\Omega}, (\widetilde{\mathcal{F}}_t)_{t\geq 0}, \widetilde{\mathcal{F}}, \widetilde{\mathbb{P}})$, construct a Lévy process $(Z_t)_{t\geq 0}$ that is independent of $(Y_t)_{t\geq 0}$ and whose characteristic exponent is ϕ_ε. We have

$$\begin{aligned}\phi_\varepsilon(\lambda) &= \int_{|x|>\varepsilon} (e^{i\lambda x} - 1 - i\lambda x \mathbf{1}_{|x|\leq 1})\nu(dx)\\ &= \int_{|x|>\varepsilon} (e^{i\lambda x} - 1)\nu(dx) - i\lambda \int_{|x|>\varepsilon} x\mathbf{1}_{|x|\leq 1}\nu(dx),\end{aligned}$$

and

$$\nu(\{x, |x| > \varepsilon\}) < +\infty.$$

Therefore, from Exercise 3.44

$$Z_t = L_t - t \int_{|x|>\varepsilon} x \mathbf{1}_{|x|\leq 1} \nu(dx),$$

where $(L_t)_{t\geq 0}$ is a compound Lévy process. We deduce that the paths of the process $(Z_t)_{t\geq 0}$ only have a finite number of jumps within any finite time interval. Moreover the size of these possible has to be larger than ε.

Since the two processes $(Y_t)_{t\geq 0}$ and $(Z_t)_{t\geq 0}$ are independent, the process

$$\widetilde{X}_t = Y_t + Z_t$$

is a Lévy process that has the same distribution as $(X_t)_{t\geq 0}$. Also, almost surely, the possible jumps of $(Y_t)_{t\geq 0}$ and $(Z_t)_{t\geq 0}$ do not intersect. Thus, every jump of $(Y_t)_{t\geq 0}$ induces a jump of $(\widetilde{X}_t)_{t\geq 0}$. Since $(X_t)_{t\geq 0}$ is assumed to be continuous, $(Y_t)_{t\geq 0}$ has no jumps. This implies

$$\nu(\{x, |x| > \varepsilon\}) = 0.$$

Since it should hold for every ε, we deduce that $\nu = 0$. □

Exercise 3.55. State and prove a multi-dimensional extension of Theorem 3.54.

Notes and comments

Sections 3.1, 3.2, 3.3. One of the first major actors of the rigorous and general theory of continuous time Markov processes as presented here is Dynkin; we refer to his book [23] for an overview of the origins of the theory. The theory of Markov processes and diffusion processes is now very rich and numerous references are available to the interested reader. We wish to mention in particular the books by Bass [5], Chung [14], and Ethier and Kurtz [24]. For the connection between pseudo-differential operators satisfying the maximum principle and Markov processes, we refer to the books by Jacob [42], [43].

Section 3.4. Lévy processes have been widely and extensively studied and several results of Chapter 2 can be extended to Lévy processes. We refer to the books by Applebaum [1], Bertoin [9] or Sato [68] for further details.

Chapter 4
Symmetric diffusion semigroups

As we have seen in Theorem 3.37, the generator of a Feller–Dynkin diffusion process is of the form

$$L = \sum_{i=1}^{n} b_i(x)\frac{\partial}{\partial x_i} + \sum_{i,j=1}^{n} \sigma_{ij}(x)\frac{\partial^2}{\partial x_i \partial x_j},$$

where b_i and σ_{ij} are continuous functions on $\mathbb{R}^n$ such that for every $x \in \mathbb{R}^n$, the matrix $(\sigma_{ij}(x))_{1\le i,j\le n}$ is symmetric and non-negative. Such second order differential operators are generally called diffusion operators.

Conversely, it is often important to know whether given a diffusion operator, is there a unique transition function that admits this operator as a generator? This problem is difficult to answer in full generality. As we will discuss it in Chapter 6, the theory of stochastic differential equations provides a fantastic probabilistic tool to study this question (see Theorem 6.15 and Proposition 6.16 in Chapter 6) but the theory of Dirichlet forms and associated symmetric diffusion semigroups that we introduce in this chapter applies in much more general situations. We assume here some basic knowledge about the theory of unbounded operators on Hilbert or Banach spaces and at some points some basic knowledge about the local regularity theory for elliptic operators. Elements of these theories are given in the appendices at the end of the book.

4.1 Essential self-adjointness, spectral theorem

Throughout the section, we consider a second order differential operator that can be written

$$L = \sum_{i,j=1}^{n} \sigma_{ij}(x)\frac{\partial^2}{\partial x_i \partial x_j} + \sum_{i=1}^{n} b_i(x)\frac{\partial}{\partial x_i},$$

where b_i and σ_{ij} are continuous functions on $\mathbb{R}^n$ and for every $x \in \mathbb{R}^n$, the matrix $(\sigma_{ij}(x))_{1\le i,j\le n}$ is a symmetric and non-negative matrix. Such operator is called a diffusion operator.

We will assume that there is Borel measure μ which is equivalent to the Lebesgue measure and that symmetrizes L in the sense that for every smooth and compactly supported functions $f, g\colon \mathbb{R}^n \to \mathbb{R}$,

$$\int_{\mathbb{R}^n} gLf d\mu = \int_{\mathbb{R}^n} fLg d\mu.$$

In what follows, as usual, we denote by $\mathcal{C}_c(\mathbb{R}^n, \mathbb{R})$ the set of smooth and compactly supported functions $f: \mathbb{R}^n \to \mathbb{R}$.

Exercise 4.1. Show that if μ is a symmetric measure for L as above, then in the sense of distributions

$$L'\mu = 0,$$

where L' is the adjoint of L in distribution sense.

Exercise 4.2. Show that if $f: \mathbb{R}^n \to \mathbb{R}$ is a smooth function and if $g \in \mathcal{C}_c(\mathbb{R}^n, \mathbb{R})$, then we still have the formula

$$\int_{\mathbb{R}^n} fLg d\mu = \int_{\mathbb{R}^n} gLf d\mu.$$

Exercise 4.3. On $\mathcal{C}_c(\mathbb{R}^n, \mathbb{R})$, let us consider the operator

$$L = \Delta + \langle \nabla U, \nabla \cdot \rangle,$$

where $U: \mathbb{R}^n \to \mathbb{R}$ is a C^1 function. Show that L is symmetric with respect to the measure

$$\mu(dx) = e^{U(x)} dx.$$

Exercise 4.4 (Divergence form operator). On $\mathcal{C}_c(\mathbb{R}^n, \mathbb{R})$, let us consider the operator

$$Lf = \operatorname{div}(\sigma \nabla f),$$

where div is the divergence operator defined on a C^1 function $\phi: \mathbb{R}^n \to \mathbb{R}^n$ by

$$\operatorname{div} \phi = \sum_{i=1}^n \frac{\partial \phi_i}{\partial x_i}$$

and where σ is a C^1 field of non-negative and symmetric matrices. Show that L is a diffusion operator which is symmetric with respect to the Lebesgue measure of $\mathbb{R}^n$.

For every smooth functions $f, g: \mathbb{R}^n \to \mathbb{R}$, let us define the so-called *carré du champ*[1], which is the symmetric first-order differential form defined by

$$\Gamma(f, g) = \frac{1}{2}\left(L(fg) - fLg - gLf\right).$$

A straightforward computation shows that

$$\Gamma(f, g) = \sum_{i,j=1}^n \sigma_{ij}(x) \frac{\partial f}{\partial x_i} \frac{\partial g}{\partial x_j},$$

so that for every smooth function f,

$$\Gamma(f, f) \geq 0.$$

[1] The literal translation from French is *square of the field.*

Exercise 4.5. (1) Show that if $f, g\colon \mathbb{R}^n \to \mathbb{R}$ are C^1 functions and $\phi_1, \phi_2\colon \mathbb{R} \to \mathbb{R}$ are also C^1, then

$$\Gamma(\phi_1(f), \phi_2(g)) = \phi_1'(f)\phi_2'(g)\Gamma(f, g).$$

(2) Show that if $f\colon \mathbb{R}^n \to \mathbb{R}$ is a C^2 function and $\phi\colon \mathbb{R} \to \mathbb{R}$ is also C^2, then

$$L\phi(f) = \phi'(f)Lf + \phi''(f)\Gamma(f, f).$$

In the sequel we shall consider the bilinear form given for $f, g \in \mathcal{C}_c(\mathbb{R}^n, \mathbb{R})$ by

$$\mathcal{E}(f, g) = \int_{\mathbb{R}^n} \Gamma(f, g)d\mu.$$

This is the so-called Dirichlet form associated to L. It is readily checked that $\mathcal{E}$ is symmetric,

$$\mathcal{E}(f, g) = \mathcal{E}(g, f),$$

and non-negative,

$$\mathcal{E}(f, f) \geq 0.$$

We observe that thanks to the symmetry of L,

$$\mathcal{E}(f, g) = -\int_{\mathbb{R}^n} fLgd\mu = -\int_{\mathbb{R}^n} gLfd\mu.$$

The operator L on its domain $\mathcal{D}(L) = \mathcal{C}_c(\mathbb{R}^n, \mathbb{R})$ is a densely defined non-positive symmetric operator on the Hilbert space $L^2_\mu(\mathbb{R}^n, \mathbb{R})$. However, it is of course not self-adjoint. Indeed, Exercise 4.2 easily shows that

$$\{f \in \mathcal{C}^\infty(\mathbb{R}^n, \mathbb{R}), \|f\|_{L^2_\mu(\mathbb{R}^n,\mathbb{R})} + \|Lf\|_{L^2_\mu(\mathbb{R}^n,\mathbb{R})} < \infty\} \subset \mathcal{D}(L^*).$$

A famous theorem of von Neumann asserts that any non-negative and symmetric operator may be extended to a self-adjoint operator. The following construction, due to Friedrichs, provides a canonical non-negative self-adjoint extension.

Theorem 4.6 (Friedrichs extension). *On the Hilbert space $L^2_\mu(\mathbb{R}^n, \mathbb{R})$, there exists a densely defined non-positive self-adjoint extension of L.*

Proof. The idea is to work with a Sobolev-type norm associated to the energy form $\mathcal{E}$. On $\mathcal{C}_c(\mathbb{R}^n, \mathbb{R})$, let us consider the norm

$$\|f\|^2_{\mathcal{E}} = \|f\|^2_{L^2_\mu(\mathbb{R}^n,\mathbb{R})} + \mathcal{E}(f, f).$$

By completing $\mathcal{C}_c(\mathbb{R}^n, \mathbb{R})$ with respect to this norm, we obtain a Hilbert space $(\mathcal{H}, \langle\cdot,\cdot\rangle_{\mathcal{E}})$. Since for $f \in \mathcal{C}_c(\mathbb{R}^n, \mathbb{R})$, $\|f\|_{L^2_\mu(\mathbb{R}^n,\mathbb{R})} \leq \|f\|_{\mathcal{E}}$, the injection map

$\iota \colon (\mathcal{C}_c(\mathbb{R}^n,\mathbb{R}), \|\cdot\|_{\mathcal{E}}) \to (L^2_\mu(\mathbb{R}^n,\mathbb{R}), \|\cdot\|_{L^2_\mu(\mathbb{R}^n,\mathbb{R})})$ is continuous and it may therefore be extended to a continuous map $\bar{\iota} \colon (\mathcal{H}, \|\cdot\|_{\mathcal{E}}) \to (L^2_\mu(\mathbb{R}^n,\mathbb{R}), \|\cdot\|_{L^2_\mu(\mathbb{R}^n,\mathbb{R})})$. Let us show that $\bar{\iota}$ is injective so that $\mathcal{H}$ may be identified with a subspace of $L^2_\mu(\mathbb{R}^n,\mathbb{R})$. So, let $f \in \mathcal{H}$ such that $\bar{\iota}(f) = 0$. We can find a sequence $f_n \in \mathcal{C}_c(\mathbb{R}^n,\mathbb{R})$ such that $\|f_n - f\|_{\mathcal{E}} \to 0$ and $\|f_n\|_{L^2_\mu(\mathbb{R}^n,\mathbb{R})} \to 0$. We have

$$\begin{aligned}
\|f\|_{\mathcal{E}} &= \lim_{m,n\to+\infty} \langle f_n, f_m\rangle_{\mathcal{E}} \\
&= \lim_{m\to+\infty}\lim_{n\to+\infty} \langle f_n, f_m\rangle_{L^2_\mu(\mathbb{R}^n,\mathbb{R})} + \mathcal{E}(f_n, f_m) \\
&= \lim_{m\to+\infty}\lim_{n\to+\infty} \langle f_n, f_m\rangle_{L^2_\mu(\mathbb{R}^n,\mathbb{R})} - \langle f_n, Lf_m\rangle_{L^2_\mu(\mathbb{R}^n,\mathbb{R})} \\
&= 0,
\end{aligned}$$

thus $f = 0$ and $\bar{\iota}$ is injective. Let us now consider the map

$$B = \bar{\iota} \cdot \bar{\iota}^* \colon L^2_\mu(\mathbb{R}^n,\mathbb{R}) \to L^2_\mu(\mathbb{R}^n,\mathbb{R}).$$

It is well defined due to the fact that since $\bar{\iota}$ is bounded, it is easily checked that

$$\mathcal{D}(\bar{\iota}^*) = L^2_\mu(\mathbb{R}^n,\mathbb{R}).$$

Moreover, B is easily seen to be symmetric, and thus self-adjoint because its domain is equal to $L^2_\mu(\mathbb{R}^n,\mathbb{R})$. Also, it is readily checked that the injectivity of $\bar{\iota}$ implies the injectivity of B. Therefore the inverse

$$A = B^{-1} \colon \mathcal{R}(\bar{\iota}\cdot\bar{\iota}^*) \subset L^2_\mu(\mathbb{R}^n,\mathbb{R}) \to L^2_\mu(\mathbb{R}^n,\mathbb{R})$$

is a densely defined self-adjoint operator on $L^2_\mu(\mathbb{R}^n,\mathbb{R})$ (see the appendix on unbounded operators on Banach spaces). Now we observe that for $f, g \in \mathcal{C}_c(\mathbb{R}^n,\mathbb{R})$,

$$\begin{aligned}
\langle f, g\rangle_{L^2_\mu(\mathbb{R}^n,\mathbb{R})} - \langle Lf, g\rangle_{L^2_\mu(\mathbb{R}^n,\mathbb{R})} &= \langle \bar{i}^{-1}(f), \bar{i}^{-1}(g)\rangle_{\mathcal{E}} \\
&= \langle (\bar{i}^{-1})^* \bar{i}^{-1} f, g\rangle_{L^2_\mu(\mathbb{R}^n,\mathbb{R})} \\
&= \langle (\bar{i}\bar{i}^*)^{-1} f, g\rangle_{L^2_\mu(\mathbb{R}^n,\mathbb{R})}.
\end{aligned}$$

Thus A and $\mathrm{Id} - L$ coincide on $\mathcal{C}_c(\mathbb{R}^n,\mathbb{R})$. By defining,

$$-\bar{L} = A - \mathrm{Id},$$

we get the required self-adjoint extension of $-L$. □

Remark 4.7. The operator $\bar{L}$, as constructed above, is called the Friedrichs extension of L. Intuitively, it is the minimal self-adjoint extension of L.

Definition 4.8. If $\bar{L}$ is the unique non-positive self-adjoint extension of L, then the operator L is said to be *essentially self-adjoint* on $\mathcal{C}_c(\mathbb{R}^n, \mathbb{R})$. In this case, there is no ambiguity and we shall write $\bar{L} = L$.

We have the following first criterion for essential self-adjointness.

Lemma 4.9. *If for some $\lambda > 0$,*

$$\operatorname{Ker}(-L^* + \lambda \mathrm{Id}) = \{0\},$$

then the operator L is essentially self-adjoint on $\mathcal{C}_c(\mathbb{R}^n, \mathbb{R})$.

Proof. We make the proof for $\lambda = 1$ and let the reader adapt it for $\lambda \neq 0$.

Let $-\tilde{L}$ be a non-negative self-adjoint extension of $-L$. We want to prove that actually $-\tilde{L} = -\bar{L}$. The assumption

$$\operatorname{Ker}(-L^* + \mathrm{Id}) = \{0\}$$

implies that $\mathcal{C}_c(\mathbb{R}^n, \mathbb{R})$ is dense in $\mathcal{D}(-L^*)$ for the norm

$$\|f\|_{\mathcal{E}}^2 = \|f\|^2_{L^2_\mu(\mathbb{R}^n,\mathbb{R})} - \langle f, L^* f\rangle_{L^2_\mu(\mathbb{R}^n,\mathbb{R})}.$$

Since $-\tilde{L}$ is a non-negative self-adjoint extension of $-L$, we have

$$\mathcal{D}(-\tilde{L}) \subset \mathcal{D}(-L^*).$$

The space $\mathcal{C}_c(\mathbb{R}^n, \mathbb{R})$ is therefore dense in $\mathcal{D}(-\tilde{L})$ for the norm $\|\cdot\|_{\mathcal{E}}$.

At that point, we use some notation introduced in the proof of the Friedrichs extension (Theorem 4.6). Since $\mathcal{C}_c(\mathbb{R}^n, \mathbb{R})$ is dense in $\mathcal{D}(-\tilde{L})$ for the norm $\|\cdot\|_{\mathcal{E}}$, we deduce that the equality

$$\langle f, g\rangle_{L^2_\mu(\mathbb{R}^n,\mathbb{R})} - \langle \tilde{L} f, g\rangle_{L^2_\mu(\mathbb{R}^n,\mathbb{R})} = \langle \bar{i}^{-1}(f), \bar{i}^{-1}(g)\rangle_{\mathcal{E}},$$

which is obviously satisfied for $f, g \in \mathcal{C}_c(\mathbb{R}^n, \mathbb{R})$, actually also holds for $f, g \in \mathcal{D}(\tilde{L})$. From the definition of the Friedrichs extension we deduce that $\bar{L}$ and $\tilde{L}$ coincide on $\mathcal{D}(\tilde{L})$. Finally, since these two operators are self-adjoint we conclude that $\bar{L} = \tilde{L}$. □

Remark 4.10. Given the fact that $-L$ is given here with the domain $\mathcal{C}_c(\mathbb{R}^n, \mathbb{R})$, the condition

$$\operatorname{Ker}(-L^* + \lambda \mathrm{Id}) = \{0\}$$

is equivalent to the fact that if $f \in L^2_\mu(\mathbb{R}^n, \mathbb{R})$ is a function that satisfies in the sense of distributions

$$-Lf + \lambda f = 0,$$

then $f = 0$.

As a corollary of the previous lemma, the following proposition provides a useful sufficient condition for essential self-adjointness that is easy to check for several diffusion operators. We recall that a diffusion operator is said to be elliptic if the matrix σ is invertible.

Proposition 4.11. *If the diffusion operator L is elliptic with smooth coefficients and if there exists an increasing sequence $h_n \in \mathcal{C}_c(\mathbb{R}^n, \mathbb{R})$, $0 \le h_n \le 1$, such that $h_n \nearrow 1$ on $\mathbb{R}^n$, and $\|\Gamma(h_n, h_n)\|_\infty \to 0$, as $n \to \infty$, then the operator L is essentially self-adjoint on $\mathcal{C}_c(\mathbb{R}^n, \mathbb{R})$.*

Proof. Let $\lambda > 0$. According to the previous lemma, it is enough to check that if $L^* f = \lambda f$ with $\lambda > 0$, then $f = 0$. As it was observed above, $L^* f = \lambda f$ is equivalent to the fact that, in sense of distributions, $Lf = \lambda f$. From the hypoellipticity of L, we therefore deduce that f is a smooth function. Now, for $h \in \mathcal{C}_c(\mathbb{R}^n, \mathbb{R})$,

$$\begin{aligned}\int_{\mathbb{R}^n} \Gamma(f, h^2 f) d\mu &= -\langle f, L(h^2 f)\rangle_{L^2_\mu(\mathbb{R}^n,\mathbb{R})} \\ &= -\langle L^* f, h^2 f\rangle_{L^2_\mu(\mathbb{R}^n,\mathbb{R})} \\ &= -\lambda\langle f, h^2 f\rangle_{L^2_\mu(\mathbb{R}^n,\mathbb{R})} \\ &= -\lambda\langle f^2, h^2\rangle_{L^2_\mu(\mathbb{R}^n,\mathbb{R})} \\ &\le 0.\end{aligned}$$

Since

$$\Gamma(f, h^2 f) = h^2\Gamma(f, f) + 2fh\Gamma(f, h),$$

we deduce that

$$\langle h^2, \Gamma(f, f)\rangle_{L^2_\mu(\mathbb{R}^n,\mathbb{R})} + 2\langle fh, \Gamma(f, h)\rangle_{L^2_\mu(\mathbb{R}^n,\mathbb{R})} \le 0.$$

Therefore, by the Cauchy–Schwarz inequality,

$$\langle h^2, \Gamma(f, f)\rangle_{L^2_\mu(\mathbb{R}^n,\mathbb{R})} \le 4\|f\|_2^2\|\Gamma(h, h)\|_\infty.$$

If we now use the sequence h_n and let $n \to \infty$, we obtain $\Gamma(f, f) = 0$ and therefore $f = 0$, as desired. □

Exercise 4.12. Let

$$L = \Delta + \langle \nabla U, \nabla \cdot\rangle,$$

where U is a smooth function on $\mathbb{R}^n$. Show that with respect to the measure $\mu(dx) = e^{U(x)}dx$, the operator L is essentially self-adjoint on $\mathcal{C}_c(\mathbb{R}^n, \mathbb{R})$.

Exercise 4.13. On $\mathbb{R}^n$, we consider the divergence form operator

$$Lf = \operatorname{div}(\sigma \nabla f),$$

where σ is a smooth field of positive and symmetric matrices that satisfies

$$a\|x\|^2 \le \langle x, \sigma x\rangle \le b\|x\|^2, \quad x \in \mathbb{R}^n,$$

for some constant $0 < a \le b$. Show that with respect to the Lebesgue measure, the operator L is essentially self-adjoint on $\mathcal{C}_c(\mathbb{R}^n, \mathbb{R})$

Exercise 4.14. On $\mathbb{R}^n$, we consider the Schrödinger-type operator

$$H = L - V,$$

where L is a diffusion operator and $V : \mathbb{R}^n \to \mathbb{R}$ is a smooth function. We write

$$\Gamma(f, g) = \frac{1}{2}(L(fg) - fLg - gLf).$$

Show the following: If there exists an increasing sequence $h_n \in \mathcal{C}_c(\mathbb{R}^n, \mathbb{R})$, $0 \le h_n \le 1$, such that $h_n \nearrow 1$ on $\mathbb{R}^n$ and $\|\Gamma(h_n, h_n)\|_\infty \to 0$ as $n \to \infty$ and if V is bounded from below, then H is essentially self-adjoint on $\mathcal{C}_c(\mathbb{R}^n, \mathbb{R})$.

From now on, we assume that the diffusion operator L is essentially self-adjoint on $\mathcal{C}_c(\mathbb{R}^n, \mathbb{R})$. Its Friedrichs extension is still denoted by L. The fact that we are now dealing with a non-negative self-adjoint operator allows us to use spectral theory in order to define e^{tL}. Indeed, we have the following so-called spectral theorem whose proof can be found in [62].

Theorem 4.15 (Spectral theorem). *Let A be a non-negative self-adjoint operator on a separable Hilbert space $\mathcal{H}$. There is a measure space (Ω, ν), a unitary map $U : L^2_\nu(\Omega, \mathbb{R}) \to \mathcal{H}$ and a non-negative real-valued measurable function λ on Ω such that*

$$U^{-1}AUf(x) = \lambda(x) f(x),$$

for $x \in \Omega$, $Uf \in \mathcal{D}(A)$. Moreover, given $f \in L^2_\nu(\Omega, \mathbb{R})$, Uf belongs to $\mathcal{D}(A)$ if only if $\int_\Omega \lambda^2 f^2 d\nu < +\infty$.

We may now apply the spectral theorem to $-L$ in order to define e^{tL}. More generally, given a Borel function $g : \mathbb{R}_{\ge 0} \to \mathbb{R}$ and the spectral decomposition of $-L$,

$$U^{-1}LUf(x) = -\lambda(x) f(x),$$

we may define an operator $g(-L)$ as being the unique operator that satisfies

$$U^{-1}g(-L)Uf(x) = (g \circ \lambda)(x) f(x).$$

We observe that $g(-L)$ is a bounded operator if g is a bounded function.

As a particular case, we define the diffusion semigroup $(\boldsymbol{P}_t)_{t\geq 0}$ on $L^2_\mu(\mathbb{R}^n,\mathbb{R})$ by the requirement

$$U^{-1}\boldsymbol{P}_t Uf(x) = e^{-t\lambda(x)}f(x).$$

This defines a family of bounded operators $\boldsymbol{P}_t : L^2_\mu(\mathbb{R}^n,\mathbb{R}) \to L^2_\mu(\mathbb{R}^n,\mathbb{R})$ satisfying the following properties which are readily checked from the spectral decomposition:

- For $f \in L^2_\mu(\mathbb{R}^n,\mathbb{R})$,

$$\|\boldsymbol{P}_t f\|_{L^2_\mu(\mathbb{R}^n,\mathbb{R})} \leq \|f\|_{L^2_\mu(\mathbb{R}^n,\mathbb{R})}.$$

- $\boldsymbol{P}_0 = \mathrm{Id}$ and for $s,t\geq 0$, $\boldsymbol{P}_s\boldsymbol{P}_t = \boldsymbol{P}_{s+t}$.
- For $f \in L^2_\mu(\mathbb{R}^n,\mathbb{R})$, the map $t \to \boldsymbol{P}_t f$ is continuous in $L^2_\mu(\mathbb{R}^n,\mathbb{R})$.
- For $f,g \in L^2_\mu(\mathbb{R}^n,\mathbb{R})$,

$$\int_{\mathbb{R}^n} (\boldsymbol{P}_t f) g\, d\mu = \int_{\mathbb{R}^n} f(\boldsymbol{P}_t g)\, d\mu.$$

We summarize the above properties by saying that $(\boldsymbol{P}_t)_{t\geq 0}$ is a self-adjoint strongly continuous contraction semigroup on $L^2_\mu(\mathbb{R}^n,\mathbb{R})$.

From the spectral decomposition, it is also easily checked that the operator L is furthermore the generator of this semigroup, that is, for $f \in \mathcal{D}(L)$,

$$\lim_{t\to 0}\left\| \frac{\boldsymbol{P}_t f - f}{t} - Lf \right\|_{L^2_\mu(\mathbb{R}^n,\mathbb{R})} = 0.$$

From the semigroup property it follows that for $t\geq 0$, $\boldsymbol{P}_t\mathcal{D}(L) \subset \mathcal{D}(L)$, and that for $f \in \mathcal{D}(L)$,

$$\frac{d}{dt}\boldsymbol{P}_t f = \boldsymbol{P}_t L f = L\boldsymbol{P}_t f,$$

the derivative on the left-hand side of the above equality being taken in $L^2_\mu(\mathbb{R}^n,\mathbb{R})$.

Exercise 4.16. Let L be an essentially self-adjoint diffusion operator on $\mathcal{C}_c(\mathbb{R}^n,\mathbb{R})$. Show that if the constant function $1 \in \mathcal{D}(L)$ and if $L1 = 0$, then

$$\boldsymbol{P}_t 1 = 1.$$

Exercise 4.17. Let L be an essentially self-adjoint diffusion operator on $\mathcal{C}_c(\mathbb{R}^n,\mathbb{R})$.

(1) Show that for every $\lambda > 0$, the range of the operator $\lambda\mathrm{Id} - L$ is dense in $L^2_\mu(\mathbb{R}^n,\mathbb{R})$.

(2) By using the spectral theorem, show that the following limit holds for the operator norm on $L^2_\mu(\mathbb{R}^n,\mathbb{R})$:

$$\boldsymbol{P}_t = \lim_{n\to+\infty}\left(\mathrm{Id} - \frac{t}{n}L\right)^{-n}.$$

Exercise 4.18. As usual, we denote by Δ the Laplace operator on $\mathbb{R}^n$:

$$\Delta = \sum_{i=1}^{n} \frac{\partial^2}{\partial x_i^2}.$$

The McDonald function with index $\nu \in \mathbb{R}$ is defined for $x \in \mathbb{R} \setminus \{0\}$ by

$$K_\nu(x) = \frac{1}{2}\left(\frac{x}{2}\right)^\nu \int_0^{+\infty} \frac{e^{-\frac{x^2}{4t}-t}}{t^{1+\nu}} dt.$$

(1) Show that for $\lambda \in \mathbb{R}^n$ and $\alpha > 0$,

$$\frac{1}{(2\pi)^{n/2}} \int_{\mathbb{R}^n} e^{i\langle\lambda,x\rangle} \left(\frac{\|x\|}{\sqrt{\alpha}}\right)^{1-\frac{n}{2}} K_{\frac{n}{2}-1}(\sqrt{\alpha}\|x\|)dx = \frac{1}{\alpha + \|\lambda\|^2}.$$

(2) Show that $K_{-\nu} = K_\nu$ for $\nu \in \mathbb{R}$.

(3) Show that

$$K_{1/2}(x) = \sqrt{\frac{\pi}{2x}} e^{-x}.$$

(4) Prove that for $f \in L^2(\mathbb{R}^n, \mathbb{R})$ and $\alpha > 0$,

$$(\alpha \mathrm{Id} - \Delta)^{-1} f(x) = \int_{\mathbb{R}^n} G_\alpha(x-y) f(y) dy,$$

where

$$G_\alpha(x) = \frac{1}{(2\pi)^{n/2}} \left(\frac{\|x\|}{\sqrt{\alpha}}\right)^{1-\frac{n}{2}} K_{\frac{n}{2}-1}(\sqrt{\alpha}\|x\|).$$

(You may use the Fourier transform to solve the partial differential equation $\alpha g - \Delta g = f$.)

(5) Prove that for $f \in L^2(\mathbb{R}^n, \mathbb{R})$,

$$\lim_{n\to+\infty} \left(\mathrm{Id} - \frac{t}{n}\Delta\right)^{-n} f = \frac{1}{(4\pi t)^{\frac{n}{2}}} \int_{\mathbb{R}^n} e^{-\frac{\|x-y\|^2}{4t}} f(y) dy,$$

the limit being taken in $L^2(\mathbb{R}^n, \mathbb{R})$. Conclude that almost everywhere

$$\boldsymbol{P}_t f(x) = \frac{1}{(4\pi t)^{\frac{n}{2}}} \int_{\mathbb{R}^n} e^{-\frac{\|x-y\|^2}{4t}} f(y) dy.$$

Exercise 4.19. (1) Show the subordination identity:

$$e^{-y|\alpha|} = \frac{y}{2\sqrt{\pi}} \int_0^{+\infty} \frac{e^{-\frac{y^2}{4t} - t\alpha^2}}{t^{3/2}} dt, \quad y > 0, \alpha \in \mathbb{R}.$$

(2) The Cauchy semigroup on $\mathbb{R}^n$ is defined by $\boldsymbol{Q}_t = e^{-t\sqrt{-\Delta}}$. Use the subordination identity and the heat semigroup on $\mathbb{R}^n$ to show that for $f \in L^2(\mathbb{R}^n, \mathbb{R})$,

$$\boldsymbol{Q}_t f(x) = \int_{\mathbb{R}^n} q(t, x - y) f(y) dy,$$

where

$$q(t, x) = \frac{\Gamma(\frac{n+1}{2})}{\pi^{\frac{n+1}{2}}} \frac{t}{(t^2 + \|x\|^2)^{\frac{n+1}{2}}}.$$

4.2 Existence and regularity of the heat kernel

In order to study the regularization properties of diffusion semigroups, we apply the theory of local elliptic regularity which is sketched in Appendix B.

Proposition 4.20. *Let L be an elliptic diffusion operator with smooth coefficients that is essentially self-adjoint with respect to a measure μ. Denote by $(\boldsymbol{P}_t)_{t \geq 0}$ the corresponding semigroup on $L^2_\mu(\mathbb{R}^n, \mathbb{R})$.*

- *If K is a compact subset of $\mathbb{R}^n$, there exists a positive constant C such that for $f \in L^2_\mu(\mathbb{R}^n, \mathbb{R})$,*

$$\sup_{x \in K} |\boldsymbol{P}_t f(x)| \leq C \Big(1 + \frac{1}{t^\kappa}\Big) \|f\|_{L^2_\mu(\mathbb{R}^n, \mathbb{R})},$$

  where κ is the smallest integer larger than $\frac{n}{4}$.
- *For $f \in L^2_\mu(\mathbb{R}^n, \mathbb{R})$, the function $(t, x) \to \boldsymbol{P}_t f(x)$ is smooth on $(0, +\infty) \times \mathbb{R}^n$.*

Proof. Let us first observe that it follows from the spectral theorem that if $f \in L^2_\mu(\mathbb{R}^n, \mathbb{R})$, then for every $k \geq 0$, $L^k \boldsymbol{P}_t f \in L^2_\mu(\mathbb{R}^n, \mathbb{R})$ and

$$\|L^k \boldsymbol{P}_t f\|_{L^2_\mu(\mathbb{R}^n, \mathbb{R})} \leq (\sup_{\lambda \geq 0} \lambda^k e^{-\lambda t}) \|f\|_{L^2_\mu(\mathbb{R}^n, \mathbb{R})}.$$

Now let K be a compact subset of $\mathbb{R}^n$. From Proposition B.11, there exists a positive constant C such that

$$(\sup_{x \in K} |\boldsymbol{P}_t f(x)|)^2 \leq C \Big(\sum_{k=0}^{\kappa} \|L^k \boldsymbol{P}_t f\|^2_{L^2_\mu(\mathbb{R}^n, \mathbb{R})} \Big).$$

Since it is immediately checked that

$$\sup_{\lambda \geq 0} \lambda^k e^{-\lambda t} = \left(\frac{k}{t}\right)^k e^{-k},$$

the bound

$$\sup_{x \in K} |\boldsymbol{P}_t f(x)| \leq C\left(1 + \frac{1}{t^\kappa}\right) \|f\|_{L^2_\mu(\mathbb{R}^n,\mathbb{R})},$$

easily follows.

We now turn to the second part. Let $f \in L^2_\mu(\mathbb{R}^n, \mathbb{R})$.

First, we fix $t > 0$. As above, from the spectral theorem we have, for every $k \geq 0$, $L^k \boldsymbol{P}_t f \in L^2_\mu(\mathbb{R}^n, \mathbb{R}) \subset \mathcal{H}_0^{\mathrm{loc}}(\Omega)$, for any bounded open set Ω. A recursive application of Lemma B.8 therefore yields $\boldsymbol{P}_t f \in \bigcap_{s>0} \mathcal{H}_s^{\mathrm{loc}}(\Omega)$, which implies from the Sobolev lemma (Theorem B.4) that $\boldsymbol{P}_t f$ is a smooth function.

Next, we prove joint continuity in the variables $(t, x) \in (0, +\infty) \times \mathbb{R}^n$. It is enough to prove that if $t_0 > 0$ and if K is a compact set on $\mathbb{R}^n$, then

$$\sup_{x \in K} |\boldsymbol{P}_t f(x) - \boldsymbol{P}_{t_0} f(x)| \xrightarrow[t \to t_0]{} 0.$$

From Proposition B.11, there exists a positive constant C such that

$$\sup_{x \in K} |\boldsymbol{P}_t f(x) - \boldsymbol{P}_{t_0} f(x)| \leq C\Big(\sum_{k=0}^{\kappa} \|L^k \boldsymbol{P}_t f - L^k \boldsymbol{P}_{t_0} f\|^2_{L^2_\mu(\mathbb{R}^n,\mathbb{R})}\Big).$$

Now, again from the spectral theorem, it is checked that

$$\lim_{t \to t_0} \sum_{k=0}^{\kappa} \|L^k \boldsymbol{P}_t f - L^k \boldsymbol{P}_{t_0} f\|^2_{L^2_\mu(\mathbb{R}^n,\mathbb{R})} = 0.$$

This gives the expected joint continuity in (t, x). The joint smoothness in (t, x) is a consequence of the second part of Proposition B.11 and the details are left to the reader. □

Remark 4.21. If the bound

$$\sup_{x \in K} |\boldsymbol{P}_t f(x)| \leq C(t) \|f\|_{L^2_\mu(\mathbb{R}^n,\mathbb{R})}$$

holds uniformly on $\mathbb{R}^n$, that is, if the operator norm

$$\|\boldsymbol{P}_t\|_{L^2_\mu(\mathbb{R}^n,\mathbb{R}) \to L^\infty_\mu(\mathbb{R}^n,\mathbb{R})} < \infty,$$

then the semigroup $(\boldsymbol{P}_t)_{t \geq 0}$ is said to be ultracontractive. The study of ultracontractive semigroups is intimately related to the beautiful theory of log-Sobolev inequalities. We refer the interested reader to the book [16].

Exercise 4.22. Let L be an elliptic diffusion operator with smooth coefficients that is essentially self-adjoint with respect to a measure μ. Let α be a multi-index. If K is a compact subset of $\mathbb{R}^n$, show that there exists a positive constant C such that for $f \in L^2_\mu(\mathbb{R}^n, \mathbb{R})$,

$$\sup_{x \in K} |\partial^\alpha \boldsymbol{P}_t f(x)| \le C\Big(1 + \frac{1}{t^{|\alpha|+\kappa}}\Big) \|f\|_{L^2_\mu(\mathbb{R}^n,\mathbb{R})},$$

where κ is the smallest integer larger than $\frac{n}{4}$.

We now prove the following fundamental theorem:

Theorem 4.23. *Let L be an elliptic and essentially self-adjoint diffusion operator. Denote by $(\boldsymbol{P}_t)_{t\ge 0}$ the corresponding semigroup on $L^2_\mu(\mathbb{R}^n, \mathbb{R})$. There is a smooth function $p(t,x,y)$, $t \in (0,+\infty)$, $x, y \in \mathbb{R}^n$, such that for every $f \in L^2_\mu(\mathbb{R}^n, \mathbb{R})$ and $x \in \mathbb{R}^n$,*

$$\boldsymbol{P}_t f(x) = \int_{\mathbb{R}^n} p(t,x,y) f(y) d\mu(y).$$

The function $p(t,x,y)$ is called the heat kernel associated to $(\boldsymbol{P}_t)_{t\ge 0}$. Furthermore, it satisfies

- $p(t,x,y) = p(t,y,x)$ *(Symmetry); and*
- $p(t+s,x,y) = \int_{\mathbb{R}^n} p(t,x,z) p(s,z,y) d\mu(z)$ *(Chapman–Kolmogorov relation).*

Proof. Let $x \in \mathbb{R}^n$ and $t > 0$. From the previous proposition, the linear form $f \to \boldsymbol{P}_t f(x)$ is continuous on $L^2_\mu(\mathbb{R}^n, \mathbb{R})$, therefore from the Riesz representation theorem, there is a function $p(t,x,\cdot) \in L^2_\mu(\mathbb{R}^n, \mathbb{R})$ such that for $f \in L^2_\mu(\mathbb{R}^n, \mathbb{R})$,

$$\boldsymbol{P}_t f(x) = \int_{\mathbb{R}^n} p(t,x,y) f(y) d\mu(y).$$

From the fact that $\boldsymbol{P}_t$ is self-adjoint on $L^2_\mu(\mathbb{R}^n, \mathbb{R})$,

$$\int_{\mathbb{R}^n} (\boldsymbol{P}_t f) g d\mu = \int_{\mathbb{R}^n} f (\boldsymbol{P}_t g) d\mu,$$

we easily deduce the symmetry property:

$$p(t,x,y) = p(t,y,x).$$

The Chapman–Kolmogorov relation $p(t+s,x,y) = \int_{\mathbb{R}^n} p(t,x,z) p(s,z,y) d\mu(z)$ stems from the semigroup property $\boldsymbol{P}_{t+s} = \boldsymbol{P}_t \boldsymbol{P}_s$. Finally, from the previous proposition the map $(t,x) \to p(t,x,\cdot) \in L^2_\mu(\mathbb{R}^n, \mathbb{R})$ is smooth on $(0,+\infty) \times \mathbb{R}^n$ for the weak topology on $L^2_\mu(\mathbb{R}^n, \mathbb{R})$. This implies that it is also smooth on $(0, +\infty) \times \mathbb{R}^n$ for the norm topology. Since from the Chapman–Kolmogorov relation

$$p(t, x, y) = \langle p(t/2, x, \cdot), p(t/2, y.\cdot)\rangle_{L^2_\mu(\mathbb{R}^n,\mathbb{R})},$$

we conclude that $(t, x, y) \to p(t, x, y)$ is smooth on $(0, +\infty) \times \mathbb{R}^n \times \mathbb{R}^n$. □

4.3 The sub-Markov property

In the previous section, we have proved that if L is a diffusion operator that is essentially self-adjoint on $\mathcal{C}_c(\mathbb{R}^n, \mathbb{R})$, then by using the spectral theorem we can define a self-adjoint strongly continuous contraction semigroup $(\boldsymbol{P}_t)_{t\geq 0}$ on $L^2_\mu(\mathbb{R}^n, \mathbb{R})$ with generator L.

In this section, we will be interested in the following additional property of the semigroup $(\boldsymbol{P}_t)_{t\geq 0}$: If $f \in L^2_\mu(\mathbb{R}^n, \mathbb{R})$ satisfies almost surely $0 \leq f \leq 1$, then almost surely $0 \leq \boldsymbol{P}_t f \leq 1, t \geq 0$. This property is called the sub-Markov property of the semigroup $(\boldsymbol{P}_t)_{t\geq 0}$. The terminology stems from the fact that it is precisely this property that makes the link with probability theory because it is equivalent to the fact that $(\boldsymbol{P}_t)_{t\geq 0}$ is associated to the transition function of a sub-Markov process.

As a first result we prove that the semigroup $(\boldsymbol{P}_t)_{t\geq 0}$ is positivity-preserving. This property may actually be seen at the level of the diffusion operator L, and relies on the following distributional inequality satisfied by L.

Lemma 4.24 (Kato inequality). *Let L be a diffusion operator on $\mathbb{R}^n$ with symmetric and invariant measure μ. Let $u \in \mathcal{C}_c(\mathbb{R}^n, \mathbb{R})$. Define*

$$\operatorname{sgn} u = \begin{cases} 0 & \text{if } u(x) = 0, \\ \frac{u(x)}{|u(x)|} & \text{if } u(x) \neq 0. \end{cases}$$

In the sense of distributions, we have the following inequality:

$$L|u| \geq (\operatorname{sgn} u) Lu.$$

Proof. If ϕ is a smooth and convex function and if u is assumed to be smooth, it is readily checked that

$$L\phi(u) = \phi'(u)Lu + \phi''(u)\Gamma(u, u) \geq \phi'(u)Lu.$$

By choosing for ϕ the function

$$\phi_\varepsilon(x) = \sqrt{x^2 + \varepsilon^2}, \quad \varepsilon > 0,$$

we deduce that for every smooth function $u \in \mathcal{C}_c(\mathbb{R}^n, \mathbb{R})$,

$$L\phi_\varepsilon(u) \geq \frac{u}{\sqrt{x^2+\varepsilon^2}} Lu.$$

As a consequence this inequality holds in the sense of distributions, that is, for every $f \in \mathcal{C}_c(\mathbb{R}^n, \mathbb{R})$, $f \geq 0$,

$$\int_{\mathbb{R}^n} f L\phi_\varepsilon(u) d\mu \geq \int_{\mathbb{R}^n} f \frac{u}{\sqrt{u^2+\varepsilon^2}} Lu d\mu$$

Letting $\varepsilon \to 0$ gives the expected result. □

We are now in a position to state and prove the positivity-preserving theorem.

Theorem 4.25. *Let L be an essentially self-adjoint diffusion operator on $\mathcal{C}_c(\mathbb{R}^n, \mathbb{R})$. If $f \in L^2_\mu(\mathbb{R}^n, \mathbb{R})$ is almost surely non-negative ($f \geq 0$), then we have $\boldsymbol{P}_t f \geq 0$ almost surely for every $t \geq 0$.*

Proof. The main idea is to prove that for $\lambda > 0$, the operator $(\lambda \mathrm{Id} - L)^{-1}$ which is well defined due to essential self-adjointness preserves the positivity of function. Then we may conclude by the fact that for $f \in L^2_\mu(\mathbb{R}^n, \mathbb{R})$, in the $L^2_\mu(\mathbb{R}^n, \mathbb{R})$-sense

$$\boldsymbol{P}_t f = \lim_{n \to +\infty} \left(\mathrm{Id} - \frac{t}{n} L \right)^{-n} f.$$

Let $\lambda > 0$. We consider on $\mathcal{C}_c(\mathbb{R}^n, \mathbb{R})$ the norm

$$\|f\|^2_\lambda = \|f\|^2_{L^2_\mu(\mathbb{R}^n,\mathbb{R})} + \lambda \mathcal{E}(f,f) = \|f\|^2_{L^2_\mu(\mathbb{R}^n,\mathbb{R})} + \lambda \int_{\mathbb{R}^n} \Gamma(f,f) d\mu$$

and denote by $\mathcal{H}_\lambda$ the completion of $\mathcal{C}_c(\mathbb{R}^n, \mathbb{R})$. Using Kato's inequality, it is obvious (multiply the inequality by u and integrate by parts, details are left to the reader) that if $u \in \mathcal{H}_\lambda$, then $|u| \in \mathcal{H}_\lambda$ and

$$\mathcal{E}(|u|, |u|) \leq \mathcal{E}(u,u). \tag{4.1}$$

Since L is essentially self-adjoint we can consider the bounded operator

$$\boldsymbol{R}_\lambda = (\mathrm{Id} - \lambda L)^{-1}$$

that goes from $L^2_\mu(\mathbb{R}^n, \mathbb{R})$ to $\mathcal{D}(L) \subset \mathcal{H}_\lambda$. For $f \in \mathcal{H}_\lambda$ and $g \in L^2_\mu(\mathbb{R}^n, \mathbb{R})$ with $g \geq 0$, we have

$$\begin{aligned}
\langle |f|, \boldsymbol{R}_\lambda g \rangle_\lambda &= \langle |f|, \boldsymbol{R}_\lambda g \rangle_{L^2_\mu(\mathbb{R}^n,\mathbb{R})} - \lambda \langle |f|, L\boldsymbol{R}_\lambda g \rangle_{L^2_\mu(\mathbb{R}^n,\mathbb{R})} \\
&= \langle |f|, (\mathrm{Id} - \lambda L)\boldsymbol{R}_\lambda g \rangle_{L^2_\mu(\mathbb{R}^n,\mathbb{R})} \\
&= \langle |f|, g \rangle_{L^2_\mu(\mathbb{R}^n,\mathbb{R})} \\
&\geq |\langle f, g \rangle_{L^2_\mu(\mathbb{R}^n,\mathbb{R})}| \\
&\geq |\langle f, \boldsymbol{R}_\lambda g \rangle_\lambda|.
\end{aligned}$$

Moreover, from inequality (4.1), for $f \in \mathcal{H}_\lambda$,

$$\| |f| \|_\lambda^2 = \| |f| \|^2_{L^2_\mu(\mathbb{R}^n,\mathbb{R})} + \lambda \mathcal{E}(|f|,|f|) \geq \|f\|^2_{L^2_\mu(\mathbb{R}^n,\mathbb{R})} + \lambda \mathcal{E}(f,f) \geq \|f\|_\lambda^2.$$

By taking $f = \boldsymbol{R}_\lambda g$ in the two above sets of inequalities, we draw the conclusion

$$|\langle \boldsymbol{R}_\lambda g, \boldsymbol{R}_\lambda g\rangle_\lambda| \leq \langle |\boldsymbol{R}_\lambda g|, \boldsymbol{R}_\lambda g\rangle_\lambda \leq \| |\boldsymbol{R}_\lambda g| \|_\lambda \|\boldsymbol{R}_\lambda g\|_\lambda \leq |\langle \boldsymbol{R}_\lambda g, \boldsymbol{R}_\lambda g\rangle_\lambda|.$$

The above inequalities are therefore equalities which implies

$$\boldsymbol{R}_\lambda g = |\boldsymbol{R}_\lambda g|.$$

As a conclusion, if $g \in L^2_\mu(\mathbb{R}^n, \mathbb{R})$ is ≥ 0, then $(\mathrm{Id} - \lambda L)^{-1} g \geq 0$ for every $\lambda > 0$. Thanks to the spectral theorem, in $L^2_\mu(\mathbb{R}^n, \mathbb{R})$ we have

$$\boldsymbol{P}_t g = \lim_{n \to +\infty} \left(\mathrm{Id} - \frac{t}{n} L\right)^{-n} g.$$

By passing to a subsequence that converges pointwise almost surely, we deduce that $\boldsymbol{P}_t g \geq 0$ almost surely. □

Exercise 4.26. Let L be an elliptic diffusion operator with smooth coefficients that is essentially self-adjoint. Denote by $p(t,x,y)$ the heat kernel of $\boldsymbol{P}_t$. Show that $p(t,x,y) \geq 0$. (Remark: It is actually possible to prove that $p(t,x,y) > 0$).

Besides the positivity-preserving property, the semigroup is a contraction on $L^\infty_\mu(\mathbb{R}^n, \mathbb{R})$. More precisely we have

Theorem 4.27. *Let L be an essentially self-adjoint diffusion operator on $\mathcal{C}_c(\mathbb{R}^n, \mathbb{R})$. If $f \in L^2_\mu(\mathbb{R}^n, \mathbb{R}) \cap L^\infty_\mu(\mathbb{R}^n, \mathbb{R})$, then $\boldsymbol{P}_t f \in L^\infty_\mu(\mathbb{R}^n, \mathbb{R})$ and*

$$\|\boldsymbol{P}_t f\|_\infty \leq \|f\|_\infty.$$

Proof. The proof is close to and relies on the same ideas as the proof of the positivity-preserving Theorem 4.25. So, we only list below the main steps and let the reader fill the details.

As before, for $\lambda > 0$, we consider on $\mathcal{C}_c(\mathbb{R}^n, \mathbb{R})$ the norm

$$\|f\|_\lambda^2 = \|f\|^2_{L^2_\mu(\mathbb{R}^n,\mathbb{R})} + \lambda \mathcal{E}(f,f) = \|f\|^2_{L^2_\mu(\mathbb{R}^n,\mathbb{R})} + \lambda \int_{\mathbb{R}^n} \Gamma(f,f) d\mu.$$

and denote by $\mathcal{H}_\lambda$ the completion of $\mathcal{C}_c(\mathbb{R}^n, \mathbb{R})$.

- The first step is to show that if $0 \leq f \in \mathcal{H}_\lambda$, then $1 \wedge f$ (minimum between 1 and f) also lies in $\mathcal{H}_\lambda$ and moreover

$$\mathcal{E}(1 \wedge f, 1 \wedge f) \leq \mathcal{E}(f,f).$$

- Let $f \in L^2_\mu(\mathbb{R}^n, \mathbb{R})$ satisfy $0 \le f \le 1$ and put $g = \boldsymbol{R}_\lambda f = (\mathrm{Id} - \lambda L)^{-1} f \in \mathcal{H}_\lambda$ and $h = 1 \wedge g$. According to the first step, $h \in \mathcal{H}_\lambda$ and $\mathcal{E}(h,h) \le \mathcal{E}(g,g)$. Now we observe that

$$\begin{aligned}
&\|g-h\|_\lambda^2 \\
&= \|g\|_\lambda^2 - 2\langle g, h\rangle_\lambda + \|h\|_\lambda^2 \\
&= \langle \boldsymbol{R}_\lambda f, f\rangle_{L^2_\mu(\mathbb{R}^n,\mathbb{R})} - 2\langle f, h\rangle_{L^2_\mu(\mathbb{R}^n,\mathbb{R})} + \|h\|^2_{L^2_\mu(\mathbb{R}^n,\mathbb{R})} + \lambda \mathcal{E}(h,h) \\
&= \langle \boldsymbol{R}_\lambda f, f\rangle_{L^2_\mu(\mathbb{R}^n,\mathbb{R})} - \|f\|^2_{L^2_\mu(\mathbb{R}^n,\mathbb{R})} + \|f-h\|^2_{L^2_\mu(\mathbb{R}^n,\mathbb{R})} + \lambda \mathcal{E}(h,h) \\
&\le \langle \boldsymbol{R}_\lambda f, f\rangle_{L^2_\mu(\mathbb{R}^n,\mathbb{R})} - \|f\|^2_{L^2_\mu(\mathbb{R}^n,\mathbb{R})} + \|f-g\|^2_{L^2_\mu(\mathbb{R}^n,\mathbb{R})} + \lambda \mathcal{E}(g,g) \\
&= 0.
\end{aligned}$$

  As a consequence $g = h$, that is, $0 \le g \le 1$.
- The previous step shows that if $f \in L^2_\mu(\mathbb{R}^n, \mathbb{R})$ satisfies $0 \le f \le 1$ then for every $\lambda > 0$, $0 \le (\mathrm{Id} - \lambda L)^{-1} f \le 1$. Thanks to spectral theorem, in $L^2_\mu(\mathbb{R}^n, \mathbb{R})$,

$$\boldsymbol{P}_t f = \lim_{n \to +\infty} \left(\mathrm{Id} - \frac{t}{n} L\right)^{-n} f.$$

  By passing to a subsequence that converges pointwise almost surely, we deduce that $0 \le \boldsymbol{P}_t f \le 1$ almost surely. □

4.4 L^p-theory: The interpolation method

In the previous section, we have seen that if L is an essentially self-adjoint diffusion operator with invariant and symmetric measure μ, then, by using the spectral theorem, we can define a self-adjoint strongly continuous contraction semigroup $(\boldsymbol{P}_t)_{t \ge 0}$ on $L^2_\mu(\mathbb{R}^n, \mathbb{R})$ with generator L.

Our goal in this section is to define $\boldsymbol{P}_t$ on $L^p_\mu(\mathbb{R}^n, \mathbb{R})$ for $1 \le p \le +\infty$. This may be done in a natural way by using the Riesz–Thorin interpolation theorem that we state below. We refer the reader to the book [63] for a proof.

Theorem 4.28 (Riesz–Thorin interpolation theorem). *Let $1 \le p_0, p_1, q_0, q_1 \le \infty$, and $\theta \in (0,1)$. Define $1 \le p, q \le \infty$ by*

$$\frac{1}{p} = \frac{1-\theta}{p_0} + \frac{\theta}{p_1}, \quad \frac{1}{q} = \frac{1-\theta}{q_0} + \frac{\theta}{q_1}.$$

If T is a linear map such that

$$\begin{aligned}
T &: L^{p_0}_\mu \to L^{q_0}_\mu, \quad \|T\|_{L^{p_0}_\mu \to L^{q_0}_\mu} = M_0, \\
T &: L^{p_1}_\mu \to L^{q_1}_\mu, \quad \|T\|_{L^{p_1}_\mu \to L^{q_1}_\mu} = M_1,
\end{aligned}$$

then, for every $f \in L^{p_0}_\mu \cap L^{p_1}_\mu$,

$$\|Tf\|_q \le M_0^{1-\theta} M_1^{\theta} \|f\|_p.$$

Hence T *extends uniquely as a bounded map from* L^p_μ *to* L^q_μ *with*

$$\|T\|_{L^p_\mu \to L^q_\mu} \le M_0^{1-\theta} M_1^{\theta}.$$

Remark 4.29. The statement that T is a linear map such that

$$T\colon L^{p_0}_\mu \to L^{q_0}_\mu, \quad \|T\|_{L^{p_0}_\mu \to L^{q_0}_\mu} = M_0,$$
$$T\colon L^{p_1}_\mu \to L^{q_1}_\mu, \quad \|T\|_{L^{p_1}_\mu \to L^{q_1}_\mu} = M_1,$$

means that there exists a map $T\colon L^{p_0}_\mu \cap L^{p_1}_\mu \to L^{q_0}_\mu \cap L^{q_1}_\mu$ with

$$\sup_{f \in L^{p_0}_\mu \cap L^{p_1}_\mu, \|f\|_{p_0} \le 1} \|Tf\|_{q_0} = M_0$$

and

$$\sup_{f \in L^{p_0}_\mu \cap L^{p_1}_\mu, \|f\|_{p_1} \le 1} \|Tf\|_{q_1} = M_1.$$

In such a case T can be uniquely extended to bounded linear maps $T_0\colon L^{p_0}_\mu \to L^{q_0}_\mu$, $T_1\colon L^{p_1}_\mu \to L^{q_1}_\mu$. With a slight abuse of notation, these two maps are both denoted by T in the theorem.

Remark 4.30. If $f \in L^{p_0}_\mu \cap L^{p_1}_\mu$ and p is defined by $\frac{1}{p} = \frac{1-\theta}{p_0} + \frac{\theta}{p_1}$, then by Hölder's inequality, $f \in L^p_\mu$ and

$$\|f\|_p \le \|f\|_{p_0}^{1-\theta} \|f\|_{p_1}^{\theta}.$$

One of the (numerous) beautiful applications of the Riesz–Thorin theorem is to construct diffusion semigroups on L^p by interpolation. More precisely, let L be an essentially self-adjoint diffusion operator. We denote by $(\boldsymbol{P}_t)_{t \ge 0}$ the self-adjoint strongly continuous semigroup associated to L constructed on L^2_μ thanks to the spectral theorem. We recall that $(\boldsymbol{P}_t)_{t \ge 0}$ satisfies the sub-Markov property: if $0 \le f \le 1$ is a function in L^2_μ, then $0 \le \boldsymbol{P}_t f \le 1$.

We now are in position to state the following theorem:

Theorem 4.31. *The space* $L^1_\mu \cap L^\infty_\mu$ *is invariant under* $\boldsymbol{P}_t$ *and* $\boldsymbol{P}_t$ *may be extended from* $L^1_\mu \cap L^\infty_\mu$ *to a contraction semigroup* $(\boldsymbol{P}_t^{(p)})_{t \ge 0}$ *on* L^p_μ *for all* $1 \le p \le \infty$*: For* $f \in L^p_\mu$,

$$\|\boldsymbol{P}_t f\|_{L^p_\mu} \le \|f\|_{L^p_\mu}.$$

These semigroups are consistent in the sense that for $f \in L^p_\mu \cap L^q_\mu$,

$$\boldsymbol{P}_t^{(p)} f = \boldsymbol{P}_t^{(q)} f.$$

Proof. If $f, g \in L^1_\mu \cap L^\infty_\mu$ which is a subset of $L^2_\mu \cap L^\infty_\mu$, then from Theorem 4.27,

$$\left| \int_{\mathbb{R}^n} (\boldsymbol{P}_t f) g d\mu \right| = \left| \int_{\mathbb{R}^n} f(\boldsymbol{P}_t g) d\mu \right| \le \|f\|_{L^1_\mu} \|\boldsymbol{P}_t g\|_{L^\infty_\mu} \le \|f\|_{L^1_\mu} \|g\|_{L^\infty_\mu}.$$

This implies

$$\|\boldsymbol{P}_t f\|_{L^1_\mu} \le \|f\|_{L^\infty_1}.$$

The conclusion then follows from the Riesz–Thorin interpolation theorem. □

Exercise 4.32. Show that if $f \in L^p_\mu$ and $g \in L^q_\mu$ with $\frac{1}{p} + \frac{1}{q} = 1$, then

$$\int_{\mathbb{R}^n} f \boldsymbol{P}^{(q)}_t g d\mu = \int_{\mathbb{R}^n} g \boldsymbol{P}^{(p)}_t f d\mu.$$

Exercise 4.33. (1) Show that for each $f \in L^1_\mu$, the L^1_μ-valued map $t \to \boldsymbol{P}^{(1)}_t f$ is continuous.

(2) Show that for each $f \in L^p_\mu$, $1 < p < 2$, the L^p_μ-valued map $t \to \boldsymbol{P}^{(p)}_t f$ is continuous.

(3) Finally, by using the reflexivity of L^p_μ, show that for each $f \in L^p_\mu$ and every $p \ge 1$, the L^p_μ-valued map $t \to \boldsymbol{P}^{(p)}_t f$ is continuous.

We mention that in general the L^∞_μ-valued map $t \to \boldsymbol{P}^{(\infty)}_t f$ is not continuous.

4.5 L^p-theory: The Hille–Yosida method

In the previous sections we learnt how to construct a semigroup from a diffusion operator by using the theory of self-adjoint unbounded operators on Hilbert spaces. In particular, we proved the following result.

Theorem 4.34. *Let L be a elliptic diffusion operator with smooth coefficients on $\mathbb{R}^n$. Suppose that*

- *there is a Borel measure μ, symmetric for L on $\mathcal{C}_c(\mathbb{R}^n, \mathbb{R})$; and*
- *there exists an increasing sequence $h_n \in \mathcal{C}_c(\mathbb{R}^n, \mathbb{R})$, $0 \le h_n \le 1$, such that $h_n \nearrow 1$ on $\mathbb{R}^n$, and $\|\Gamma(h_n, h_n)\|_\infty \to 0$ as $n \to \infty$.*

Then by using the spectral theorem we can define a self-adjoint strongly continuous contraction semigroup $(\boldsymbol{P}_t)_{t \ge 0}$ on $L^2_\mu(\mathbb{R}^n, \mathbb{R})$ with generator L. Moreover, $\boldsymbol{P}_t$ is a sub-Markov semigroup.

In this section we will work under the assumptions of the above theorem. Our goal will be to define, for $1 < p < +\infty$, $\boldsymbol{P}_t$ on $L^p_\mu(\mathbb{R}^n, \mathbb{R})$. This can be done by using the interpolation method described in the previous section but also by using so-called Hille–Yosida theory of contraction semigroups on Banach spaces.

4.5.1 Semigroups on Banach spaces

Let $(B, \|\cdot\|)$ be a Banach space (which for us will later be $L^p_\mu(\mathbb{R}^n, \mathbb{R})$, $1 < p < +\infty$). We first have the following basic definition.

Definition 4.35. A family of bounded operators $(T_t)_{t\geq 0}$ on B is called a *strongly continuous contraction semigroup* if

- $T_0 = \mathrm{Id}$ and for $s, t \geq 0$, $T_{s+t} = T_s T_t$;
- for each $x \in B$, the map $t \to T_t x$ is continuous;
- for each $x \in B$ and $t \geq 0$, $\|T_t x\| \leq \|x\|$.

Now let us recall that a densely defined linear operator

$$A\colon \mathcal{D}(A) \subset B \to B$$

is said to be closed if for every sequence $x_n \in \mathcal{D}(A)$ that converges to $x \in B$ and such that $Ax_n \to y \in B$, we have $x \in \mathcal{D}(A)$ and $y = Ax$. We observe that any densely defined operator A may be extended to a closed operator. Indeed, let us consider on $\mathcal{D}(A)$ the norm

$$\|x\|_A = \|x\| + \|Ax\|,$$

and let us complete $\mathcal{D}(A)$ to obtain a Banach space $\mathcal{B}$. Since $\|x\| \leq \|x\|_A$, we may identify $\mathcal{B}$ as a subset of B. Moreover $\|Ax\| \leq \|x\|_A$, therefore A can be extended to an operator $\bar{A}\colon \mathcal{B} \to B$, and it is readily checked that $\bar{A}$ with domain $\mathcal{B}$ is a closed extension of A. The operator $\bar{A}$ is called the closure of A.

In this situation, we have the following proposition which extends Theorem 3.31 and whose proof is pretty much the same and therefore left to the reader.

Proposition 4.36. *Let $(T_t)_{t\geq 0}$ be a strongly continuous contraction semigroup on B. There exists a closed and densely defined operator*

$$A\colon \mathcal{D}(A) \subset B \to B,$$

where

$$\mathcal{D}(A) = \left\{ x \in B,\ \lim_{t\to 0} \frac{T_t x - x}{t} \ \textit{exists} \right\}$$

such that for $x \in \mathcal{D}(A)$,

$$\lim_{t\to 0} \left\| \frac{T_t x - x}{t} - Ax \right\| = 0.$$

The operator A is called the generator of the semigroup $(T_t)_{t\geq 0}$. We also say that A generates $(T_t)_{t\geq 0}$.

Remark 4.37. We observe that the proof of the above result does not involve the contraction property of $(T_t)_{t\geq 0}$, so that it may be extended to strongly continuous semigroups.

The following important theorem is due to Hille and Yosida and provides, through spectral properties, a characterization of closed operators that are generators of contraction semigroups.

Let $A\colon \mathcal{D}(A) \subset B \to B$ be a densely defined closed operator. A constant $\lambda \in \mathbb{R}$ is said to be in the spectrum of A if the operator $\lambda\mathrm{Id} - A$ is not bijective. In that case, it is a consequence of the closed graph theorem[2] that if λ is not in the spectrum of A, then the operator $\lambda\mathrm{Id} - A$ has a bounded inverse. The spectrum of an operator A shall be denoted by $\rho(A)$.

Theorem 4.38 (Hille–Yosida theorem). *A necessary and sufficient condition that a densely defined closed operator A generates a strongly continuous contraction semigroup is that*

- $\rho(A) \subset (-\infty, 0]$; *and*
- $\|(\lambda\mathrm{Id} - A)^{-1}\| \leq \frac{1}{\lambda}$ *for all* $\lambda > 0$.

Proof. Let us first assume that A generates a strongly continuous contraction semigroup $(T_t)_{t\geq 0}$. Let $\lambda > 0$. We want to prove that $\lambda\mathrm{Id} - A$ is a bijective operator $\mathcal{D}(A) \to B$.

The formal Laplace transform formula

$$\int_0^{+\infty} e^{-\lambda t} e^{tA} dt = (\lambda\mathrm{Id} - A)^{-1},$$

suggests that the operator

$$\boldsymbol{R}_\lambda = \int_0^{+\infty} e^{-\lambda t} T_t dt$$

is the inverse of $\lambda\mathrm{Id} - A$. First, let us observe that $\boldsymbol{R}_\lambda$ is well defined as a Riemann integral since $t \to T_t$ is continuous and $\|T_t\| \leq 1$. We now show that for $x \in B$, $\boldsymbol{R}_\lambda x \in \mathcal{D}(A)$. For $h > 0$,

$$\begin{aligned}\frac{T_h - \mathrm{Id}}{h}\boldsymbol{R}_\lambda x &= \int_0^{+\infty} e^{-\lambda t}\frac{T_h - \mathrm{Id}}{h} T_t x dt \\ &= \int_0^{+\infty} e^{-\lambda t}\frac{T_{h+t} - T_t}{h} x dt\end{aligned}$$

[2] An everywhere defined operator between two Banach spaces $A\colon B_1 \to B_2$ is bounded if and only if it is closed.

$$
\begin{aligned}
&= e^{\lambda h} \int_h^{+\infty} e^{-\lambda s} \frac{T_s - T_{s-h}}{h} x ds \\
&= \frac{e^{\lambda h}}{h} \Big(\boldsymbol{R}_\lambda x - \int_0^h e^{-\lambda s} T_s x ds - \int_h^{+\infty} e^{-\lambda s} T_{s-h} x ds \Big) \\
&= \frac{e^{\lambda h} - 1}{h} \boldsymbol{R}_\lambda x - \frac{e^{\lambda h}}{h} \int_0^h e^{-\lambda s} T_s x ds.
\end{aligned}
$$

By letting $h \to 0$, we deduce that $\boldsymbol{R}_\lambda x \in \mathcal{D}(A)$ and moreover

$$
A\boldsymbol{R}_\lambda x = \lambda \boldsymbol{R}_\lambda x - x.
$$

Therefore we proved

$$
(\lambda \mathrm{Id} - A)\boldsymbol{R}_\lambda = \mathrm{Id}.
$$

Furthermore, it is readily checked that, since A is closed, for $x \in \mathcal{D}(A)$,

$$
\begin{aligned}
A\boldsymbol{R}_\lambda x &= A \int_0^{+\infty} e^{-\lambda t} T_t x dt \\
&= \int_0^{+\infty} e^{-\lambda t} A T_t x dt = \int_0^{+\infty} e^{-\lambda t} T_t A x dt = \boldsymbol{R}_\lambda A x.
\end{aligned}
$$

We therefore conclude that

$$
(\lambda \mathrm{Id} - A)\boldsymbol{R}_\lambda = \boldsymbol{R}_\lambda(\lambda \mathrm{Id} - A) = \mathrm{Id}.
$$

Thus,

$$
\boldsymbol{R}_\lambda = (\lambda \mathrm{Id} - A)^{-1},
$$

and it is clear that

$$
\|\boldsymbol{R}_\lambda\| \le \frac{1}{\lambda}.
$$

Let us now assume that A is a densely defined closed operator such that

- $\rho(A) \subset (-\infty, 0]$;
- $\|(\lambda \mathrm{Id} - A)^{-1}\| \le \frac{1}{\lambda}$ for all $\lambda > 0$.

The idea is to consider the following sequence of bounded operators:

$$
A_n = -n\mathrm{Id} + n^2(n\mathrm{Id} - A)^{-1}.
$$

From this sequence it is easy to define a contraction semigroup and then to show that $A_n \to A$. We will then define a contraction semigroup associated to A as the limit of the contraction semigroups associated to A_n.

First, for $x \in \mathcal{D}(A)$, we have

$$
A_n x = n(n\mathrm{Id} - A)^{-1} A x \xrightarrow[n \to +\infty]{} 0.
$$

Now, since A_n is a bounded operator, we may define a semigroup $(T_t^n)_{t\geq 0}$ through the formula

$$T_t^n = \sum_{k=0}^{+\infty} \frac{t^k A_n^k}{k!}.$$

At this point let us observe that we also have

$$T_t^n = e^{-nt} \sum_{k=0}^{+\infty} \frac{n^{2k} t^k (n\mathrm{Id} - A)^{-k}}{k!}.$$

As a consequence we have

$$\|T_t^n\| \leq e^{-nt} \sum_{k=0}^{+\infty} \frac{n^{2k} \|(n\mathrm{Id} - A)^{-1}\|^k}{k!} \leq e^{-nt} \sum_{k=0}^{+\infty} \frac{n^k t^k}{k!} \leq 1$$

and $(T_t^n)_{t\geq 0}$ is therefore a contraction semigroup. The strong continuity is also easily checked:

$$\|T_{t+h}^n - T_t^n\| = \|T_t^n (T_h^n - \mathrm{Id})\| \leq \|T_h^n - \mathrm{Id})\| \leq \sum_{k=1}^{+\infty} \frac{h^k \|A_n\|^k}{k!} \underset{h\to 0}{\longrightarrow} 0.$$

We now prove that for fixed $t \geq 0$, $x \in \mathcal{D}(A)$, $(T_t^n x)_{n\geq 1}$ is a Cauchy sequence. We have

$$\begin{aligned}
\|T_t^n x - T_t^m x\| &= \left\| \int_0^t \frac{d}{ds}(T_s^n T_{t-s}^m x) ds \right\| \\
&= \left\| \int_0^t T_s^n T_{t-s}^m (A_n x - A_m x) ds \right\| \\
&\leq \int_0^t \|A_n x - A_m x\| ds \\
&\leq t \|A_n x - A_m x\|.
\end{aligned}$$

Therefore for $x \in \mathcal{D}(A)$, $(T_t^n x)_{n\geq 1}$ is a Cauchy sequence and we can define

$$T_t x = \lim_{n\to+\infty} T_t^n x.$$

Since $\mathcal{D}(A)$ is dense and the family $(T_t^n)_{n\geq 1}$ uniformly bounded, the above limit actually exists for every $x \in B$, so that $(T_t)_{t\geq 0}$ is well defined on B. It is clear that $(T_t)_{t\geq 0}$ is a strongly continuous semigroup, inheriting this property from $(T_t^n)_{t\geq 0}$ (the details are left to the reader here).

It remains to show that the generator of $(T_t)_{t\geq 0}$, call it $\tilde{A}$, is equal to A. For every $t \geq 0$, $x \in \mathcal{D}(A)$ and $n \geq 1$,

$$T_t^n x = x + \int_0^t T_s^n A x ds,$$

therefore

$$T_t^n x = x + \int_0^t T_s^n A x ds.$$

Hence $\mathcal{D}(A) \subset \mathcal{D}(\tilde{A})$ and for $x \in \mathcal{D}(A)$, $\tilde{A}x = Ax$. Finally, since for $\lambda > 0$, $(\lambda \mathrm{Id} - A)\mathcal{D}(A) = B = (\lambda \mathrm{Id} - \tilde{A})\mathcal{D}(\tilde{A})$, it follows that $\mathcal{D}(A) = \mathcal{D}(\tilde{A})$. □

Exercise 4.39. By using the proof of Theorem 4.38, show the following fact: If A_1 and A_2 are the generators of contraction semigroups $(T_t^1)_{t\geq 0}$ and $(T_t^2)_{t\geq 0}$, then for $x \in B$, the two following statements are equivalent:

- $(\lambda \mathrm{Id} - A_1)^{-1}x = (\lambda \mathrm{Id} - A_2)^{-1}x$ for all $\lambda > 0$;
- $T_t^1 x = T_t^2 x$ for all $t \geq 0$.

As powerful as it is, it is difficult to directly apply the Hille–Yosida theorem to the theory of diffusion semigroups. We shall need a corollary of it that is more suited to this case.

Definition 4.40. A densely defined operator on a Banach space B is called *dissipative* if for each $x \in \mathcal{D}(A)$, we can find an element ϕ of the dual space B^* such that

- $\|\phi\| = \|x\|$;
- $\phi(x) = \|x\|^2$;
- $\phi(Ax) \leq 0$.

With this new definition in hands, we have the following corollary of the Hille–Yosida theorem:

Corollary 4.41. *A closed operator A on a Banach space B is the generator of a strongly continuous contraction semigroup if and only if the following holds.*

- *A is dissipative.*
- *For $\lambda > 0$, the range of the operator $\lambda \mathrm{Id} - A$ is B.*

Proof. Let us first assume that A is the generator of a contraction semigroup $(T_t)_{t\geq 0}$. From the Hahn–Banach theorem, there exists $\phi \in B^*$ such that $\|\phi\| = \|x\|$ and $\phi(x) = \|x\|^2$. We have, at $t = 0$,

$$\frac{d}{dt}\phi(T_t x) = \phi(Ax),$$

but

$$|\phi(T_t x)| \leq \|\phi\| \|T_t x\| \leq \|\phi\| \|x\| \leq \|x\|^2 \leq \phi(x),$$

thus, at $t = 0$,

$$\frac{d}{dt}\phi(T_t x) \leq 0,$$

and we conclude that

$$\phi(Ax) \leq 0.$$

The fact that for $\lambda > 0$, the range of the operator $\lambda \mathrm{Id} - A$ is B is a straightforward consequence of Theorem 4.38.

Let us now assume that A is a densely defined closed operator such that

- A is dissipative;
- for $\lambda > 0$, the range of the operator $\lambda \mathrm{Id} - A$ is B.

Let $x \in \mathcal{D}(A)$ and let $\phi \in B^*$ such that

- $\|\phi\| = \|x\|$;
- $\phi(x) = \|x\|^2$;
- $\phi(Ax) \leq 0$.

For $\lambda > 0$,

$$\lambda \|x\|^2 \leq \lambda \phi(x) - \phi(Ax) \leq \phi((\lambda \mathrm{Id} - A)x) \leq \|x\| \|(\lambda \mathrm{Id} - A)x\|.$$

Thus,

$$\|(\lambda \mathrm{Id} - A)x\| \geq \lambda \|x\|.$$

This implies that the range $\mathcal{R}_\lambda$ of the operator $\lambda \mathrm{Id} - A$ is closed and that this operator has a bounded inverse from $\mathcal{R}_\lambda$ to $\mathcal{D}(A)$ with norm lower than $\frac{1}{\lambda}$. Since $\mathcal{R}_\lambda = B$, the proof is complete. □

4.5.2 Applications to diffusion semigroups

After this digression on the theory of contraction semigroups on Banach spaces, we now come back to the case of diffusion semigroups. Let us recall that we consider an elliptic diffusion operator L on $\mathbb{R}^n$ with smooth coefficients such that the following holds:

- There is a Borel measure μ, symmetric and invariant for L on $\mathcal{C}_c(\mathbb{R}^n, \mathbb{R})$.
- There exists an increasing sequence $h_n \in \mathcal{C}_c(\mathbb{R}^n, \mathbb{R})$, $0 \leq h_n \leq 1$, such that $h_n \nearrow 1$ on $\mathbb{R}^n$, and $\|\Gamma(h_n, h_n)\|_\infty \to 0$ as $n \to \infty$.

Let $1 < p < +\infty$. We want to apply the previous theorems for the operator L on the Banach space $L^p_\mu(\mathbb{R}^n, \mathbb{R})$. We denote by $L^{(p)}$ the closure in $L^p_\mu(\mathbb{R}^n, \mathbb{R})$ of the operator L, densely defined on $\mathcal{C}_c(\mathbb{R}^n, \mathbb{R})$.

First, we have the following lemma.

Lemma 4.42. *Let $1 < p \le q < +\infty$ and $\lambda > 0$. If $f \in L^p_\mu(\mathbb{R}^n, \mathbb{R}) + L^q_\mu(\mathbb{R}^n, \mathbb{R})$ and satisfies in the sense of distributions*

$$Lf = \lambda f,$$

then $f = 0$.

Proof. We use an idea already present in the proof of Proposition 4.11. Let us first observe that if $f \in L^p_\mu(\mathbb{R}^n, \mathbb{R}) + L^q_\mu(\mathbb{R}^n, \mathbb{R})$ satisfies in the sense of distributions

$$Lf = \lambda f,$$

then it is actually a smooth function, due to the ellipticity of L. Now let $h \in \mathcal{C}_c(\mathbb{R}^n, \mathbb{R})$, $h \ge 0$, and let ϕ be a smooth non-negative function to be made precise later. We have

$$\int_{\mathbb{R}^n} \Gamma(f, h^2\phi(f)f)d\mu = -\int_{\mathbb{R}^n} fL(h^2\phi(f)f)d\mu = -\lambda \int_{\mathbb{R}^n} h^2\phi(f)f^2 d\mu \le 0.$$

But

$$\begin{aligned}\Gamma(f, h^2\phi(f)f) &= h^2\Gamma(f, \phi(f)f) + 2\phi(f)fh\Gamma(f,h)\\ &= h^2(\phi'(f)f + \phi(f))\Gamma(f,f) + 2\phi(f)fh\Gamma(f,h),\end{aligned}$$

therefore

$$\int_{\mathbb{R}^n} h^2(\phi'(f)f + \phi(f))\Gamma(f,f) + 2\phi(f)fh\Gamma(f,h)d\mu \le 0,$$

which implies that

$$\int_{\mathbb{R}^n} h^2(\phi'(f)f + \phi(f))\Gamma(f,f)d\mu \le -2\int_{\mathbb{R}^n} \phi(f)fh\Gamma(f,h)d\mu. \tag{4.2}$$

From the Cauchy–Schwarz inequality, we have

$$\begin{aligned}&\left|\int_{\mathbb{R}^n} \phi(f)fh\Gamma(f,h)d\mu\right|\\ &\quad\le 2\sqrt{\|\Gamma(h,h)\|_\infty}\sqrt{\int_{\mathbb{R}^n} \phi(f)h^2\Gamma(f,f)d\mu}\sqrt{\int_K \phi(f)f^2 d\mu},\end{aligned} \tag{4.3}$$

where K is any compact set containing the support of h. Let us now assume that we may chose the function ϕ in such a way that for every $x \in \mathbb{R}$,

$$\phi'(x)x + \phi(x) \ge C\phi(x)$$

with $C > 0$. In that case, from (4.2) and (4.3) we have

$$\int_{\mathbb{R}^n} \phi(f) h^2 \Gamma(f, f) d\mu \leq \frac{4}{C^2} \|\Gamma(h,h)\|_\infty \int_K \phi(f) f^2 d\mu. \tag{4.4}$$

We now proceed to the choice for the function ϕ. For $0 < \varepsilon < 1$, let

$$\phi_\varepsilon(x) = \begin{cases} |x|^{p-2}, & |x| \geq 1, \\ (x^2 + \varepsilon^2)^{\frac{q}{2}-1}, & |x| \leq 1 - \varepsilon. \end{cases}$$

We then extend ϕ_ε in a smooth way to $\mathbb{R}$ in such a way that there is a constant $C > 0$, independent of ε, such that for every $x \in \mathbb{R}$,

$$\phi'_\varepsilon(x) x + \phi_\varepsilon(x) \geq C \phi_\varepsilon(x).$$

The proof of the possibility of such an extension is left to the reader.

By using the inequality (4.4) for ϕ_ε, we obtain

$$\int_{\mathbb{R}^n} \phi_\varepsilon(f) h^2 \Gamma(f, f) d\mu \leq \frac{4}{C^2} \|\Gamma(h,h)\|_\infty \int_K \phi_\varepsilon(f) f^2 d\mu.$$

By letting $\varepsilon \to 0$, we draw the conclusion that

$$\int_{\mathbb{R}^n} \phi_0(f) h^2 \Gamma(f, f) d\mu \leq \frac{4}{C^2} \|\Gamma(h,h)\|_\infty \int_K \phi_0(f) f^2 d\mu,$$

where

$$\phi_0(x) = \begin{cases} |x|^{p-2}, & |x| \geq 1, \\ |x|^{q-2}, & |x| \leq 1. \end{cases}$$

Due to the assumption $f \in L^p_\mu(\mathbb{R}^n, \mathbb{R}) + L^q_\mu(\mathbb{R}^n, \mathbb{R})$, the function ϕ_0 is integrable on $\mathbb{R}^n$, thus

$$\int_{\mathbb{R}^n} \phi_0(f) h^2 \Gamma(f, f) d\mu \leq \frac{4}{C^2} \|\Gamma(h,h)\|_\infty \int_{\mathbb{R}^n} \phi_0(f) f^2 d\mu.$$

By using the previous inequality with an increasing sequence $h_n \in \mathcal{C}_c(\mathbb{R}^n, \mathbb{R})$, $0 \leq h_n \leq 1$, such that $h_n \nearrow 1$ on $\mathbb{R}^n$, and $\|\Gamma(h_n, h_n)\|_\infty \to 0$, as $n \to \infty$, and letting $n \to +\infty$,

$$\int_{\mathbb{R}^n} \phi_0(f) \Gamma(f, f) d\mu = 0.$$

As a consequence, since μ charges the open sets, if $f \neq 0$, then $\Gamma(f, f) = 0$. Since $f = \frac{1}{\lambda} L f$, we conclude that $f = 0$. □

We are now in a position to prove the main result of this section. For $1 < p < +\infty$, denote by $L^{(p)}$ the closure in $L^p_\mu(\mathbb{R}^n, \mathbb{R})$ of the operator L, originally only densely defined on $\mathcal{C}_c(\mathbb{R}^n, \mathbb{R})$.

Proposition 4.43. *The operator $L^{(p)}$ is the generator of a strongly continuous contraction semigroup $(\boldsymbol{P}_t^{(p)})_{t\geq 0}$ on $L^p_\mu(\mathbb{R}^n,\mathbb{R})$. Moreover, for $1 < p \leq q < +\infty$ and $f \in L^p_\mu(\mathbb{R}^n,\mathbb{R}) \cap L^q_\mu(\mathbb{R}^n,\mathbb{R})$,*

$$\boldsymbol{P}_t^{(p)} f = \boldsymbol{P}_t^{(q)} f.$$

Proof. We apply Corollary 4.41. We have to show that

- $L^{(p)}$ is dissipative;
- for $\lambda > 0$, the range of the operator $\lambda \mathrm{Id} - L^{(p)}$ is $L^p_\mu(\mathbb{R}^n,\mathbb{R})$.

We first check that $L^{(p)}$ is dissipative. Since it is easily seen that the closure of a dissipative operator is dissipative, we have to check that L is dissipative on $\mathcal{C}_c(\mathbb{R}^n,\mathbb{R})$.

Let $f \in \mathcal{C}_c(\mathbb{R}^n,\mathbb{R})$ and let

$$g = \frac{1}{\|f\|^{p/q}_{L^p_\mu(\mathbb{R}^n,\mathbb{R})}} |f|^{p-2} f,$$

where $\frac{1}{p} + \frac{1}{q} = 1$. We have

- $\|g\|_{L^q_\mu(\mathbb{R}^n,\mathbb{R})} = \|f\|_{L^p_\mu(\mathbb{R}^n,\mathbb{R})}$;
- $\int_{\mathbb{R}^n} g f d\mu = \|f\|^2_{L^p_\mu(\mathbb{R}^n,\mathbb{R})}$.

Because $L^q_\mu(\mathbb{R}^n,\mathbb{R})$ is the dual space of $L^p_\mu(\mathbb{R}^n,\mathbb{R})$, it remains to prove that $\int_{\mathbb{R}^n} gLf d\mu \leq 0$. For $\varepsilon > 0$, let

$$\Phi_\varepsilon(x) = (x^2 + \varepsilon^2)^{\frac{p}{2}-1} x.$$

We have

$$\int_{\mathbb{R}^n} \Phi_\varepsilon(f) L f d\mu = -\int_{\mathbb{R}^n} \Gamma(\Phi_\varepsilon(f), f) d\mu = -\int_{\mathbb{R}^n} \Phi'_\varepsilon(f) \Gamma(f,f) d\mu \leq 0.$$

From the Lebesgue dominated convergence theorem applied when $\varepsilon \to 0$, this implies

$$\int_{\mathbb{R}^n} |f|^{p-2} f L f d\mu \leq 0.$$

As a conclusion, L is dissipative on $\mathcal{C}_c(\mathbb{R}^n,\mathbb{R})$ and $L^{(p)}$ is dissipative on its domain.

We now show that for $\lambda > 0$, the range of the operator $\lambda \mathrm{Id} - L^{(p)}$ is $L^p_\mu(\mathbb{R}^n,\mathbb{R})$. If not, we could find a non-zero $g \in L^q_\mu(\mathbb{R}^n,\mathbb{R})$, $\frac{1}{p} + \frac{1}{q} = 1$, such that, in the sense of distributions,

$$Lg = \lambda g.$$

But, according to Lemma 4.42, the above equation implies that $g = 0$. The range of the operator $\lambda \mathrm{Id} - L^{(p)}$ is therefore $L^p_\mu(\mathbb{R}^n,\mathbb{R})$.

We conclude from Corollary 4.41 that $L^{(p)}$ is the generator of a contraction semigroup $(\boldsymbol{P}_t^{(p)})_{t\geq 0}$ on $L^p_\mu(\mathbb{R}^n, \mathbb{R})$.

We finally show that for $1 < p \leq q < +\infty$, $(\boldsymbol{P}_t^{(p)})_{t\geq 0}$ and $(\boldsymbol{P}_t^{(q)})_{t\geq 0}$ coincide on $L^p_\mu(\mathbb{R}^n, \mathbb{R}) \cap L^q_\mu(\mathbb{R}^n, \mathbb{R})$. Let $f \in L^p_\mu(\mathbb{R}^n, \mathbb{R}) \cap L^q_\mu(\mathbb{R}^n, \mathbb{R})$. For $\lambda > 0$, the function

$$g = (\lambda \mathrm{Id} - L^{(p)})^{-1} f - (\lambda \mathrm{Id} - L^{(q)})^{-1} f$$

is a function that belongs to $L^p_\mu(\mathbb{R}^n, \mathbb{R}) + L^q_\mu(\mathbb{R}^n, \mathbb{R})$ and that satisfies in the sense of distributions

$$Lg = \lambda g.$$

From Lemma 4.42, we get $g = 0$. As a consequence for every $\lambda > 0$,

$$(\lambda \mathrm{Id} - L^{(p)})^{-1} f = (\lambda \mathrm{Id} - L^{(q)})^{-1} f.$$

This implies that for $t \geq 0$, $\boldsymbol{P}_t^{(p)} f = \boldsymbol{P}_t^{(q)} f$ (See Exercise 4.39). □

Exercise 4.44. Show that on $L^2_\mu(\mathbb{R}^n, \mathbb{R})$, the semigroup $(\boldsymbol{P}_t^{(2)})_{t\geq 0}$ coincides with the semigroup $(\boldsymbol{P}_t)_{t\geq 0}$ that was constructed with the spectral theorem. (You may use Exercise 4.39.)

Remark 4.45. Let us observe that the Hille–Yosida methods fail in the limiting cases $p = 1$, $p = +\infty$.

Since $(\boldsymbol{P}_t^{(p)})_{t\geq 0}$ and $(\boldsymbol{P}_t^{(q)})_{t\geq 0}$ coincide on $L^p_\mu(\mathbb{R}^n, \mathbb{R}) \cap L^q_\mu(\mathbb{R}^n, \mathbb{R})$, there is no ambiguity and we can safely remove the superscript (p) from $(\boldsymbol{P}_t^{(p)})_{t\geq 0}$. So that we can use only the notation $(\boldsymbol{P}_t)_{t\geq 0}$, but will make precise on which space it is defined if there is ambiguity. The following duality lemma is then easy to prove.

Lemma 4.46. *Let $f \in L^p_\mu(\mathbb{R}^n, \mathbb{R})$ and let $g \in L^q_\mu(\mathbb{R}^n, \mathbb{R})$, with $\frac{1}{p} + \frac{1}{q} = 1$. For $t \geq 0$,*

$$\int_{\mathbb{R}^n} (\boldsymbol{P}_t f) g \, d\mu = \int_{\mathbb{R}^n} (\boldsymbol{P}_t g) f \, d\mu.$$

Proof. Let us first assume that $g \in \mathcal{C}_c(\mathbb{R}^n, \mathbb{R})$. The linear form $\Phi \colon L^p_\mu(\mathbb{R}^n, \mathbb{R}) \to \mathbb{R}$, $f \to \int_{\mathbb{R}^n} (\boldsymbol{P}_t f) g \, d\mu - \int_{\mathbb{R}^n} (\boldsymbol{P}_t g) f \, d\mu$ is continuous and vanishes on the dense subspace $\mathcal{C}_c(\mathbb{R}^n, \mathbb{R})$, it is therefore zero on $L^p_\mu(\mathbb{R}^n, \mathbb{R})$. This proves that for $f \in L^p_\mu(\mathbb{R}^n, \mathbb{R})$ and $g \in \mathcal{C}_c(\mathbb{R}^n, \mathbb{R})$,

$$\int_{\mathbb{R}^n} (\boldsymbol{P}_t f) g \, d\mu = \int_{\mathbb{R}^n} (\boldsymbol{P}_t g) f \, d\mu.$$

The same density argument applied on the linear form $g \to \int_{\mathbb{R}^n} (\boldsymbol{P}_t f) g \, d\mu - \int_{\mathbb{R}^n} (\boldsymbol{P}_t g) f \, d\mu$ provides the expected result. □

4.6 Diffusion semigroups as solutions of a parabolic Cauchy problem

In this section, we connect the semigroup associated to a diffusion operator L to the following parabolic Cauchy problem:

$$\frac{\partial u}{\partial t} = Lu, \quad u(0,x) = f(x).$$

As before, let L be a elliptic diffusion operator with smooth coefficients on $\mathbb{R}^n$ such that

- there is a Borel measure μ, symmetric for L on $\mathcal{C}_c(\mathbb{R}^n, \mathbb{R})$;
- there exists an increasing sequence $h_n \in \mathcal{C}_c(\mathbb{R}^n, \mathbb{R})$, $0 \le h_n \le 1$, such that $h_n \nearrow 1$ on $\mathbb{R}^n$, and $\|\Gamma(h_n, h_n)\|_\infty \to 0$, as $n \to \infty$.

Proposition 4.47. *Let $f \in L^p_\mu(\mathbb{R}^n, \mathbb{R})$, $1 \le p \le \infty$, and let*

$$u(t,x) = P_t f(x), \quad t \ge 0, x \in \mathbb{R}^n.$$

Then u is smooth on $(0, +\infty) \times \mathbb{R}^n$ and is a solution of the Cauchy problem

$$\frac{\partial u}{\partial t} = Lu, \quad u(0,x) = f(x).$$

Proof. For $\phi \in \mathcal{C}_c((0,+\infty) \times \mathbb{R}^n, \mathbb{R})$, we have

$$\begin{aligned}
&\int_{\mathbb{R}^n \times \mathbb{R}} \Big(\Big(-\frac{\partial}{\partial t} - L\Big)\phi(t,x)\Big) u(t,x)\,d\mu(x)dt \\
&= \int_{\mathbb{R}}\int_{\mathbb{R}^n} \Big(\Big(-\frac{\partial}{\partial t} - L\Big)\phi(t,x)\Big) P_t f(x) dx dt \\
&= \int_{\mathbb{R}}\int_{\mathbb{R}^n} P_t\Big(\Big(-\frac{\partial}{\partial t} - L\Big)\phi(t,x)\Big) f(x) dx dt \\
&= \int_{\mathbb{R}}\int_{\mathbb{R}^n} -\frac{\partial}{\partial t}(P_t\phi(t,x) f(x)) dx dt \\
&= 0.
\end{aligned}$$

Therefore u is a weak solution of the equation $\frac{\partial u}{\partial t} = Lu$. Since we already know that u is smooth it is also a strong solution. □

We now address the uniqueness question.

Proposition 4.48. *Let $v(x,t)$ be a non-negative function such that*

$$\frac{\partial v}{\partial t} \le Lv, \quad v(x,0) = 0,$$

and such that for every $t > 0$,

$$\|v(\cdot, t)\|_{L^p_\mu(\mathbb{R}^n,\mathbb{R})} < +\infty,$$

where $1 < p < +\infty$. Then $v(x,t) = 0$.

Proof. Let $x_0 \in \mathbb{R}^n$ and $h \in \mathcal{C}_c(\mathbb{R}^n, \mathbb{R})$. Since v is a subsolution with the zero initial data, for any $\tau \in (0, T)$,

$$\begin{aligned}
\int_0^\tau \int_{\mathbb{R}^n} h^2(x) v^{p-1}(x,t) L v(x,t) d\mu(x) dt &\geq \int_0^\tau \int_{\mathbb{R}^n} h^2(x) v^{p-1} \frac{\partial v}{\partial t} d\mu(x) dt \\
&= \frac{1}{p} \int_0^\tau \frac{\partial}{\partial t} \Big(\int_{\mathbb{R}^n} h^2(x) v^p d\mu(x) \Big) dt \\
&= \frac{1}{p} \int_{\mathbb{R}^n} h^2(x) v^p(x,\tau) d\mu(x).
\end{aligned}$$

On the other hand, integrating by parts yields

$$\begin{aligned}
&\int_0^\tau \int_{\mathbb{R}^n} h^2(x) v^{p-1}(x,t) L v(x,t) d\mu(x) dt \\
&\quad = - \int_0^\tau \int_{\mathbb{R}^n} 2 h v^{p-1} \Gamma(h, v) d\mu dt - \int_0^\tau \int_{\mathbb{R}^n} h^2 (p-1) v^{p-2} \Gamma(v) d\mu dt.
\end{aligned}$$

Observing that

$$\begin{aligned}
0 &\leq \bigg(\sqrt{\frac{2}{p-1} \Gamma(h)} v - \sqrt{\frac{p-1}{2} \Gamma(v)} h \bigg)^2 \\
&\leq \frac{2}{p-1} \Gamma(h) v^2 + 2 \Gamma(h, v) h v + \frac{p-1}{2} \Gamma(v) h^2,
\end{aligned}$$

we obtain the following estimate:

$$\begin{aligned}
&\int_0^\tau \int_{\mathbb{R}^n} h^2(x) v^{p-1}(x,t) L v(x,t) d\mu(x) dt \\
&\quad \leq \int_0^\tau \int_{\mathbb{R}^n} \frac{2}{p-1} \Gamma(h) v^p d\mu dt - \int_0^\tau \int_{\mathbb{R}^n} \frac{p-1}{2} h^2 v^{p-2} \Gamma(v) d\mu dt \\
&\quad = \int_0^\tau \int_{\mathbb{R}^n} \frac{2}{p-1} \Gamma(h) v^p d\mu dt - \frac{2(p-1)}{p^2} \int_0^\tau \int_{\mathbb{R}^n} h^2 \Gamma(v^{p/2}) d\mu dt.
\end{aligned}$$

Combining with the previous conclusion we obtain

$$\begin{aligned}
&\int_{\mathbb{R}^n} h^2(x) v^p(x,\tau) d\mu(x) + \frac{2(p-1)}{p} \int_0^\tau \int_{\mathbb{R}^n} h^2 \Gamma(v^{p/2}) d\mu dt \\
&\quad \leq \frac{2p}{(p-1)} \|\Gamma(h)\|_\infty^2 \int_0^\tau \int_{\mathbb{R}^n} v^p d\mu dt.
\end{aligned}$$

By using the previous inequality with an increasing sequence $h_n \in \mathcal{C}_c(\mathbb{R}^n, \mathbb{R})$, $0 \leq h_n \leq 1$, such that $h_n \nearrow 1$ on $\mathbb{R}^n$, and $\|\Gamma(h_n, h_n)\|_\infty \to 0$ as $n \to \infty$, and letting $n \to +\infty$, we obtain $\int_{\mathbb{R}^n} v^p(x,\tau)d\mu(x) = 0$ thus $v = 0$. □

As a consequence of this result, any solution in $L^p_\mu(\mathbb{R}^n, \mathbb{R})$, $1 < p < +\infty$ of the heat equation $\frac{\partial u}{\partial t} = Lu$ is uniquely determined by its initial condition, and is therefore of the form $u(t,x) = P_t f(x)$ (apply the above lemma with the subsolution $|u|$). We stress that without further conditions, this result fails when $p = 1$ or $p = +\infty$.

4.7 The Dirichlet semigroup

In this section we will construct the Dirichlet semigroup associated to a symmetric diffusion operator which is given on some a relatively compact domain of $\mathbb{R}^n$. A later application of this construction will be the proof of the existence of a continuous Markov process associated with the given symmetric diffusion operator.

Let L be an elliptic diffusion operator on $\mathbb{R}^n$ with smooth coefficients. We assume that L, defined on $\mathcal{C}_c(\mathbb{R}^n, \mathbb{R})$, is symmetric with respect to a measure with smooth and positive density μ. Let $\Omega \subset \mathbb{R}^n$ be a non-empty open set whose closure $\bar{\Omega}$ is compact and whose boundary $\partial\Omega$ is smooth. We recall that the following Green identity holds for L (see for instance [26]):

Proposition 4.49 (Green identity). *There exists a never vanishing first order differential operator ν defined on $\partial\Omega$ such that for every C^2 functions $f, g : \bar{\Omega} \to \mathbb{R}$,*

$$\int_\Omega fLg d\mu = -\int_\Omega \Gamma(f,g)d\mu + \int_{\partial\Omega} f(\nu g)d\mu_S,$$

where, as usual, $\Gamma(f,g) = \frac{1}{2}(L(fg) - fLg - gLf)$ and μ_S is the surface measure induced by μ on $\partial\Omega$

The basic result to construct the Dirichlet semigroup associated to L on Ω is the following fact.

Proposition 4.50. *The operator L is essentially self-adjoint on the set*

$$\mathcal{C}_0^\infty(\bar{\Omega}) = \{u : \bar{\Omega} \to \mathbb{R},\ u \text{ smooth on } \bar{\Omega},\ u = 0 \text{ on } \partial\Omega\}.$$

Proof. According to Lemma 4.9, it is enough to prove that

$$\operatorname{Ker}(-L^* + \mathrm{Id}) = \{0\}.$$

Let $f \in \operatorname{Ker}(-L^* + \mathrm{Id}) = \{0\}$. Since $\mathcal{C}_0^\infty(\bar{\Omega})$ contains the space of smooth and compactly supported functions inside Ω, we have in the sense of distributions

$Lf = f$. By ellipticity of L, this implies that f is smooth in Ω. From Green's identity we now have for $g \in \mathcal{C}_0^\infty(\bar{\Omega})$,

$$\int_\Omega fLg d\mu - \int_\Omega gLf d\mu + \int_{\partial\Omega} f(\nu g) d\mu_S.$$

Since $f \in \mathcal{D}(L^*)$, the square of the right-hand side must be controlled, for every g, by $\|g\|^2_{L^2_\mu(\bar{\Omega},\mathbb{R})}$. This is only possible when $f = 0$ on $\partial\Omega$. Thus, $f \in \mathcal{C}_0^\infty(\bar{\Omega})$. Since $Lf = f$, from Green's identity we have

$$\int_\Omega f^2 d\mu = \int_\Omega fLf d\mu = -\int_\Omega \Gamma(f, f) d\mu \le 0.$$

This implies that $f = 0$ and, as a conclusion, L is essentially self-adjoint on $\mathcal{C}_0^\infty(\bar{\Omega})$. □

The self-adjoint extension of the previous proposition is called the Dirichlet extension of L on Ω and will be denoted by L_Ω. The semigroup generated by L_Ω is called the Dirichlet semigroup and denoted by $(\boldsymbol{P}_t^\Omega)_{t\ge0}$. By using the same Sobolev embedding techniques as in the previous sections (see Theorems 4.20 and 4.23), it is seen that the semigroup $\boldsymbol{P}_t^\Omega$ has a smooth heat kernel $p^\Omega(t, x, y)$. It is then easy to see that for $f \in L^2_\mu(\bar{\Omega}, \mathbb{R})$, $\boldsymbol{P}_t^\Omega f \in \mathcal{C}_0^\infty(\bar{\Omega})$, $t > 0$. In particular, if x or y are on $\partial\Omega$, then $p^\Omega(t, x, y) = 0$, $t > 0$. This also implies that $\boldsymbol{P}_t^\Omega 1 \ne 1$, that is, $\boldsymbol{P}_t^\Omega$ is a strictly sub-Markov semigroup.

It is important to observe that L_Ω is not the unique self-adjoint extension of L on the space of smooth and compactly supported functions inside Ω. The Neumann extension presented in the following section is also particularly important.

It turns out that the compactness of $\bar{\Omega}$ implies the compactness of the semigroup $\boldsymbol{P}_t^\Omega$ (see Appendix A for further details on compact operators, in particular see page 260 for the definition of trace class and Hilbert–Schmidt operators).

Proposition 4.51. *For $t > 0$ the operator $\boldsymbol{P}_t^\Omega$ is a compact operator on the Hilbert space $L^2_\mu(\bar{\Omega}, \mathbb{R})$. Moreover, it is a trace class operator and*

$$\mathrm{Tr}(\boldsymbol{P}_t^\Omega) = \int_\Omega p^\Omega(t, x, x) d\mu(x).$$

Proof. From the existence of the heat kernel we have

$$\boldsymbol{P}_t^\Omega f(x) = \int_\Omega p^\Omega(t, x, y) f(y) d\mu(y).$$

But from the compactness of $\bar{\Omega}$ and the continuity of $p(t, \cdot, \cdot)$ on $\bar{\Omega} \times \bar{\Omega}$, we have

$$\int_\Omega \int_\Omega p^\Omega(t, x, y)^2 d\mu(x) d\mu(y) < +\infty.$$

Therefore, from Theorem A.12 in Appendix A

$$\boldsymbol{P}_t^{\Omega} \colon L^2_{\mu}(\bar{\Omega}, \mathbb{R}) \to L^2_{\mu}(\bar{\Omega}, \mathbb{R})$$

is a Hilbert–Schmidt operator. It is thus in particular a compact operator.

Since $\boldsymbol{P}_t^{\Omega} = \boldsymbol{P}_{t/2}^{\Omega} \boldsymbol{P}_{t/2}^{\Omega}$, $\boldsymbol{P}_t^{\Omega}$ is a product of two Hilbert–Schmidt operators. It is therefore a trace class operator and

$$\operatorname{Tr}(\boldsymbol{P}_t^{\Omega}) = \int_{\Omega} \int_{\Omega} p^{\Omega}(t/2, x, y) p^{\Omega}(t/2, y, x) d\mu(x) d\mu(y).$$

We then conclude by applying the Chapman–Kolmogorov relation that

$$\operatorname{Tr}(\boldsymbol{P}_t^{\Omega}) = \int_{\Omega} p^{\Omega}(t, x, x) d\mu(x). \qquad \square$$

We have the following expansion of the Dirichlet heat kernel:

Theorem 4.52. *There exists a complete orthonormal basis $(\phi_n)_{n \geq 1}$ of $L^2_{\mu}(\bar{\Omega}, \mathbb{R})$, consisting of eigenfunctions of $-L$, with $\phi_n \in \mathcal{C}_0^{\infty}(\bar{\Omega})$ having an eigenvalue λ_n with finite multiplicity and*

$$0 < \lambda_1 \leq \lambda_2 \leq \cdots \leq \lambda_n \leq \cdots \nearrow +\infty.$$

Moreover, for $t > 0$, $x, y \in \bar{\Omega}$,

$$p^{\Omega}(t, x, y) = \sum_{n=1}^{+\infty} e^{-\lambda_n t} \phi_n(x) \phi_n(y),$$

where the convergence is absolute and uniform for each $t > 0$.

Proof. Let $t > 0$. From the Hilbert–Schmidt theorem (see Theorem A.8 in Appendix A), for the non-negative self-adjoint compact operator $\boldsymbol{P}_t^{\Omega}$ there exists a complete orthonormal basis $(\phi_n(t))_{n \geq 1}$ of $L^2_{\mu}(\bar{\Omega}, \mathbb{R})$ and a non-increasing sequence $\alpha_n(t) \geq 0$, $\alpha_n(t) \searrow 0$ such that

$$\boldsymbol{P}_t^{\Omega} \phi_n(t) = \alpha_n(t) \phi_n(t).$$

The semigroup property $\boldsymbol{P}_{t+s}^{\Omega} = \boldsymbol{P}_t^{\Omega} \boldsymbol{P}_s^{\Omega}$ implies that for $k \in \mathbb{N}$, $k \geq 1$,

$$\phi_n(k) = \phi_n(1), \alpha_n(k) = \alpha_n(1)^k.$$

The same result is then seen to hold for $k \in \mathbb{Q}$, $k > 0$ and finally for $k \in \mathbb{R}$, due to the strong continuity of the semigroup. Since the map $t \to \|\boldsymbol{P}_t^{\Omega}\|_{L^2_{\mu}(\bar{\Omega}, \mathbb{R})}$ is decreasing, we deduce that $\alpha_n(1) \leq 1$. Thus, there is a $\lambda_n \geq 0$ such that

$$\alpha_n(1) = e^{-\lambda_n}.$$

As a conclusion, there exists a complete orthonormal basis $(\phi_n)_{n\geq 1}$ of $L^2_\mu(\bar{\Omega}, \mathbb{R})$, and a sequence λ_n satisfying

$$0 \leq \lambda_1 \leq \lambda_2 \leq \cdots \nearrow +\infty,$$

such that

$$\boldsymbol{P}_t \phi_n = e^{-\lambda_n t} \phi_n.$$

If $f \in L^2_\mu(\bar{\Omega}, \mathbb{R})$ is such that $\boldsymbol{P}^\Omega_t f = f$, it is straightforward that $f \in \mathcal{D}(L_\Omega)$ and that $L_\Omega f = 0$, so that $f = 0$. Therefore we have $\lambda_1 > 0$.

Since $\boldsymbol{P}^\Omega_t \phi_n = e^{-\lambda_n t}\phi_n$, by differentiating as $t \to 0$ in $L^2_\mu(\bar{\Omega}, \mathbb{R})$, we obtain furthermore that $\phi_n \in \mathcal{D}(L_\Omega)$ and that $L_\Omega \phi_n = -\lambda_n \phi_n$. By ellipticity of L, we deduce that $\phi_n \in \mathcal{C}^\infty_0(\bar{\Omega})$.

The family $(x, y) \to \phi_n(x)\phi_m(y)$ forms an orthonormal basis of $L^2_{\mu\otimes\mu}(\bar{\Omega} \times \bar{\Omega}, \mathbb{R})$. We therefore have a decomposition in $L^2_{\mu\otimes\mu}(\bar{\Omega} \times \bar{\Omega}, \mathbb{R})$,

$$p^\Omega(t, x, y) = \sum_{m,n\in\mathbb{N}} c_{mn}\phi_m(x)\phi_n(y).$$

Since $p^\Omega(t, \cdot, \cdot)$ is the kernel of the symmetric semigroup $\boldsymbol{P}^\Omega_t$, it is then straightforward that for $m \neq n$, $c_{mn} = 0$ and that $c_{nn} = e^{-\lambda_n t}$. Therefore in $L^2_\mu(\bar{\Omega}, \mathbb{R})$,

$$p^\Omega(t, x, y) = \sum_{n=1}^{+\infty} e^{-\lambda_n t}\phi_n(x)\phi_n(y).$$

The continuity of p^Ω together with the positivity of $\boldsymbol{P}^\Omega_t$ imply, via Mercer's theorem, that the above series is actually absolutely and uniformly convergent for $t > 0$. □

As we stressed it in the statement of the theorem, in the decomposition

$$p^\Omega(t, x, y) = \sum_{n=1}^{+\infty} e^{-\lambda_n t}\phi_n(x)\phi_n(y)$$

the eigenvalue λ_n is repeated according to its multiplicity. It is often useful to rewrite this decomposition in the form

$$p^\Omega(t, x, y) = \sum_{n=1}^{+\infty} e^{-\alpha_n t} \sum_{k=1}^{d_n} \phi^n_k(x)\phi^n_k(y),$$

where the eigenvalue α_n is not repeated, that is,

$$0 < \alpha_1 < \alpha_2 < \cdots.$$

In this decomposition, d_n is the dimension of the eigenspace $\mathcal{V}_n$ corresponding to the eigenvalue α_n and $(\phi_k^n)_{1\le k\le d_n}$ is an orthonormal basis of $\mathcal{V}_n$. If we write

$$\mathcal{K}_n(x,y)=\sum_{k=1}^{d_n}\phi_k^n(x)\phi_k^n(y),$$

then $\mathcal{K}_n$ is called the reproducing kernel of the eigenspace $\mathcal{V}_n$. It satisfies the following properties whose proofs are left to the reader:

Proposition 4.53. *$\mathcal{K}_n$ does not depend on the choice of the basis $(\phi_k^n)_{1\le k\le d_n}$.*

If $f\in\mathcal{V}_n$, then $\int_{\mathbb{M}}\mathcal{K}_n(x,y)f(y)d\mu(y)=f(x)$.

We finally observe that from the very definition of the reproducing kernels, we have

$$p^{\Omega}(t,x,y)=\sum_{n=1}^{+\infty}e^{-\alpha_n t}\mathcal{K}_n(x,y).$$

Exercise 4.54. (1) Let $f\in L^2_\mu(\bar{\Omega},\mathbb{R})$. Show that uniformly on $\bar{\Omega}$, when $t\to+\infty$, then $\boldsymbol{P}_t^{\Omega}f\to 0$.

(2) Let $f\in\mathcal{C}_0^\infty(\bar{\Omega})$. Show that uniformly on $\bar{\Omega}$, when $t\to 0$, then $\boldsymbol{P}_t^{\Omega}f\to f$.

Exercise 4.55. Show that if K is a compact subset of Ω, then

$$\lim_{t\to 0}\sup_{x\in K}\frac{1}{t}(\boldsymbol{P}_t^{\Omega}1_{B(x,\varepsilon)^c\cap\Omega})(x)=0.$$

4.8 The Neumann semigroup

In this section we consider another semigroup which is canonically associated to diffusion operators on sub-domains of $\mathbb{R}^n$: the Neumann semigroup. The difference with respect to the Dirichlet semigroup is the boundary condition.

The framework is identical to the framework of the previous section. Let L be an elliptic diffusion operator on $\mathbb{R}^n$ with smooth coefficients. We assume that L, defined on $\mathcal{C}_c(\mathbb{R}^n,\mathbb{R})$, is symmetric with respect to a measure with smooth and positive density μ. Let $\Omega\subset\mathbb{R}^n$ be a non-empty open set whose closure $\bar{\Omega}$ is compact and whose boundary $\partial\Omega$ is smooth. We list the main results concerning the Neumann semigroup but leave the arguments as an exercise to the reader since the proofs are almost identical to the ones of the previous section.

Proposition 4.56. *The operator L is essentially self-adjoint on the set*

$$\mathcal{C}_\nu^\infty(\bar{\Omega})=\{u\colon\bar{\Omega}\to\mathbb{R},\ u\text{ smooth on }\bar{\Omega},\ \nu u=0\text{ on }\partial\Omega\},$$

where, as before, ν denotes the vector field in the Green formula Proposition 4.49. *The self-adjoint extension of L on that set is called the Neumann extension of L and will be denoted by $L_{\Omega,\nu}$.*

The semigroup generated by $L_{\Omega,\nu}$ is called the Neumann semigroup and we will denote it by $(\boldsymbol{P}_t^{\Omega,\nu})_{t\geq 0}$. The semigroup $\boldsymbol{P}_t^{\Omega,\nu}$ has a smooth heat kernel $p^{\Omega,\nu}(t,x,y)$ and for $f \in L^2_\mu(\bar{\Omega},\mathbb{R})$, $\boldsymbol{P}_t^{\Omega} f \in \mathcal{C}^\infty_\nu(\bar{\Omega})$, $t > 0$. The main difference to the Dirichlet semigroup is that the Neumann semigroup is a Markov semigroup, that is, $\boldsymbol{P}_t^{\Omega,\nu} 1 = 1$. This comes from the fact that $1 \in \mathcal{D}(L_{\Omega,\nu})$.

Proposition 4.57. *For $t > 0$ the operator $\boldsymbol{P}_t^{\Omega,\nu}$ is a compact operator on the Hilbert space $L^2_\mu(\bar{\Omega},\mathbb{R})$. Moreover, it is a trace class operator and*

$$\mathrm{Tr}(\boldsymbol{P}_t^{\Omega,\nu}) = \int_\Omega p^{\Omega,\nu}(t,x,x)\,d\mu(x).$$

Theorem 4.58. *There exists a complete orthonormal basis $(\psi_n)_{n\geq 1}$ of $L^2_\mu(\bar{\Omega},\mathbb{R})$, consisting of eigenfunctions of $-L$, with $\psi_n \in \mathcal{C}^\infty_\nu(\bar{\Omega})$ having an eigenvalue λ_n with finite multiplicity and $\psi_1 = \frac{1}{\sqrt{\mu(\Omega)}}$,*

$$0 = \lambda_1 < \lambda_2 \leq \cdots \leq \lambda_n \leq \cdots \nearrow +\infty.$$

Moreover, for $t > 0$, $x, y \in \bar{\Omega}$,

$$p^{\Omega,\nu}(t,x,y) = \sum_{n=1}^{+\infty} e^{-\lambda_n t}\psi_n(x)\psi_n(y),$$

where the convergence is absolute and uniform for each $t > 0$.

Exercise 4.59. (1) Let $f \in L^2_\mu(\bar{\Omega},\mathbb{R})$. Show that uniformly on $\bar{\Omega}$, when $t \to +\infty$, $\boldsymbol{P}_t^{\Omega,\nu} f \to \frac{1}{\mu(\Omega)}\int_\Omega f\,d\mu$.

(2) Let $f \in \mathcal{C}^\infty_\nu(\bar{\Omega})$. Show that uniformly on $\bar{\Omega}$, when $t \to 0$, $\boldsymbol{P}_t^{\Omega,\nu} f \to f$.

Exercise 4.60. Show that if K is a compact subset of Ω,

$$\lim_{t\to 0}\sup_{x\in K}\frac{1}{t}(\boldsymbol{P}_t^{\Omega,\nu} 1_{\mathrm{B}(x,\varepsilon)^c\cap\Omega})(x) = 0.$$

4.9 Symmetric diffusion processes

Our goal in this section will be to associate to a symmetric and elliptic diffusion operator L a somehow canonical continuous Markov process. In general, as we have seen, L does not generate a Markov semigroup but a sub-Markov semigroup. The corresponding process will thus be a sub-Markov process (see the end of Section 3.1 in Chapter 3 for a definition).

Let L be an elliptic diffusion operator on $\mathbb{R}^n$ with smooth coefficients. We assume that L, defined on $\mathcal{C}_c(\mathbb{R}^n, \mathbb{R})$, is symmetric with respect to a measure μ which has a smooth and positive density with respect to the Lebesgue measure.

Proposition 4.61 (Neumann process). *Let $\Omega \subset \mathbb{R}^n$ be a non-empty open set whose closure $\bar{\Omega}$ is compact and whose boundary $\partial\Omega$ is smooth. For every $x \in \Omega$, there exists a continuous Markov process $(X_t^x)_{t\geq 0}$ with semigroup $(P_t^{\Omega,\nu})_{t\geq 0}$ (Neumann semigroup) such that $X_0^x = x$. Moreover, for every $f \in \mathcal{C}_c(\Omega, \mathbb{R})$, the process*

$$f(X_t^x) - \int_0^t Lf(X_s^x)ds$$

is a martingale.

Proof. We can use the same arguments as in the proofs of Theorem 3.28, Proposition 3.30 and Proposition 3.36. □

Proposition 4.62 (Dirichlet process). *Let $\Omega \subset \mathbb{R}^n$ be a non-empty open set whose closure $\bar{\Omega}$ is compact and whose boundary $\partial\Omega$ is smooth. For every $x \in \Omega$, there exists a continuous sub-Markov process $(X_t^x)_{t\geq 0}$ with semigroup $(P_t^{\Omega})_{t\geq 0}$ (Dirichlet semigroup) and extinction time $\boldsymbol{e}_\Omega(x) = \inf\{t \geq 0, X_t^x \in \partial\Omega\}$. Moreover, for every $f \in \mathcal{C}_c(\Omega, \mathbb{R})$, the process*

$$f(X_t^x) - \int_0^t Lf(X_s^x)ds, \quad t < \boldsymbol{e}_\Omega(x)$$

is a martingale.

Proof. Let $\mathcal{O}$ be a non-empty open set whose closure $\bar{\mathcal{O}}$ is compact and whose boundary $\partial\mathcal{O}$ is smooth. We also assume that $\Omega \subset \mathcal{O}$. Let $x \in \Omega$ and let $(Y_t^x)_{t\geq 0}$ be the Neumann process in $\mathcal{O}$ started at x. Let $T_x = \inf\{t \geq 0, Y_t^x \in \partial\Omega\}$ and consider the killed process

$$X_t^x = \begin{cases} Y_t^x, & t \leq T_x. \\ \star, & t > T_x. \end{cases}$$

It is easily seen that $(X_t^x)_{t\geq 0}$ is a sub-Markov process with semigroup

$$\boldsymbol{Q}_t f(x) = \mathbb{E}(f(X_t^x)1_{t<T_x}).$$

Moreover, for every $f \in \mathcal{C}_c(\mathcal{O}, \mathbb{R})$, the process

$$f(Y_t^x) - \int_0^t Lf(Y_s^x)ds$$

is a martingale. Let now f be a smooth function on Ω such f and Lf vanish on $\partial\Omega$. From the Doob stopping theorem, we have

$$\mathbb{E}(f(Y^x_{t\wedge T_x}) - \int_0^{t\wedge T_x} Lf(Y^x_s)ds) = f(x).$$

But,

$$\mathbb{E}(f(Y^x_{t\wedge T_x})) = \mathbb{E}(f(X^x_t)1_{t<T_x})$$

and

$$\mathbb{E}\Big(\int_0^{t\wedge T_x} Lf(Y^x_s)ds\Big) = \mathbb{E}\Big(\int_0^t Lf(X^x_s)1_{s<T_x}ds\Big),$$

thus we obtain

$$\boldsymbol{Q}_t f(x) = f(x) + \int_0^t \boldsymbol{Q}_s Lf(x)ds.$$

Applying this with an eigenfunction Ψ_n of the Dirichlet Laplacian L_Ω, we get that

$$\boldsymbol{Q}_t \Psi_n = \Psi_n - \lambda_n \int_0^t \boldsymbol{Q}_s \Psi_n ds.$$

This implies that $\boldsymbol{Q}_t \Psi_n = e^{-\lambda_n t}\Psi_n$. Using then the spectral decomposition of the Dirichlet semigroup $\boldsymbol{P}^\Omega_t$ we easily conclude that, actually, $\boldsymbol{Q}_t = \boldsymbol{P}^\Omega_t$. □

With these two preliminaries at hand, we are now in a position to prove the main result of this section.

Theorem 4.63. *For every $x \in \mathbb{R}^n$, there exists a continuous sub-Markov process $(X^x_t)_{t\geq 0}$ such that $X^x_0 = x$, a.s. and for every $f \in \mathcal{C}_c(\mathbb{R}^n, \mathbb{R})$, the process*

$$f(X^x_t) - \int_0^t Lf(X^x_s)ds, \quad t < \boldsymbol{e}(x),$$

is a martingale. If L is moreover essentially self-adjoint on $\mathcal{C}_c(\mathbb{R}^n, \mathbb{R})$, then $(X^x_t)_{t\geq 0}$ admits the semigroup generated by L as a transition semigroup.

Proof. Let $x \in \mathbb{R}^n$. Let us denote by $\mathrm{B}_n = \mathrm{B}(x, n)$ the open ball with center x and radius n. Consider now independent continuous sub-Markov processes $(X^{y,n}_t)_{t\geq 0, y\in \mathrm{B}_n}$, $n \geq 1$, such that the transition semigroup of $(X^{y,n}_t)_{t\geq 0}$ is the Dirichlet semigroup $\boldsymbol{P}^{\mathrm{B}_n}_t$ and $X^{y,n}_0 = y$. Let

$$\mathrm{T}_1 = \inf\{t \geq 0, X^{x,1}_t \in \partial\, \mathrm{B}_1\}$$

and

$$X^1_{\partial \mathrm{B}_1} = \lim_{t\to \mathrm{T}_1} X^{x,1}_t.$$

We then define by induction

$$\mathrm{T}_n = \inf\{t \geq 0, X_t^{X^{n-1}_{\partial \mathrm{B}_{n-1}}, n} \in \partial \mathrm{B}_n\}$$

and

$$X^n_{\partial \mathrm{B}_n} = \lim_{t \to \mathrm{T}_n} X_t^{X^{n-1}_{\partial \mathrm{B}_{n-1}}, n}.$$

We finally define the sub-Markov process $(X_t^x)_{t \geq 0}$ as follows: For $0 \leq t < \mathrm{T}_1$, $X_t^x = X_t^{1,x}$, and for $\mathrm{T}_n \leq t < \mathrm{T}_n$, $X_t = X_t^{n+1, X^n_{\partial \mathrm{B}_n}}$. By the very construction of $(X_t^x)_{t \geq 0}$, it is clear that it is a continuous sub-Markov process such that for every $n \geq 1$,

$$\mathrm{T}_n = \inf\{t \geq 0, X_t^x \in \partial \mathrm{B}_n\},$$

and such that the killed process defined by

$$\tilde{X}_t = \begin{cases} X_t^x, & t < \mathrm{T}_n, \\ \star, & t \geq \mathrm{T}_n, \end{cases}$$

is a sub-Markov process with semigroup $\boldsymbol{P}_t^{\mathrm{B}_n}$. In particular, for every $f \in \mathcal{C}_c(\mathrm{B}_n, \mathbb{R})$, the process

$$f(X_t^x) - \int_0^t Lf(X_s^x)ds, \quad t < \mathrm{T}_n,$$

is a martingale. This easily implies that for every $f \in \mathcal{C}_c(\mathbb{R}^n, \mathbb{R})$, the process

$$f(X_t^x) - \int_0^t Lf(X_s^x)ds, \quad t < \boldsymbol{e}(x), \tag{4.5}$$

is a martingale. This proves the first part of the theorem.

Let us now assume that L is moreover essentially self-adjoint on $\mathcal{C}_c(\mathbb{R}^n, \mathbb{R})$. Let us denote by $\boldsymbol{Q}_t$ the transition semigroup of $(X_t^x)_{t \geq 0}$. The construction of $(X_t^x)_{t \geq 0}$ shows that $\boldsymbol{Q}_t$ is limit in $L^2_\mu(\mathbb{R}^n, \mathbb{R})$ of the self-adjoint operators $\boldsymbol{P}_t^{\mathrm{B}_n}$. This implies that $\boldsymbol{Q}_t$ is a contraction semigroup in $L^2_\mu(\mathbb{R}^n, \mathbb{R})$. Let $\bar{L}$ be the generator of this semigroup. By using the martingale (4.5), we easily see that the domain of $\bar{L}$ contains $\mathcal{C}_c(\mathbb{R}^n, \mathbb{R})$ and that for $f \in \mathcal{C}_c(\mathbb{R}^n, \mathbb{R})$,

$$\bar{L}f = Lf.$$

The operator $\bar{L}$ is therefore the unique self-adjoint extension of L. □

Notes and comments

Section 4.1. For a detailed account on the theory of self-adjointness on Hilbert spaces we refer the reader to the books by Reed and Simon, vol. I [62] and vol. II [63]. The criterion for essential self-adjointness given in Proposition 4.11 appears in a note by Bakry [3] and relies on the proof by Strichartz [75] that the Laplace–Beltrami operator on a complete Riemannian manifold is essentially self-adjoint.

Section 4.2. The existence of the heat kernel can also be proved by using the parametrix method, see the book by Friedman [28]. Our method that relies on a Sobolev embedding theorem has the advantage to extend to more general operators like subelliptic operators.

Section 4.3. The fact that the Kato inequality for the generator is equivalent to the positivity-preserving property of the corresponding contraction semigroup is pointed out by Simon in [72] and relies on the theory of Dirichlet forms developed by Beurling and Deny in [10]. Dirichlet forms provide an abstract framework which allows one to generalize some of the results of this section to much more general situations, see for instance Sturm [78] for the analysis of Dirichlet forms on metric spaces. We refer the reader to the book by Fukushima–Oshima–Takeda [30] for an extensive account on the general theory of Dirichlet forms and to the Chapter 1 of the book by E. B. Davies [16] for an introduction.

Section 4.4. The Riesz–Thorin theorem was first proved by Riesz in [65] but the idea to use complex methods to prove the theorem goes back to his student Thorin. An extension of this theorem to analytic families is due to Stein [74].

Section 4.5. The Hille–Yosida theorem (Theorem 4.38) was independently proved by Hille in [36], [37] and Yosida in [81], [82]. The construction of the heat semigroup by using Hille–Yosida theory follows the approach of Strichartz [75] but is extended here in the context of general elliptic operators.

Section 4.6. The connection between partial differential equations and semigroup theory goes back to Hadamard [32], [33], [34] who observed the semigroup property for solutions of the Cauchy problem. However semigroup theory was not applied systematically to partial differential equations until Hille and Yosida developed the analytical tools in the late 1940s. One of the seminal fundamental papers applying semigroup techniques to diffusion equations is the paper by Feller [25].

Section 4.9. By using the theory of Dirichlet forms, it is possible to greatly generalize Theorem 4.63. We refer the interested reader to the book by Fukushima–Oshima–Takeda [30] for the construction of continuous Markov processes associated with local Dirichlet forms. We also mention the reference [13].

Chapter 5
Itô calculus

The main goal of the chapter is to construct an integral with respect to the Brownian motion. This integral may not be defined by a usual Riemann–Stieltjes integral because Brownian paths are almost surely of infinite variation. However, by using the full strength of probability methods, it is possible to develop a natural and fruitful integration theory for stochastic processes. Once the stochastic integral is constructed, we will prove the change of variable for this integral: The Döblin–Itô formula. Several applications of the Döblin–Itô formula are then investigated.

5.1 Variation of the Brownian paths

We first study some properties of the Brownian paths that will be useful to develop an integral with respect to Brownian motion. As a first step, we prove that Brownian paths almost surely have an infinite variation.

If

$$\Delta_n[0,t] = \{0 = t_0^n \le t_1^n \le \dots \le t_n^n = t\}$$

is a subdivision of the time interval $[0,t]$, we denote by

$$|\Delta_n[0,t]| = \max\{|t_{k+1}^n - t_k^n|,\ k = 0,\dots,n-1\},$$

the mesh of this subdivision.

Proposition 5.1. *Let $(B_t)_{t\ge 0}$ be a standard Brownian motion. Let $t \ge 0$. For every sequence $\Delta_n[0,t]$ of subdivisions such that*

$$\lim_{n\to+\infty} |\Delta_n[0,t]| = 0,$$

the following convergence takes place in L^2 (and thus in probability):

$$\lim_{n\to+\infty} \sum_{k=1}^{n} (B_{t_k^n} - B_{t_{k-1}^n})^2 = t.$$

As a consequence, almost surely, Brownian paths have an infinite variation on the time interval $[0,t]$.

Proof. Let us write

$$V_n = \sum_{k=1}^{n} (B_{t_k^n} - B_{t_{k-1}^n})^2.$$

Thanks to the stationarity and the independence of Brownian increments, we have

$$
\begin{aligned}
\mathbb{E}((V_n - t)^2) &= \mathbb{E}(V_n^2) - 2t\mathbb{E}(V_n) + t^2 \\
&= \sum_{j,k=1}^{n} \mathbb{E}((B_{t_j^n} - B_{t_{j-1}^n})^2 (B_{t_k^n} - B_{t_{k-1}^n})^2) - t^2 \\
&= \sum_{k=1}^{n} \mathbb{E}((B_{t_j^n} - B_{t_{j-1}^n})^4) \\
&\quad + 2 \sum_{1 \le j < k \le n}^{n} \mathbb{E}((B_{t_j^n} - B_{t_{j-1}^n})^2 (B_{t_k^n} - B_{t_{k-1}^n})^2) - t^2 \\
&= \sum_{k=1}^{n} (t_k^n - t_{k-1}^n)^2 \mathbb{E}(B_1^4) + 2 \sum_{1 \le j < k \le n}^{n} (t_j^n - t_{j-1}^n)(t_k^n - t_{k-1}^n) - t^2 \\
&= 3 \sum_{k=1}^{n} (t_j^n - t_{j-1}^n)^2 + 2 \sum_{1 \le j < k \le n}^{n} (t_j^n - t_{j-1}^n)(t_k^n - t_{k-1}^n) - t^2 \\
&= 2 \sum_{k=1}^{n} (t_k^n - t_{k-1}^n)^2 \\
&\le 2t |\Delta_n[0,t]| \xrightarrow[n \to +\infty]{} 0.
\end{aligned}
$$

Let us now prove that, as a consequence of this convergence, the paths of the process $(B_t)_{t \ge 0}$ almost surely have an infinite variation on the time interval $[0, t]$. It suffices to prove that there exists a sequence of subdivisions $\Delta_n[0, t]$ such that almost surely

$$\lim_{n \to +\infty} \sum_{k=1}^{n} |B_{t_k^n} - B_{t_{k-1}^n}| = +\infty.$$

Assume by contradiction that the supremum on all the subdivisions of the time interval $[0, t]$ of the sums

$$\lim_{n \to +\infty} \sum_{k=1}^{n} |B_{t_k^n} - B_{t_{k-1}^n}|$$

is bounded from above by some positive M. From the above result, since the convergence in probability implies the existence of an almost surely convergent subsequence, we can find a sequence of subdivisions $\Delta_n[0, t]$ whose mesh tends to 0 and such that almost surely

$$\lim_{n \to +\infty} \sum_{k=1}^{n} (B_{t_k^n} - B_{t_{k-1}^n})^2 = t.$$

We then get

$$\sum_{k=1}^{n}(B_{t_k^n}-B_{t_{k-1}^n})^2 \le M \sup_{1\le k\le n}|B_{t_k^n}-B_{t_{k-1}^n}| \xrightarrow[n\to+\infty]{} 0,$$

which is clearly absurd. □

Exercise 5.2. Let $(B_t)_{t\ge 0}$ be a Brownian motion. Show that for $t \ge 0$ almost surely

$$\lim_{n\to+\infty}\sum_{k=1}^{2^n}(B_{\frac{kt}{2^n}}-B_{\frac{(k-1)t}{2^n}})^2 = t.$$

5.2 Itô integral

Since a Brownian motion $(B_t)_{t\ge 0}$ does not have absolutely continuous paths, we can not directly use the theory of Riemann–Stieltjes integrals to give a sense to integrals like $\int_0^t \Theta_s dB_s$ for every continuous stochastic process $(\Theta_s)_{s\ge 0}$. However, if $(\Theta_s)_{s\ge 0}$ is regular enough in the Hölder sense, then $\int_0^t \Theta_s dB_s$ can still be constructed as a limit of Riemann sums by using the so-called Young integral. In the sequel, we shall denote by $\mathcal{C}^\alpha(I)$ the space of α-Hölder continuous functions that are defined on an interval I.

Theorem 5.3 (Young integral). *Let $f \in \mathcal{C}^\beta([0,T])$ and let $g \in \mathcal{C}^\gamma([0,T])$. If $\beta+\gamma > 1$, then for every subdivision $\Delta_n[0,T]$ whose mesh tends to 0, the Riemann sums*

$$\sum_{i=0}^{n-1} f(t_i^n)(g(t_{i+1}^n)-g(t_i^n))$$

converge, when $n\to\infty$, to a limit which is independent of the subdivision t_i^n. This limit is denoted by $\int_0^T f dg$ and called the Young integral of f with respect to g.

As a consequence of the previous result, we can therefore use Young's integral to give a sense to the integral $\int_0^t \Theta_s dB_s$ as soon as the stochastic process $(\Theta_s)_{s\ge 0}$ has γ-Hölder paths with $\gamma > 1/2$. This is not satisfying enough, since for instance the integral $\int_0^t B_s dB_s$ is not even well defined. The alternative to using Riemann sums and studying its almost sure convergence is to take advantage of the quadratic variation of the Brownian motion paths and use the full power of probabilistic methods.

In what follows, we consider a Brownian motion $(B_t)_{t\ge 0}$ which is defined on a filtered probability space $(\Omega, (\mathcal{F}_t)_{t\ge 0}, \mathcal{F}, \mathbb{P})$. $(B_t)_{t\ge 0}$ is assumed to be adapted to the filtration $(\mathcal{F}_t)_{t\ge 0}$. We also assume that $(\Omega, (\mathcal{F}_t)_{t\ge 0}, \mathcal{F}, \mathbb{P})$ satisfies the usual conditions as they were defined in Definition 1.43. These assumptions imply in particular the following facts that we record here for later use:

(1) A limit (in L^2 or in probability) of adapted processes is still adapted.

(2) A modification of a progressively measurable process is still a progressively measurable process.

Exercise 5.4. Let $(B_t)_{t\geq 0}$ be a standard Brownian motion. We denote by $(\mathcal{F}_t^B)_{t\geq 0}$ its natural filtration: $\mathcal{F}_\infty^B = \sigma(B_u, u \geq 0)$ and by $\mathcal{N}$ the null sets of $\mathcal{F}_\infty^B$. Show that the filtration $(\sigma(\mathcal{F}_t^B, \mathcal{N}))_{t\geq 0}$ satisfies the usual conditions.

We denote by $L^2(\Omega, (\mathcal{F}_t)_{t\geq 0}, \mathbb{P})$ the set of processes $(u_t)_{t\geq 0}$ that are progressively measurable with respect to the filtration $(\mathcal{F}_t)_{t\geq 0}$ and such that

$$\mathbb{E}\Big(\int_0^{+\infty} u_s^2 ds\Big) < +\infty.$$

Exercise 5.5. Show that the space $L^2(\Omega, (\mathcal{F}_t)_{t\geq 0}, \mathbb{P})$ endowed with the norm

$$\|u\|^2 = \mathbb{E}\Big(\int_0^{+\infty} u_s^2 ds\Big)$$

is a Hilbert space.

We now denote by $\mathcal{E}$ the set of processes $(u_t)_{t\geq 0}$ that may be written as:

$$u_t = \sum_{i=0}^{n-1} F_i 1_{(t_i, t_{i+1}]}(t),$$

where $0 \leq t_0 \leq \cdots \leq t_n$ and where F_i is a random variable that is measurable with respect to $\mathcal{F}_{t_i}$ and such that $\mathbb{E}(F_i^2) < +\infty$. The set $\mathcal{E}$ is often called the set of simple previsible processes. We first observe that it is straightforward to check that

$$\mathcal{E} \subset L^2(\Omega, (\mathcal{F}_t)_{t\geq 0}, \mathbb{P}).$$

The following theorem provides the basic definition of the so-called Itô integral.

Theorem 5.6 (Itô integral). *There is a unique linear map*

$$\mathcal{I}\colon L^2(\Omega, (\mathcal{F}_t)_{t\geq 0}, \mathbb{P}) \to L^2(\Omega, \mathcal{F}, \mathbb{P})$$

such that

(1) *for* $u = \sum_{i=0}^{n-1} F_i 1_{(t_i, t_{i+1}]} \in \mathcal{E}$,

$$\mathcal{I}(u) = \sum_{i=0}^{n-1} F_i (B_{t_{i+1}} - B_{t_i});$$

(2) *for* $u \in L^2(\Omega, (\mathcal{F}_t)_{t\geq 0}, \mathbb{P})$,

$$\mathbb{E}\left(\mathcal{I}(u)^2\right) = \mathbb{E}\Big(\int_0^{+\infty} u_s^2 ds\Big).$$

The map $\mathcal{I}$ *is called the Itô integral and for* $u \in L^2(\Omega, (\mathcal{F}_t)_{t\geq 0}, \mathbb{P})$, *we will use the notation*

$$\mathcal{I}(u) = \int_0^{+\infty} u_s dB_s.$$

Proof. Since $L^2(\Omega, (\mathcal{F}_t)_{t\geq 0}, \mathbb{P})$ endowed with the norm

$$\|u\|^2 = \mathbb{E}\Big(\int_0^{+\infty} u_s^2 ds\Big)$$

is a Hilbert space, from the isometries extension theorem we just have to prove that

(1) for $u = \sum_{i=0}^{n-1} F_i 1_{(t_i,t_{i+1}]} \in \mathcal{E}$,

$$\mathbb{E}\Big(\Big(\sum_{i=0}^{n-1} F_i(B_{t_{i+1}} - B_{t_i})\Big)^2\Big) = \mathbb{E}\Big(\int_0^{+\infty} u_s^2 ds\Big),$$

(2) the set $\mathcal{E}$ is dense in $L^2(\Omega, (\mathcal{F}_t)_{t\geq 0}, \mathbb{P})$.

Let $u = \sum_{i=0}^{n-1} F_i 1_{(t_i,t_{i+1}]} \in \mathcal{E}$. Due to the independence of the Brownian motion increments, we have

$$\begin{aligned}
\mathbb{E}\Big(\Big(\sum_{i=0}^{n-1} F_i(B_{t_{i+1}} - B_{t_i})\Big)^2\Big) &= \mathbb{E}\Big(\sum_{i,j=0}^{n-1} F_i F_j(B_{t_{i+1}} - B_{t_i})(B_{t_{j+1}} - B_{t_j})\Big) \\
&= \mathbb{E}\Big(\sum_{i=0}^{n-1} F_i^2(B_{t_{i+1}} - B_{t_i})^2\Big) \\
&\quad + 2\mathbb{E}\Big(\sum_{0\leq i<j\leq n-1} F_i F_j(B_{t_{i+1}} - B_{t_i})(B_{t_{j+1}} - B_{t_j})\Big) \\
&= \mathbb{E}\Big(\sum_{i=0}^{n-1} F_i^2(t_{i+1} - t_i)\Big) \\
&= \mathbb{E}\Big(\int_0^{+\infty} u_s^2 ds\Big).
\end{aligned}$$

Let us now prove that $\mathcal{E}$ is dense in $L^2(\Omega, (\mathcal{F}_t)_{t\geq 0}, \mathbb{P})$. We proceed in several steps. As a first step, let us observe that the set of progressively measurable bounded processes is dense in $L^2(\Omega, (\mathcal{F}_t)_{t\geq 0}, \mathbb{P})$. Indeed, for $u \in L^2(\Omega, (\mathcal{F}_t)_{t\geq 0}, \mathbb{P})$, the sequence $(u_t \mathbf{1}_{[0,n]}(|u_t|))_{t\geq 0}$ converges to u.

As a second step, we remark that if $u \in L^2(\Omega, (\mathcal{F}_t)_{t\geq 0}, \mathbb{P})$ is a bounded process, then u is a limit of bounded processes that are in $L^2(\Omega, (\mathcal{F}_t)_{t\geq 0}, \mathbb{P})$ and such that almost all paths are supported in a fixed compact set (consider the sequence $(u_t \mathbf{1}_{[0,n]}(t)_{t\geq 0})$.

Thirdly, if $u \in L^2(\Omega, (\mathcal{F}_t)_{t\geq 0}, \mathbb{P})$ is a bounded process such that almost all paths are supported in a fixed compact set, then the sequence $\left(\frac{1}{n}\int_{t-\frac{1}{n}}^t u_s ds \mathbf{1}_{(\frac{1}{n},+\infty)}(t)\right)_{t\geq 0}$ is seen to converge toward u. Therefore, u is a limit of left continuous and bounded processes that are in $L^2(\Omega, (\mathcal{F}_t)_{t\geq 0}, \mathbb{P})$ and such that almost every paths are supported in a fixed compact set.

Finally, it suffices to prove that if $u \in L^2(\Omega, (\mathcal{F}_t)_{t\geq 0}, \mathbb{P})$ is a left continuous and bounded process such that almost every paths are supported in a fixed compact set, then u is a limit of processes that belong to $\mathcal{E}$. This may be proved by considering the sequence

$$u_t^n = \sum_{i=0}^{+\infty} u_{\frac{i}{n}} 1_{(\frac{i}{n}, \frac{i+1}{n}]}(t). \qquad \square$$

Exercise 5.7. Let $u, v \in L^2(\Omega, (\mathcal{F}_t)_{t\geq 0}, \mathbb{P})$, show that

$$\mathbb{E}\left(\int_0^{+\infty} u_s dB_s\right) = 0$$

and

$$\mathbb{E}\left(\int_0^{+\infty} u_s dB_s \int_0^{+\infty} v_s dB_s\right) = \mathbb{E}\left(\int_0^{+\infty} u_s v_s ds\right).$$

Associated with Itô's integral, we can construct an integral process. Its fundamental property is that it is a continuous martingale.

Proposition 5.8. *Let $u \in L^2(\Omega, (\mathcal{F}_t)_{t\geq 0}, \mathbb{P})$. The process*

$$\left(\int_0^t u_s dB_s\right)_{t\geq 0} = \left(\int_0^{+\infty} u_s 1_{[0,t]}(s) dB_s\right)_{t\geq 0}$$

is a martingale with respect to the filtration $(\mathcal{F}_t)_{t\geq 0}$ that admits a continuous modification.

Proof. We first prove the martingale property. If

$$u_t = \sum_{i=0}^{n-1} F_i 1_{(t_i, t_{i+1}]}(t)$$

is in $\mathcal{E}$, then for every $t \geq s$,

$$\begin{aligned}\mathbb{E}\Big(\int_0^t u_v dB_v \mid \mathcal{F}_s\Big) &= \mathbb{E}\Big(\sum_{i=0}^{n-1} F_i (B_{t_{i+1}\wedge t} - B_{t_i \wedge t}) \mid \mathcal{F}_s\Big) \\ &= \sum_{i=0}^{n-1} F_i (B_{t_{i+1}\wedge s} - B_{t_i \wedge s}) \\ &= \int_0^s u_v dB_v.\end{aligned}$$

Thus if $u \in \mathcal{E}$, the process

$$\Big(\int_0^t u_s dB_s\Big)_{t \geq 0} = \Big(\int_0^{+\infty} u_s 1_{[0,t]}(s) dB_s\Big)_{t \geq 0}$$

is a martingale with respect to the filtration $(\mathcal{F}_t)_{t\geq 0}$. Since $\mathcal{E}$ is dense in $L^2(\Omega, (\mathcal{F}_t)_{t\geq 0}, \mathbb{P})$, and since it is easily checked that a limit in $L^2(\Omega, (\mathcal{F}_t)_{t\geq 0}, \mathbb{P})$ of martingales is still a martingale, we deduce the expected result.

We now prove the existence of a continuous version.

If $u \in \mathcal{E}$, the continuity of the integral process easily stems from the continuity of the Brownian paths. Let $u \in L^2(\Omega, (\mathcal{F}_t)_{t\geq 0}, \mathbb{P})$ and let u^n be a sequence in $\mathcal{E}$ that converges to u. From Doob's inequality, we have for $m, n \geq 0$ and $\varepsilon > 0$,

$$\begin{aligned}\mathbb{P}\Big(\sup_{t\geq 0}\Big|\int_0^t (u_s^n - u_s^m) dB_s\Big| \geq \varepsilon\Big) &\leq \frac{\mathbb{E}(|\int_0^{+\infty}(u_s^n - u_s^m) dB_s|^2)}{\varepsilon^2} \\ &\leq \frac{\mathbb{E}(\int_0^{+\infty}(u_s^n - u_s^m)^2 ds)}{\varepsilon^2}.\end{aligned}$$

There exists thus a sequence $(n_k)_{k\geq 0}$ such that

$$\mathbb{P}\Big(\sup_{t\geq 0}\Big|\int_0^t (u_s^{n_{k+1}} - u_s^{n_k}) dB_s\Big| \geq \frac{1}{2^k}\Big) \leq \frac{1}{2^k}.$$

From the Borel–Cantelli lemma, the sequence of processes $(\int_0^t u_s^{n_k} dB_s)_{t\geq 0}$ converges then almost surely uniformly to the process $(\int_0^t u_s dB_s)_{t\geq 0}$ which is therefore continuous. □

As a straightforward consequence of the previous proposition and Doob inequalities, we obtain

Proposition 5.9. *Let* $u \in L^2(\Omega, (\mathcal{F}_t)_{t\geq 0}, \mathbb{P})$.

(1) *For every* $\lambda > 0$,

$$\mathbb{P}\left(\sup_{t \geq 0} \left| \int_0^t u_s dB_s \right| \geq \lambda \right) \leq \frac{\mathbb{E}(\int_0^{+\infty} u_s^2 ds)}{\lambda^2};$$

(2)

$$\mathbb{E}\left(\left(\sup_{t \geq 0} \left| \int_0^t u_s dB_s \right| \right)^2 \right) \leq 4\mathbb{E}\left(\int_0^{+\infty} u_s^2 ds \right).$$

Once again, we stress the fact that Itô's integral is not pathwise in the sense that for $u \in L^2(\Omega, (\mathcal{F}_t)_{t \geq 0}, \mathbb{P})$, the Riemann sums

$$\sum_{k=0}^{n-1} u_{\frac{kt}{n}} \left(B_{\frac{(k+1)t}{n}} - B_{\frac{kt}{n}} \right),$$

need not to almost surely converge to $\int_0^t u_s dB_s$. However the following proposition shows that under continuity assumptions we have a convergence in probability.

Proposition 5.10. *Let* $u \in L^2(\Omega, (\mathcal{F}_t)_{t \geq 0}, \mathbb{P})$ *be a continuous process. Let* $t \geq 0$. *For every sequence of subdivisions* $\Delta_n[0, t]$ *such that*

$$\lim_{n \to +\infty} \mid \Delta_n[0, t] \mid = 0,$$

the following convergence holds in probability:

$$\lim_{n \to +\infty} \sum_{k=0}^{n-1} u_{t_k^n} (B_{t_{k+1}^n} - B_{t_k^n}) = \int_0^t u_s dB_s.$$

Proof. Let us first assume that u is bounded almost surely. We have

$$\sum_{k=0}^{n-1} u_{t_k^n} (B_{t_{k+1}^n} - B_{t_k^n}) = \int_0^t u_s^n dB_s,$$

where $u_s^n = \sum_{k=0}^{n-1} u_{t_k^n} 1_{(t_k^n, t_{k+1}^n]}(s)$. The Itô isometry and the Lebesgue dominated convergence theorem shows then that $\int_0^t u_s^n dB_s$ converges to $\int_0^t u_s dB_s$ in L^2 and therefore in probability. For general u's we can use a localization procedure. For $N \geq 0$, consider the random time

$$T_N = \inf\{t \geq 0, |u_t| \geq N\}.$$

We have for every $\varepsilon > 0$,

$$\mathbb{P}\Big(\Big|\sum_{k=0}^{n-1} u_{t_k^n}(B_{t_{k+1}^n} - B_{t_k^n}) - \int_0^t u_s dB_s\Big| \geq \varepsilon\Big)$$

$$\leq \mathbb{P}(T_N \leq t) + \mathbb{P}\Big(\Big|\sum_{k=0}^{n-1} u_{t_k^n}(B_{t_{k+1}^n} - B_{t_k^n}) - \int_0^t u_s dB_s\Big| \geq \varepsilon, T_N \geq t\Big)$$

$$\leq \mathbb{P}(T_N \leq t) + \mathbb{P}\Big(\Big|\sum_{k=0}^{n-1} u_{t_k^n} 1_{|u_{t_k^n}| \leq M}(B_{t_{k+1}^n} - B_{t_k^n}) - \int_0^t u_s 1_{|u_s| \leq M} dB_s\Big| \geq \varepsilon\Big).$$

This easily implies the convergence in probability. □

Exercise 5.11. (1) Show that for $t \geq 0$,

$$\int_0^t B_s dB_s = \frac{1}{2}(B_t^2 - t).$$

What is surprising in this formula?

(2) Show that when $n \to +\infty$, the sequence

$$\sum_{k=0}^{n-1} B_{\frac{(k+1)t}{n}}\big(B_{\frac{(k+1)t}{n}} - B_{\frac{kt}{n}}\big),$$

converges in probability to a random variable that shall be computed.

Exercise 5.12. Show that if $f : \mathbb{R}_{\geq 0} \to \mathbb{R}$ is locally square integrable, that is, $\int_0^t f^2(x)dx < +\infty$, $t \geq 0$, then the process $(\int_0^t f(s)dB_s)_{t \geq 0}$ is a Gaussian process. Compute its mean and its covariance.

5.3 Square integrable martingales and quadratic variations

It turns out that stochastic integrals may be defined for other stochastic processes than Brownian motions. The key properties that were used in the above approach were the martingale property and the square integrability of the Brownian motion.

As above, we consider a filtered probability space $(\Omega, (\mathcal{F}_t)_{t \geq 0}, \mathcal{F}, \mathbb{P})$ that satisfies the usual conditions. A martingale $(M_t)_{t \geq 0}$ defined on this space is said to be square integrable if for every $t \geq 0$, $\mathbb{E}(M_t^2) < +\infty$.

For instance, if $(B_t)_{t \geq 0}$ is a Brownian motion on $(\Omega, (\mathcal{F}_t)_{t \geq 0}, \mathcal{F}, \mathbb{P})$ and if $(u_t)_{t \geq 0}$ is a process which is progressively measurable with respect to the filtration $(\mathcal{F}_t)_{t \geq 0}$ such that for every $t \geq 0$, $\mathbb{E}\big(\int_0^t u_s^2 ds\big) < +\infty$, then the process

$$M_t = \int_0^t u_s dB_s, \quad t \geq 0,$$

is a square integrable martingale.

The most important theorem concerning continuous square integrable martingales is that they admit a quadratic variation. Before proving this theorem, we state a preliminary lemma.

Lemma 5.13. *Let $(M_t)_{0\le t\le T}$ be a continuous martingale such that*

$$\sup_{\Delta_n[0,T]} \sum_{k=0}^{n-1} |M_{t_{k+1}^n} - M_{t_k^n}| < +\infty.$$

Then $(M_t)_{0\le t\le T}$ is constant.

Proof. We may assume $M_0 = 0$. For $N \ge 0$, let us consider the stopping time

$$T_N = \inf\Big\{s \in [0,T], |M_s| \ge N, \sup_{\Delta_n[0,s]} \sum_{k=0}^{n-1} |M_{t_{k+1}^n} - M_{t_k^n}| \ge N\Big\} \wedge T.$$

The stopped process $(M_{t\wedge T_N})_{0\le t\le T}$ is a martingale and therefore for $s \le t$,

$$\mathbb{E}((M_{t\wedge T_N} - M_{s\wedge T_N})^2) = \mathbb{E}(M_{t\wedge T_N}^2) - \mathbb{E}(M_{s\wedge T_N}^2).$$

Consider now a sequence of subdivisions $\Delta_n[0,T]$ whose mesh tends to 0. By summing up the above inequality on the subdivision, we obtain

$$\begin{aligned}
\mathbb{E}(M_{T_N}^2) &= \sum_{k=0}^{n-1} (M_{t_k^n \wedge T_N} - M_{t_{k-1}^n \wedge T_N})^2 \\
&\le \sup |M_{t_k^n \wedge T_N} - M_{t_{k-1}^n \wedge T_N}| \mathbb{E}\Big(\sum_{k=0}^{n-1} |M_{t_k^n \wedge T_N} - M_{t_{k-1}^n \wedge T_N}|\Big) \\
&\le N \sup |M_{t_k^n \wedge T_N} - M_{t_{k-1}^n \wedge T_N}|.
\end{aligned}$$

By letting $n \to +\infty$, we get $\mathbb{E}(M_{T_N}^2) = 0$. This implies that $M_{T_N} = 0$. Letting now $N \to \infty$, we conclude that $M_T = 0$. □

Theorem 5.14 (Quadratic variation of a martingale). *Let $(M_t)_{t\ge 0}$ be a martingale on $(\Omega, (\mathcal{F}_t)_{t\ge 0}, \mathcal{F}, \mathbb{P})$ which is continuous and square integrable and such that $M_0 = 0$. There is a unique continuous and increasing process denoted by $(\langle M\rangle_t)_{t\ge 0}$ that satisfies the following properties:*

(1) $\langle M\rangle_0 = 0$;

(2) *the process $(M_t^2 - \langle M\rangle_t)_{t\ge 0}$ is a martingale.*

Actually for every $t \geq 0$ and for every sequence of subdivisions $\Delta_n[0,t]$ such that

$$\lim_{n\to+\infty} |\Delta_n[0,t]| = 0,$$

the following convergence takes place in probability:

$$\lim_{n\to+\infty} \sum_{k=1}^{n} (M_{t_k^n} - M_{t_{k-1}^n})^2 = \langle M \rangle_t.$$

The process $(\langle M \rangle_t)_{t\geq 0}$ is called the quadratic variation process of $(M_t)_{t\geq 0}$.

Proof. We first assume that the martingale $(M_t)_{t\geq 0}$ is bounded and prove that if $\Delta_n[0,t]$ is a sequence of subdivisions of the interval $[0,t]$ such that

$$\lim_{n\to+\infty} |\Delta_n[0,t]| = 0,$$

then the limit

$$\lim_{n\to+\infty} \sum_{k=1}^{n} (M_{t_k^n} - M_{t_{k-1}^n})^2$$

exists in L^2 and thus in probability.

Toward this goal, we introduce some notations. If $\Delta[0,T]$ is a subdivision of the time interval $[0,T]$ and if $(X_t)_{t\geq 0}$ is a stochastic process, then we write

$$S_t^{\Delta[0,T]}(X) = \sum_{i=0}^{k-1} (X_{t_{i+1}} - X_{t_i})^2 + (X_t - X_{t_k})^2,$$

where k is such that $t_k \leq t < t_{k+1}$.

An easy computation on conditional expectations shows that if $(X_t)_{t\geq 0}$ is a martingale, then the process

$$X_t^2 - S_t^{\Delta[0,T]}(X), \quad t \leq T,$$

is also a martingale. Also, if $\Delta[0,T]$ and $\Delta'[0,T]$ are two subdivisions of the time interval $[0,T]$, we will denote by $\Delta \vee \Delta'[0,T]$ the subdivision obtained by putting together the points $\Delta[0,T]$ and the points of $\Delta'[0,T]$. Let now $\Delta_n[0,T]$ be a sequence of subdivisions of $[0,T]$ such that

$$\lim_{n\to+\infty} |\Delta_n[0,T]| = 0.$$

Let us show that the sequence $S_T^{\Delta_n[0,T]}(M)$ is a Cauchy sequence in L^2. Since the process $S^{\Delta_n[0,T]}(M) - S^{\Delta_p[0,T]}(M)$ is a martingale (as a difference of two martingales), we deduce that

$$\begin{aligned}
&\mathbb{E}((S_T^{\Delta_n[0,T]}(M) - S_T^{\Delta_p[0,T]}(M))^2) \\
&\quad = \mathbb{E}(S_T^{\Delta_n \vee \Delta_p[0,T]}(S^{\Delta_n[0,T]}(M) - S^{\Delta_p[0,T]}(M))) \\
&\quad \le 2(\mathbb{E}(S_T^{\Delta_n \vee \Delta_p[0,T]}(S^{\Delta_n[0,T]}(M))) + \mathbb{E}(S_T^{\Delta_n \vee \Delta_p[0,T]}(S^{\Delta_p[0,T]}(M)))).
\end{aligned}$$

Let us denote by s_k the points of the subdivision $\Delta_n \vee \Delta_p[0,T]$ and for fixed s_k, we denote by t_l the point of $\Delta_n[0,T]$ which is the closest to s_k and such that $t_l \le s_k \le t_{l+1}$. We have

$$\begin{aligned}
S_{s_{k+1}}^{\Delta_n[0,T]}(M) - S_{s_k}^{\Delta_n[0,T]}(M) &= (M_{s_{k+1}} - M_{t_l})^2 - (M_{s_k} - M_{t_l})^2 \\
&= (M_{s_{k+1}} - M_{s_k})(M_{s_{k+1}} + M_{s_k} - 2M_{t_l}).
\end{aligned}$$

Therefore, from the Cauchy–Schwarz inequality, we have

$$\begin{aligned}
&\mathbb{E}(S_T^{\Delta_n \vee \Delta_p[0,T]}(S^{\Delta_n[0,T]}(M))) \\
&\quad \le \mathbb{E}(\sup_k (M_{s_{k+1}} + M_{s_k} - 2M_{t_l})^4)^{1/2} \mathbb{E}((S_T^{\Delta_n \vee \Delta_p[0,T]}(M))^2)^{1/2}.
\end{aligned}$$

Since the martingale M is assumed to be continuous, when $n, p \to +\infty$,

$$\mathbb{E}(\sup_k (M_{s_{k+1}} + M_{s_k} - 2M_{t_l})^4) \to 0.$$

Thus, in order to conclude, it suffices to prove that $\mathbb{E}((S_T^{\Delta_n \vee \Delta_p[0,T]}(M))^2)$ is bounded. This is a consequence of the fact that M is assumed to be bounded and we let the reader work out the details of the argument. Therefore, in the L^2 sense the following convergence holds:

$$\langle M \rangle_t = \lim_{n \to +\infty} \sum_{k=1}^{n} (M_{t_k^n} - M_{t_{k-1}^n})^2.$$

The process $(M_t^2 - \langle M \rangle_t)_{t \ge 0}$ is seen to be a martingale because for every n and $T \ge 0$, the process

$$M_t^2 - S_t^{\Delta_n[0,T]}(M), \quad t \le T,$$

is a martingale. Let us now show that the obtained process $\langle M \rangle$ is a continuous process. From Doob's inequality, for $n, p \ge 0$ and $\varepsilon > 0$,

$$\begin{aligned}
&\mathbb{P}(\sup_{0 \le t \le T} (S_t^{\Delta_n[0,T]}(M) - S_t^{\Delta_p[0,T]}(M)) > \varepsilon) \\
&\quad \le \frac{\mathbb{E}((S_T^{\Delta_n[0,T]}(M) - S_T^{\Delta_p[0,T]}(M))^2)}{\varepsilon^2}.
\end{aligned}$$

From the Borel–Cantelli lemma, there exists therefore a sequence n_k such that the sequence of continuous stochastic processes $(S_t^{\Delta_{n_k}[0,T]}(M))_{0 \le t \le T}$ almost surely uniformly converges to the process $(\langle M \rangle_t)_{0 \le t \le T}$. This proves the existence of a continuous version for $\langle M \rangle$. Finally, to prove that $\langle M \rangle$ is increasing, it is enough to consider a an increasing sequence of subdivisions whose mesh tends to 0. Let us now prove that $\langle M \rangle$ is the unique process such that $M^2 - \langle M \rangle$ is a martingale. Let A and A' be two continuous and increasing stochastic processes such that $A_0 = A'_0 = 0$ and such that $(M_t^2 - A_t)_{t \ge 0}$ and $(M_t^2 - A'_t)_{t \ge 0}$ are martingales. The process $(N_t)_{t \ge 0} = (A_t - A'_t)_{t \ge 0}$ is then seen to be a martingale that has a bounded variation. From the previous lemma, this implies that $(N_t)_{t \ge 0}$ is constant and therefore equal to 0 due to its initial condition.

We now turn to the case where $(M_t)_{t \ge 0}$ is not necessarily bounded. Let us introduce the sequence of stopping times:

$$T_N = \inf\{t \ge 0, |M_t| \ge N\}.$$

According to the previous arguments, for every $N \ge 0$, there is an increasing process A^N such that $(M_{t \wedge T_N}^2 - A_t^N)_{t \ge 0}$ is a martingale. By the uniqueness of this process, it is clear that $A_{t \wedge T_N}^{N+1} = A_t^N$, therefore we can define a process A_t by requiring that $A_t(\omega) = A_t^N(\omega)$ provided that $T_N(\omega) \ge t$. By using convergence theorems, it is then checked that $(M_t^2 - A_t)_{t \ge 0}$ is a martingale.

Finally, let $\Delta_n[0,t]$ be a sequence of subdivisions whose mesh tends to 0. We have for every $\varepsilon > 0$,

$$\begin{aligned} &\mathbb{P}\Big(\Big|A_t - \sum_{k=1}^{n} (M_{t_k^n} - M_{t_{k-1}^n})^2\Big| \ge \varepsilon\Big) \\ &\qquad \le \mathbb{P}(T_N \le t) + \mathbb{P}\Big(\Big|A_t^N - \sum_{k=1}^{n} (M_{t_k^n \wedge T_N} - M_{t_{k-1}^n \wedge T_N})^2\Big| \ge \varepsilon\Big). \end{aligned}$$

This easily implies the announced convergence in probability of the quadratic variations to A_t. □

Exercise 5.15. Let $(M_t)_{t \ge 0}$ be a square integrable martingale on a filtered probability space $(\Omega, (\mathcal{F}_t)_{t \ge 0}, \mathcal{F}, \mathbb{P})$. Assume that $M_0 = 0$. If $\Delta[0,T]$ is a subdivision of the time interval $[0,T]$ and if $(X_t)_{t \ge 0}$ is a stochastic process, we write

$$S_t^{\Delta[0,T]}(X) = \sum_{i=0}^{k-1} (X_{t_{i+1}} - X_{t_i})^2 + (X_t - X_{t_k})^2,$$

where k is such that $t_k \le t < t_{k+1}$. Let $\Delta_n[0,T]$ be a sequence of subdivisions of $[0,T]$ such that

$$\lim_{n \to +\infty} |\Delta_n[0,T]| = 0.$$

Show that the following convergence holds in probability:

$$\lim_{n\to+\infty} \sup_{0\le t\le T} |S_t^{\Delta[0,T]}(M) - \langle M\rangle_t| = 0.$$

Thus, in the previous theorem, the convergence is actually uniform on compact intervals.

We have already pointed out that stochastic integrals with respect to Brownian motion provide an example of square integrable martingale, they therefore have a quadratic variation. The next proposition explicitly computes this variation.

Proposition 5.16. *Let $(B_t)_{t\ge0}$ be a Brownian motion on a filtered probability space $(\Omega, (\mathcal{F}_t)_{t\ge0}, \mathcal{F}, \mathbb{P})$ that satisfies the usual conditions. Let $(u_t)_{t\ge0}$ be a progressively measurable process such that for every $t \ge 0$, $\mathbb{E}(\int_0^t u_s^2 ds) < +\infty$. For $t \ge 0$,*

$$\left\langle \int_0^{\cdot} u_s dB_s \right\rangle_t = \int_0^t u_s^2 ds.$$

Proof. Since the process $(\int_0^t u_s^2 ds)_{t\ge0}$ is continuous, increasing and equals 0 when $t = 0$, we just need to prove that

$$\left(\int_0^t u_s dB_s\right)^2 - \int_0^t u_s^2 ds$$

is a martingale.

If $u \in \mathcal{E}$ is a simple process, it is easily seen that for $t \ge s$,

$$\begin{aligned}
\mathbb{E}\left(\left(\int_0^t u_v dB_v\right)^2 \mid \mathcal{F}_s\right) &= \mathbb{E}\left(\left(\int_0^s u_v dB_v + \int_s^t u_v dB_v\right)^2 \mid \mathcal{F}_s\right) \\
&= \mathbb{E}\left(\left(\int_0^s u_v dB_v\right)^2 \mid \mathcal{F}_s\right) + \mathbb{E}\left(\left(\int_s^t u_v dB_v\right)^2 \mid \mathcal{F}_s\right) \\
&= \left(\int_0^s u_v dB_v\right)^2 + \mathbb{E}\left(\int_s^t u_v^2 dv \mid \mathcal{F}_s\right).
\end{aligned}$$

We may then conclude by using the density of $\mathcal{E}$ in $L^2(\Omega, (\mathcal{F}_t)_{t\ge0}, \mathbb{P})$. □

As a straightforward corollary of Proposition 5.14, we immediately obtain:

Corollary 5.17. *Let $(M_t)_{t\ge0}$ and $(N_t)_{t\ge0}$ be two continuous square integrable martingales on $(\Omega, (\mathcal{F}_t)_{t\ge0}, \mathcal{F}, \mathbb{P})$ such that $M_0 = N_0 = 0$. There is a unique continuous process $(\langle M, N\rangle_t)_{t\ge0}$ with bounded variation that satisfies*

(1) $\langle M, N\rangle_0 = 0$;

(2) *the process $(M_t N_t - \langle M, N\rangle_t)_{t\ge0}$ is a martingale.*

Moreover, for $t \geq 0$ and for every sequence $\Delta_n[0,t]$ such that

$$\lim_{n\to+\infty} |\Delta_n[0,t]| = 0,$$

the following convergence holds in probability:

$$\lim_{n\to+\infty} \sum_{k=1}^{n} (M_{t_k^n} - M_{t_{k-1}^n})(N_{t_k^n} - N_{t_{k-1}^n}) = \langle M, N\rangle_t.$$

The process $(\langle M, N\rangle_t)_{t\geq 0}$ is called the quadratic covariation process of $(M_t)_{t\geq 0}$ and $(N_t)_{t\geq 0}$.

Proof. We may actually just use the formula

$$\langle M, N\rangle = \frac{1}{4}(\langle M+N\rangle - \langle M-N\rangle)$$

as a definition of the covariation and then check that the above properties are indeed satisfied due to Proposition 5.14. □

Exercise 5.18. Let $(B_t^1)_{t\geq 0}$ and $(B_t^2)_{t\geq 0}$ be two independent Brownian motions. Show that

$$\langle B^1, B^2\rangle_t = 0.$$

In the same way that a stochastic integral with respect to Brownian motion was constructed, a stochastic integral with respect to square integrable martingales may be defined. We shall not repeat this construction, since it was done in the Brownian motion case, but we point out the main results without proofs and leave the proofs as an exercise to the reader.

Let $(M_t)_{t\geq 0}$ be a continuous square integrable martingale on a filtered probability space $(\Omega, (\mathcal{F}_t)_{t\geq 0}, \mathcal{F}, \mathbb{P})$ that satisfies the usual conditions. We assume that $M_0 = 0$. Let us denote by $\mathcal{L}_M^2(\Omega, (\mathcal{F}_t)_{t\geq 0}, \mathbb{P})$ the set of processes $(u_t)_{t\geq 0}$ that are progressively measurable with respect to the filtration $(\mathcal{F}_t)_{t\geq 0}$ and such that

$$\mathbb{E}\Big(\int_0^{+\infty} u_s^2 d\langle M\rangle_s\Big) < +\infty.$$

We still denote by $\mathcal{E}$ the set of simple and predictable processes, that is, the set of processes $(u_t)_{t\geq 0}$ that may be written as

$$u_t = \sum_{i=0}^{n-1} F_i \mathbf{1}_{(t_i, t_{i+1}]}(t),$$

where $0 \le t_0 \le \cdots \le t_n$ and where F_i is a random variable that is measurable with respect to $\mathcal{F}_{t_i}$ and such that $\mathbb{E}(F_i^2) < +\infty$. We define an equivalence relation $\mathcal{R}$ on the set $\mathcal{L}_M^2(\Omega, (\mathcal{F}_t)_{t\ge 0}, \mathbb{P})$ as follows:

$$u\mathcal{R}v \iff \mathbb{E}\Big(\int_0^{+\infty} (u_s - v_s)^2 d\langle M\rangle_s\Big) = 0$$

and denote by

$$L_M^2(\Omega, (\mathcal{F}_t)_{t\ge 0}, \mathbb{P}) = \mathcal{L}_M^2(\Omega, (\mathcal{F}_t)_{t\ge 0}, \mathbb{P})/\mathcal{R}$$

the set of equivalence classes. It is easy to check that $L_M^2(\Omega, (\mathcal{F}_t)_{t\ge 0}, \mathbb{P})$ endowed with the norm

$$\|u\|^2 = \mathbb{E}\Big(\int_0^{+\infty} u_s^2 d\langle M\rangle_s\Big),$$

is a Hilbert space.

The following results are then proved similarly as for the Brownian motion.

Theorem 5.19. *There exists a unique linear map*

$$\mathcal{I}_M \colon L_M^2(\Omega, (\mathcal{F}_t)_{t\ge 0}, \mathbb{P}) \to L^2(\Omega, \mathcal{F}, \mathbb{P})$$

such that the following holds:

- *For* $u = \sum_{i=0}^{n-1} F_i 1_{(t_i, t_{i+1}]} \in \mathcal{E}$,

$$\mathcal{I}(u) = \sum_{i=0}^{n-1} F_i (M_{t_{i+1}} - M_{t_i}).$$

- *For* $u \in L_M^2(\Omega, (\mathcal{F}_t)_{t\ge 0}, \mathbb{P})$,

$$\mathbb{E}(\mathcal{I}_M(u)^2) = \mathbb{E}\Big(\int_0^{+\infty} u_s^2 d\langle M\rangle_s\Big).$$

The map $\mathcal{I}_M$ is called the Itô integral with respect to the continuous and square integrable martingale $(M_t)_{t\ge 0}$. We write for $u \in L_M^2(\Omega, (\mathcal{F}_t)_{t\ge 0}, \mathbb{P})$,

$$\mathcal{I}_M(u) = \int_0^{+\infty} u_s dM_s.$$

In the case where $(M_t)_{t\ge 0}$ is itself a stochastic integral with respect to Brownian motion, we can express the stochastic integral with respect to $(M_t)_{t\ge 0}$ as an integral with respect to the underlying Brownian motion.

Proposition 5.20. *Let us assume that* $M_t = \int_0^t \Theta_s dB_s$ *where* $(B_t)_{t\geq 0}$ *is a Brownian motion on* $(\Omega, (\mathcal{F}_t)_{t\geq 0}, \mathcal{F}, \mathbb{P})$ *and where* $(\Theta_t)_{t\geq 0}$ *is a progressively measurable process such that for every* $t \geq 0$, $\mathbb{E}(\int_0^t \Theta_s^2 ds) < +\infty$. *Then* $u \in L^2_M(\Omega, (\mathcal{F}_t)_{t\geq 0}, \mathbb{P})$ *if and only if* $u\Theta \in L^2_B(\Omega, (\mathcal{F}_t)_{t\geq 0}, \mathbb{P})$ *and in this case, for every* $t \geq 0$,

$$\int_0^t u_s dM_s = \int_0^t u_s \Theta_s dB_s.$$

It should come as no surprise, that stochastic integrals with respect to martingales are still martingales. This is confirmed in the next proposition.

Proposition 5.21. *Let* $(u_t)_{t\geq 0}$ *be a stochastic process which is progressively measurable with respect to the filtration* $(\mathcal{F}_t)_{t\geq 0}$ *and such that for every* $t \geq 0$, $\mathbb{E}(\int_0^t u_s^2 d\langle M\rangle_s) < +\infty$. *The process*

$$\left(\int_0^t u_s dM_s\right)_{t\geq 0} = \left(\int_0^{+\infty} u_s 1_{[0,t]}(s) dM_s\right)_{t\geq 0}$$

is a square integrable martingale with respect to the filtration $(\mathcal{F}_t)_{t\geq 0}$ *that admits a continuous modification and its quadratic variation process is*

$$\left\langle \int_0^\cdot u_s dM_s \right\rangle_t = \int_0^t u_s^2 d\langle M\rangle_s.$$

As in the Brownian motion case, the stochastic integral with respect to a martingale is not an almost sure limit of Riemann sums, it is however a limit in probability.

Proposition 5.22. *Let* $u \in L^2_M(\Omega, (\mathcal{F}_t)_{t\geq 0}, \mathbb{P})$ *be a continuous stochastic process. Let* $t \geq 0$. *For every sequence of subdivisions* $\Delta_n[0,t]$ *such that*

$$\lim_{n\to+\infty} |\Delta_n[0,t]| = 0,$$

the following convergence holds in probability:

$$\lim_{n\to+\infty} \sum_{k=0}^{n-1} u_{t_k^n}(M_{t_{k+1}^n} - M_{t_k^n}) = \int_0^t u_s dM_s.$$

5.4 Local martingales, semimartingales and integrators

The goal of this section is to extend the domain of definition of the Itô integral with respect to Brownian motion. The idea is to use the fruitful concept of localization. We will then finally be interested in the widest class of processes for which it is possible to define a stochastic integral satisfying natural probabilistic properties. This will lead to the natural notion of semimartingales.

As before, we consider here a Brownian motion $(B_t)_{t\geq 0}$ that is defined on a filtered probability space $(\Omega, (\mathcal{F}_t)_{t\geq 0}, \mathcal{F}, \mathbb{P})$ that satisfies the usual conditions.

Definition 5.23. We define the space $L^2_{\mathrm{loc}}(\Omega, (\mathcal{F}_t)_{t\geq 0}, \mathbb{P})$ to be the set of the processes $(u_t)_{t\geq 0}$ that are progressively measurable with respect to the filtration $(\mathcal{F}_t)_{t\geq 0}$ and such that for every $t \geq 0$,

$$\mathbb{P}\Big(\int_0^t u_s^2 ds < +\infty\Big) = 1.$$

We first have the following fact whose proof is left as an exercise to the reader:

Lemma 5.24. *Let $u \in L^2_{\mathrm{loc}}(\Omega, (\mathcal{F}_t)_{t\geq 0}, \mathbb{P})$. There exists an increasing family of stopping times $(T_n)_{n\geq 0}$ for the filtration $(\mathcal{F}_t)_{t\geq 0}$ such that the following holds.*

(1) *Almost surely,*

$$\lim_{n\to+\infty} T_n = +\infty.$$

(2)

$$\mathbb{E}\Big(\int_0^{T_n} u_s^2 ds\Big) < +\infty.$$

Thanks to this lemma, it is now easy to naturally define $\int_0^t u_s dB_s$ for $u \in L^2_{\mathrm{loc}}(\Omega, (\mathcal{F}_t)_{t\geq 0}, \mathbb{P})$. Indeed, let $u \in L^2_{\mathrm{loc}}(\Omega, (\mathcal{F}_t)_{t\geq 0}, \mathbb{P})$ and let $t \geq 0$. According to the previous lemma, let us now consider an increasing sequence of stopping times $(T_n)_{n\geq 0}$ such that

(1) almost surely,

$$\lim_{n\to+\infty} T_n = +\infty;$$

(2)

$$\mathbb{E}\Big(\int_0^{T_n} u_s^2 ds\Big) < +\infty.$$

Since

$$\mathbb{E}\Big(\int_0^{T_n} u_s^2 ds\Big) < +\infty,$$

the stochastic integral

$$\int_0^{T_n} u_s dB_s = \int_0^{+\infty} u_s 1_{[0,T_n]}(s) dB_s$$

exists. We may therefore define in a unique way a stochastic process

$$\Big(\int_0^t u_s dB_s\Big)_{t\geq 0}$$

with the following properties:

(1)
$$\left(\int_0^t u_s dB_s\right)_{t\geq 0}$$
is a continuous stochastic process adapted to the filtration $(\mathcal{F}_t)_{t\geq 0}$;

(2) the stochastic process
$$\left(\int_0^{t\wedge T_n} u_s dB_s\right)_{t\geq 0}$$
is a uniformly integrable martingale with respect to the filtration $(\mathcal{F}_t)_{t\geq 0}$ (because it is bounded in L^2).

This leads to the following definition:

Definition 5.25 (Local martingale). A stochastic process $(M_t)_{t\geq 0}$ is called a *local martingale* (with respect to the filtration $(\mathcal{F}_t)_{t\geq 0}$) if there is a sequence of stopping times $(T_n)_{n\geq 0}$ such that

(1) the sequence $(T_n)_{n\geq 0}$ is increasing and almost surely satisfies $\lim_{n\to+\infty} T_n = +\infty$;

(2) for $n \geq 1$, the process $(M_{t\wedge T_n})_{t\geq 0}$ is a uniformly integrable martingale with respect to the filtration $(\mathcal{F}_t)_{t\geq 0}$.

Thus, as an example, if $u \in L^2_{\mathrm{loc}}(\Omega, (\mathcal{F}_t)_{t\geq 0}, \mathbb{P})$ then the process $(\int_0^t u_s dB_s)_{t\geq 0}$ is a local martingale. Of course, any martingale turns out to be a local martingale. But, as we will see later, in general the converse is not true.

The following exercise gives a useful criterion to prove that a given local martingale is actually a martingale.

Exercise 5.26. Let $(M_t)_{t\geq 0}$ be a continuous local martingale such that for $t \geq 0$,
$$\mathbb{E}(\sup_{s\leq t} |M_s|) < +\infty.$$
Show that $(M_t)_{t\geq 0}$ is a martingale. As a consequence, bounded local martingales necessarily are martingales.

Exercise 5.27. Show that a positive local martingale is a supermartingale.

It is interesting to observe that if $(M_t)_{t\geq 0}$ is a local martingale, then the sequence of stopping times may explicitly be chosen in such a way that the resulting stopped martingales enjoy nice properties.

Lemma 5.28. *Let $(M_t)_{t\geq 0}$ be a continuous local martingale on $(\Omega, (\mathcal{F}_t)_{t\geq 0}, \mathcal{F}, \mathbb{P})$ such that $M_0 = 0$. Let*
$$T_n = \inf\{t \geq 0, |M_t| \geq n\}.$$
Then, for $n \in \mathbb{N}$, the process $(M_{t\wedge T_n})_{t\geq 0}$ is a bounded martingale.

Proof. Let $(S_n)_{n\geq 0}$ be a sequence of stopping times such that

(1) the sequence $(S_n)_{n\geq 0}$ is increasing and almost surely $\lim_{n\to+\infty} S_n = +\infty$;
(2) for every $n \geq 1$, the process $(M_{t\wedge S_n})_{t\geq 0}$ is a uniformly integrable martingale with respect to the filtration $(\mathcal{F}_t)_{t\geq 0}$.

For $t \geq s$ and $k, n \geq 0$, we have

$$\mathbb{E}(M_{t\wedge S_k\wedge T_n} \mid \mathcal{F}_s) = M_{s\wedge S_k\wedge T_n}.$$

Letting $k \to +\infty$ leads then to the expected result. □

Since bounded martingales are of course square integrable, we easily deduce from the previous lemma that the following result holds:

Theorem 5.29. *Suppose that $(M_t)_{t\geq 0}$ is a continuous local martingale on $(\Omega, (\mathcal{F}_t)_{t\geq 0}, \mathcal{F}, \mathbb{P})$ such that $M_0 = 0$. Then there is a unique continuous increasing process $(\langle M\rangle_t)_{t\geq 0}$ such that*

(1) $\langle M\rangle_0 = 0$;
(2) *the process $(M_t^2 - \langle M\rangle_t)_{t\geq 0}$ is a local martingale.*

Furthermore, for every $t \geq 0$ and every sequence of subdivisions $\Delta_n[0,t]$ such that

$$\lim_{n\to+\infty} |\Delta_n[0,t]| = 0,$$

the following limit holds in probability:

$$\lim_{n\to+\infty} \sum_{k=1}^{n} (M_{t_k^n} - M_{t_{k-1}^n})^2 = \langle M\rangle_t.$$

The process $(\langle M\rangle_t)_{t\geq 0}$ is called the quadratic variation of the local martingale. Moreover, if u is a progressively measurable process such that for every $t \geq 0$,

$$\mathbb{P}\Big(\int_0^t u_s^2 d\langle M\rangle_s < +\infty\Big) = 1,$$

then we may define a stochastic integral $(\int_0^t u_s dM_s)_{t\geq 0}$ such that the stochastic process $(\int_0^t u_s dM_s)_{t\geq 0}$ is a continuous local martingale.

At that point, we already almost found the widest class of stochastic processes with respect to which it was possible to naturally construct a stochastic integral. To go further in that direction, let us first observe that if we add a bounded variation process to a local martingale, then we obtain a process with respect to which a stochastic integral is naturally defined.

More precisely, if $(X_t)_{t\ge 0}$ may be written in the form

$$X_t = X_0 + A_t + M_t,$$

where $(A_t)_{t\ge 0}$ is a bounded variation process and where $(M_t)_{t\ge 0}$ is a continuous local martingale on $(\Omega, (\mathcal{F}_t)_{t\ge 0}, \mathcal{F}, \mathbb{P})$ such that $M_0 = 0$. Then, if u is a progressively measurable process such that for $t \ge 0$,

$$\mathbb{P}\Big(\int_0^t u_s^2 d\langle M\rangle_s < +\infty\Big) = 1,$$

we may define a stochastic integral by

$$\Big(\int_0^t u_s dX_s\Big)_{t\ge 0} = \Big(\int_0^t u_s dA_s + \int_0^t u_s dM_s\Big)_{t\ge 0},$$

where $\int_0^t u_s dA_s$ is simply understood as the Riemann–Stieltjes integral with respect to the process $(A_t)_{t\ge 0}$.

The class of stochastic processes that we obtained is called the class of semimartingales and, as we will see later, is the most relevant one:

Definition 5.30 (Semimartingale). Let $(X_t)_{t\ge 0}$ be an adapted continuous stochastic process on the filtered probability space $(\Omega, (\mathcal{F}_t)_{t\ge 0}, \mathcal{F}, \mathbb{P})$. We say that $(X_t)_{t\ge 0}$ is a *semimartingale with respect to the filtration* $(\mathcal{F}_t)_{t\ge 0}$ if $(X_t)_{t\ge 0}$ may be written as

$$X_t = X_0 + A_t + M_t,$$

where $(A_t)_{t\ge 0}$ is a bounded variation process and $(M_t)_{t\ge 0}$ is a continuous local martingale such that $M_0 = 0$. If it exists, then the previous decomposition is unique.

Remark 5.31. It is possible to prove that if $(X_t)_{t\ge 0}$ is a semimartingale with respect to a filtration $(\mathcal{F}_t)_{t\ge 0}$, then it is also a semimartingale in its own natural filtration (see [61]).

Exercise 5.32. Let $(M_t)_{t\ge 0}$ be a continuous local martingale on the filtered probability space $(\Omega, (\mathcal{F}_t)_{t\ge 0}, \mathcal{F}, \mathbb{P})$. Show that $(M_t^2)_{t\ge 0}$ is a semimartingale.

Since a bounded variation process has a zero quadratic variation, it is easy to prove the following theorem:

Proposition 5.33. *Let*

$$X_t = X_0 + A_t + M_t, \quad t \ge 0,$$

be a continuous adapted semimartingale. For every $t \geq 0$ and every sequence of subdivisions $\Delta_n[0,t]$ such that

$$\lim_{n\to+\infty} |\Delta_n[0,t]| = 0,$$

the following limit holds in probability:

$$\lim_{n\to+\infty} \sum_{k=1}^{n} (X_{t_k^n} - X_{t_{k-1}^n})^2 = \langle M\rangle_t.$$

We therefore call $\langle M\rangle$ the quadratic variation of X and write $\langle X\rangle = \langle M\rangle$.

Exercise 5.34. Let $(X_t)_{t\geq 0}$ be a continuous semimartingale on the filtered probability space $(\Omega, (\mathcal{F}_t)_{t\geq 0}, \mathcal{F}, \mathbb{P})$. If $\Delta[0,T]$ is a subdivision of the time interval $[0,T]$, we write

$$S_t^{\Delta[0,T]}(X) = \sum_{i=0}^{k-1} (X_{t_{i+1}} - X_{t_i})^2 + (X_t - X_{t_k})^2,$$

where k is such that $t_k \leq t < t_{k+1}$. Let $\Delta_n[0,T]$ be a sequence of subdivisions of $[0,T]$ such that

$$\lim_{n\to+\infty} |\Delta_n[0,T]| = 0.$$

Show that the following limit holds in probability:

$$\lim_{n\to+\infty} \sup_{0\leq t\leq T} |S_t^{\Delta[0,T]}(X) - \langle X\rangle_t| = 0.$$

Exercise 5.35. Let $(X_t)_{t\geq 0}$ be a continuous semimartingale on $(\Omega, (\mathcal{F}_t)_{t\geq 0}, \mathcal{F}, \mathbb{P})$. Let u^n be a sequence of locally bounded and adapted processes almost surely converging toward 0 such that $u^n \leq u$, where u is a locally bounded process. Show that for $T \geq 0$, the following limit holds in probability:

$$\lim_{n\to+\infty} \sup_{0\leq t\leq T} \left| \int_0^t u_s^n \, dX_s \right| = 0.$$

It already has been observed that in the Brownian case, though the stochastic integral is not an almost sure limit of Riemann sums, it is however a limit in probability of such sums. This may be extended to semimartingales in the following way.

Proposition 5.36. *Let u be a continuous and adapted process, let $(X_t)_{t\geq 0}$ be a continuous and adapted semimartingale and let $t \geq 0$. For every sequence of subdivisions $\Delta_n[0,t]$ such that*

$$\lim_{n\to+\infty} |\Delta_n[0,t]| = 0,$$

the following limit holds in probability:

$$\lim_{n\to+\infty}\sum_{k=0}^{n-1} u_{t_k^n}(X_{t_{k+1}^n}-X_{t_k^n})=\int_0^t u_s dX_s.$$

Exercise 5.37 (Backward and Stratonovitch integrals). Let $(X_t)_{t\geq 0}, (Y_t)_{t\geq 0}$ be continuous semimartingales.

(1) Show that for every sequence of subdivisions $\Delta_n[0,t]$ such that

$$\lim_{n\to+\infty} \mid \Delta_n[0,t] \mid = 0,$$

the following limit exists in probability:

$$\lim_{n\to+\infty}\sum_{k=0}^{n-1} \frac{X_{t_k^n}+X_{t_{k+1}^n}}{2}(Y_{t_{k+1}^n}-Y_{t_k^n}).$$

This limit is called the Stratonovitch integral and is denoted by $\int_0^t X_s \circ dY_s$.

(2) Show the formula

$$\int_0^t X_s \circ dY_s = \int_0^t X_s dY_s + \frac{1}{2}\langle X, Y\rangle_t.$$

(3) With the same assumptions as above, show that the following limit exists in probability:

$$\lim_{n\to+\infty}\sum_{k=0}^{n-1} X_{t_{k+1}^n}(Y_{t_{k+1}^n}-Y_{t_k^n}).$$

The limit is called the backward stochastic integral. Find a formula relating the backward integral to Stratonovitch's.

As we already suggested, the class of semimartingales is actually the widest class of stochastic processes with respect to which we may define a stochastic integral that enjoys natural probabilistic properties. Let us explain more precisely what the previous statement means.

We denote by $\mathcal{E}_b$ the set of processes $(u_t)_{t\geq 0}$ with

$$u_t = \sum_{i=1}^{N} F_i 1_{(S_i,T_i]}(t),$$

where $0 \leq S_1 \leq T_1 \leq \cdots \leq S_N \leq T_N$ are bounded stopping times and where the F_i's are random variable that are bounded and measurable with respect to $\mathcal{F}_{S_i}$.

If $(X_t)_{t\geq 0}$ is a continuous and adapted process and if $u \in \mathcal{E}_b$, then we naturally define

$$\int_0^t u_s dX_s = \sum_{i=1}^N F_i(X_{T_i\wedge t} - X_{S_i\wedge t}).$$

We have the following theorem whose proof goes back to Meyer and can be found in the book [61] by Protter:

Theorem 5.38. *Let $(X_t)_{t\geq 0}$ be a continuous and adapted process. The process $(X_t)_{t\geq 0}$ is a semimartingale if and only if for every sequence u^n in $\mathcal{E}_b$ that almost surely converges to 0, we have for every $t \geq 0$ and $\varepsilon > 0$,*

$$\lim_{n\to+\infty} \mathbb{P}\left(\left|\int_0^t u_s^n dX_s\right| > \varepsilon\right) = 0.$$

If the previous continuity in probability is not asked, then integrals in the rough paths sense of Lyons (see Chapter 7) may be useful in applications. This integral is a natural extension of Young's integral which coincides with the Stratonovitch integral (see Exercise 5.37) for semimartingales but which does not enjoy any nice probabilistic property.

5.5 Döblin–Itô formula

The Döblin–Itô formula is certainly the most important and useful formula of stochastic calculus. It is the change of variable formula for stochastic integrals. It is a very simple formula whose specificity is the appearance of a quadratic variation term. This reflects the fact that semimartingales have a finite quadratic variation.

Due to its importance, we first provide a heuristic argument on how to derive Itô formula. Let $f : \mathbb{R} \to \mathbb{R}$ be a smooth function and $x : \mathbb{R} \to \mathbb{R}$ be a C^1 path. We have the following heuristic computation:

$$\begin{aligned} f(x_{t+dt}) &= f(x_t + (x_{t+dt} - x_t)) \\ &= f(x_t) + f'(x_t)(x_{t+dt} - x_t) \\ &= f(x_t) + f'(x_t)dx_t. \end{aligned}$$

This suggests, by summation, the following (correct) formula:

$$f(x_t) = f(x_0) + \int_0^t f'(x_s)dx_s.$$

Let us now try to consider a Brownian motion $(B_t)_{t\geq 0}$ instead of the smooth path x and let us try to adapt the previous computation to this case. Since Brownian motion has quadratic variation which is not zero, $\langle B\rangle_t = t$, we need to go at the order 2 in the Taylor expansion of f. This leads to the following heuristic computation:

$$\begin{aligned} f(B_{t+dt}) &= f(B_t + (B_{t+dt} - B_t)) \\ &= f(B_t) + f'(B_t)(B_{t+dt} - B_t) + \frac{1}{2} f''(B_t)((B_{t+dt} - B_t))^2 \\ &= f(B_t) + f'(B_t)dB_t + \frac{1}{2} f''(B_t)dt. \end{aligned}$$

By summation, we are therefore led to the formula

$$f(B_t) = f(0) + \int_0^t f'(B_s)dB_s + \frac{1}{2}\int_0^t f''(B_s)ds,$$

which is, as we will see later, perfectly correct.

In what follows, we consider a filtered probability space $(\Omega, (\mathcal{F}_t)_{t\geq 0}, \mathcal{F}, \mathbb{P})$ that satisfies the usual conditions. Our starting point to prove Döblin–Itô formula is the following formula which is known as the integration by parts formula for semimartingales:

Theorem 5.39 (Integration by parts formula). *Let $(X_t)_{t\geq 0}$ and $(Y_t)_{t\geq 0}$ be two continuous semimartingales. Then the process $(X_tY_t)_{t\geq 0}$ is a continuous semimartingale and we have*

$$X_tY_t = X_0Y_0 + \int_0^t X_s dY_s + \int_0^t Y_s dX_s + \langle X, Y\rangle_t, \quad t \geq 0.$$

Proof. By bilinearity of the multiplication, we may assume that $X = Y$. Also by considering, if needed, $X - X_0$ instead of X, we may assume that $X_0 = 0$.

Let $t \geq 0$. For every sequence $\Delta_n[0,t]$ such that

$$\lim_{n\to+\infty} |\Delta_n[0,t]| = 0,$$

we have

$$\sum_{k=1}^n (X_{t_k^n} - X_{t_{k-1}^n})^2 = X_t^2 - 2\sum_{k=1}^n X_{t_{k-1}^n}(X_{t_k^n} - X_{t_{k-1}^n}).$$

By letting $n \to \infty$, and using Propositions 5.33 and 5.36 we therefore obtain the following identity which yields the expected result:

$$X_t^2 = 2\int_0^t X_s dX_s + \langle X\rangle_t. \qquad \square$$

Exercise 5.40. Let

$$L = \sum_{i=1}^{n} b_i(x)\frac{\partial}{\partial x_i} + \frac{1}{2}\sum_{i,j=1}^{n} a_{ij}(x)\frac{\partial^2}{\partial x_i \partial x_j},$$

be a diffusion operator. Let us assume that there exists a diffusion process $(X_t)_{t\geq 0}$ with generator L. Show that for $f \in \mathcal{C}_c(\mathbb{R}^n, \mathbb{R})$, the quadratic variation of the martingale

$$M_t^f = f(X_t) - \int_0^t Lf(X_s)ds$$

is given by

$$\langle M^f \rangle_t = \int_0^t \Gamma(f)(X_s)^2 ds,$$

where $\Gamma(f) = \frac{1}{2}(Lf^2 - 2fLf)$ is the *carré du champ.*

We are now in a position to prove the Döblin–Itô formula in its simpler form.

Theorem 5.41 (Döblin–Itô formula I). *Let $(X_t)_{t\geq 0}$ be a continuous and adapted semimartingale and let $f : \mathbb{R} \to \mathbb{R}$ be a function which is twice continuously differentiable. The process $(f(X_t))_{t\geq 0}$ is a semimartingale and the following change of variable formula holds:*

$$f(X_t) = f(X_0) + \int_0^t f'(X_s)dX_s + \frac{1}{2}\int_0^t f''(X_s)d\langle X \rangle_s. \tag{5.1}$$

Proof. We assume that the semimartingale $(X_t)_{t\geq 0}$ is bounded. If this is not the case, we may apply the following arguments to the semimartingale $(X_{t\wedge T_n})_{t\geq 0}$, where $T_n = \inf\{t \geq 0, X_t \geq n\}$ and then let $n \to \infty$.

Let $\mathcal{A}$ be the set of two times continuously differentiable functions f for which the formula (5.1) holds. It is straightforward that $\mathcal{A}$ is a vector space.

Let us show that $\mathcal{A}$ is also an algebra, that is, $\mathcal{A}$ is stable under multiplication. Let $f, g \in \mathcal{A}$. By using the integration by parts formula with the semimartingales $(f(X_t))_{t\geq 0}$ and $(g(X_t))_{t\geq 0}$, we obtain

$$f(X_t)g(X_t) = f(X_0)g(X_0) + \int_0^t f(X_s)dg(X_s) + \int_0^t g(X_s)df(X_s) + \langle f(X), g(X)\rangle_t.$$

The terms of the this sum may be treated separately in the following way. Since

$f, g \in \mathcal{A}$, we get:

$$\int_0^t f(X_s)dg(X_s) = \int_0^t f(X_s)g'(X_s)dX_s + \frac{1}{2}\int_0^t f(X_s)g''(X_s)d\langle X\rangle_s,$$
$$\int_0^t g(X_s)df(X_s) = \int_0^t g(X_s)f'(X_s)dX_s + \frac{1}{2}\int_0^t g(X_s)f''(X_s)d\langle X\rangle_s,$$
$$\langle f(X), g(X)\rangle_t = \int_0^t f'(X_s)g'(X_s)d\langle X\rangle_s.$$

Therefore,

$$\begin{aligned} f(X_t)g(X_t) &= f(X_0)g(X_0) + \int_0^t f(X_s)g'(X_s)dX_s + \int_0^t g(X_s)f'(X_s)dX_s \\ &\quad + \frac{1}{2}\int_0^t f(X_s)g''(X_s)d\langle X\rangle_s + \int_0^t f'(X_s)g'(X_s)d\langle X\rangle_s \\ &\quad + \frac{1}{2}\int_0^t g(X_s)f''(X_s)d\langle X\rangle_s \\ &= f(X_0)g(X_0) + \int_0^t (fg)'(X_s)dX_s + \frac{1}{2}\int_0^t (fg)''(X_s)d\langle X\rangle_s. \end{aligned}$$

We deduce that $fg \in \mathcal{A}$.

As a conclusion, $\mathcal{A}$ is an algebra of functions. Since $\mathcal{A}$ contains the function $x \to x$, we deduce that $\mathcal{A}$ actually contains every polynomial function. Now in order to show that every function f which is twice continuously differentiable is actually in $\mathcal{A}$, we first observe that since X is assumed to be bounded, it take its values in a compact set.

It is then possible to find a sequence of polynomials P_n such that, on this compact set, P_n uniformly converges toward f, P_n' uniformly converges toward f' and P_n'' uniformly converges toward f''. We may then conclude by using the result of Exercise 5.35. □

As a particular case of the previous formula, if we apply this formula with X a Brownian motion, we get the formula that was already pointed out at the beginning of the section: If $f : \mathbb{R} \to \mathbb{R}$ is twice continuously differentiable function, then

$$f(B_t) = f(0) + \int_0^t f'(B_s)dB_s + \frac{1}{2}\int_0^t f''(B_s)ds.$$

It is easy to derive the following variations of the Döblin–Itô formula:

Theorem 5.42 (Döblin–Itô formula II). *Let $(X_t)_{t\geq 0}$ be a continuous and adapted semimartingale, and let $(A_t)_{t\geq 0}$ be an adapted bounded variation process. If $f : \mathbb{R} \times \mathbb{R} \to \mathbb{R}$ is a function that is once continuously differentiable with respect* *to its first variable and that is twice continuously differentiable with respect to its second variable, then, for* $t \geq 0$,

$$f(A_t, X_t) = f(A_0, X_0) + \int_0^t \frac{\partial f}{\partial t}(A_s, X_s) dA_s + \int_0^t \frac{\partial f}{\partial x}(A_s, X_s) dX_s + \frac{1}{2} \int_0^t \frac{\partial^2 f}{\partial x^2}(A_s, X_s) d\langle X \rangle_s.$$

Theorem 5.43 (Döblin–Itô formula III). *Let* $(X_t^1)_{t\geq 0}, \dots, (X_t^n)_{t\geq 0}$ *be* n *adapted and continuous semimartingales and let* $f\colon \mathbb{R}^n \to \mathbb{R}$ *be a twice continuously differentiable function. We have*

$$f(X_t^1, \dots, X_t^n) = f(X_0^1, \dots, X_0^n) + \sum_{i=1}^n \int_0^t \frac{\partial f}{\partial x_i}(X_s^1, \dots, X_s^n) dX_s^i + \frac{1}{2} \sum_{i,j=1}^n \int_0^t \frac{\partial^2 f}{\partial x_i \partial x_j}(X_s^1, \dots, X_s^n) d\langle X^i, X^j \rangle_s.$$

Exercise 5.44. Let $f\colon \mathbb{R}_{\geq 0} \times \mathbb{R} \to \mathbb{C}$ be a function that is once continuously differentiable with respect to its first variable and twice continuously differentiable with respect to its second variable and that satisfies

$$\frac{1}{2}\frac{\partial^2 f}{\partial x^2} + \frac{\partial f}{\partial t} = 0.$$

Show that if $(M_t)_{t\geq 0}$ is a continuous local martingale, then $(f(\langle M \rangle_t, M_t))_{t\geq 0}$ is a continuous local martingale. Deduce that for $\lambda \in \mathbb{C}$, the process

$$\left(\exp(\lambda M_t - \frac{1}{2}\lambda^2 \langle M \rangle_t) \right)_{t\geq 0}$$

is a local martingale.

Exercise 5.45. The Hermite polynomial of order n is defined by

$$H_n(x) = (-1)^n \frac{1}{n!} e^{\frac{x^2}{2}} \frac{d^n}{dx^n} e^{-\frac{x^2}{2}}.$$

(1) Compute H_0, H_1, H_2, H_3.
(2) Show that if $(B_t)_{t\geq 0}$ is a Brownian motion, then the process $(t^{n/2} H_n(\frac{B_t}{\sqrt{t}}))_{t\geq 0}$ is a martingale.
(3) Show that

$$t^{n/2} H_n\left(\frac{B_t}{\sqrt{t}}\right) = \int_0^t \int_0^{t_n} \dots \int_0^{t_2} dB_{t_1} \dots dB_{t_n}.$$

5.6 Recurrence and transience of the Brownian motion in higher dimensions

The purpose of the next sections is to illustrate through several applications the power of the stochastic integration theory and particularly of the Döblin–Itô formula. We start with a study of the multi-dimensional Brownian motion. As already pointed out, a multi-dimensional stochastic process $(B_t)_{t\geq 0} = (B_t^1, \dots, B_t^n)_{t\geq 0}$ is called a Brownian motion if the processes $(B_t^1)_{t\geq 0}, \dots, (B_t^n)_{t\geq 0}$ are independent Brownian motions. In the sequel we denote by Δ the Laplace operator on $\mathbb{R}^n$, that is,

$$\Delta = \sum_{i=1}^{n} \frac{\partial^2}{\partial x_i^2}.$$

The following result is an easy consequence of the Döblin–Itô formula.

Proposition 5.46. *Let $f : \mathbb{R}_{\geq 0} \times \mathbb{R}^n \to \mathbb{R}$ be a function that is once continuously differentiable with respect to its first variable and twice continuously differentiable with respect to its second variable and let $(B_t)_{t\geq 0} = (B_t^1, \dots, B_t^n)_{t\geq 0}$ be an n-dimensional Brownian motion. The process*

$$X_t = f(t, B_t) - \Big(\int_0^t \frac{1}{2} \Delta f(s, B_s) + \frac{\partial f}{\partial t}(s, B_s) ds \Big)$$

is a local martingale. If moreover f is such that

$$\sum_{i=1}^{n} \Big(\frac{\partial f}{\partial x_i}(t, x) \Big)^2 \leq \phi(t) e^{K\|x\|}$$

for some continuous function ϕ and some constant $K \in \mathbb{R}$, then $(X_t)_{t\geq 0}$ is a martingale.

In particular, if f is a harmonic function, i.e., $\Delta f = 0$, and if $(B_t)_{t\geq 0}$ is a multi-dimensional Brownian motion, then the process $(f(B_t))_{t\geq 0}$ is a local martingale. As we will see later, this nice fact has many consequences. A first nice application is the study of recurrence or transience of the multi-dimensional Brownian motion paths. As we have seen before, the one-dimensional Brownian motion is recurrent: it reaches any value with probability 1. In higher dimensions, the situation is more subtle.

Let $(B_t)_{t\geq 0} = (B_t^1, \dots, B_t^n)_{t\geq 0}$ be an n-dimensional Brownian motion with $n \geq 2$. For $a > 0$ and $x \in \mathbb{R}^n$, we consider the stopping time

$$T_a^x = \inf\{t \geq 0, \|B_t + x\| = a\}.$$

Proposition 5.47. *For $a < \|x\| < b$,*

$$\mathbb{P}(T_a^x < T_b^x) = \begin{cases} \frac{\ln b - \ln \|x\|}{\ln b - \ln a}, & n = 2, \\ \frac{\|x\|^{2-n} - b^{2-n}}{a^{2-n} - b^{2-n}}, & n \geq 3. \end{cases}$$

Proof. For $a < \|x\| < b$, we consider the function

$$f(x) = \Psi(\|x\|) = \begin{cases} \ln \|x\|, & n = 2, \\ \|x\|^{2-n}, & n \geq 3. \end{cases}$$

A straightforward computation shows that $\Delta f = 0$. The process

$$(f(B_{t \wedge T_a^x \wedge T_b^x}))_{t \geq 0}$$

is therefore a martingale, which implies $\mathbb{E}(f(B_{T_a^x \wedge T_b^x})) = f(x)$. This yields

$$\Psi(a)\mathbb{P}(T_a^x < T_b^x) + \Psi(b)\mathbb{P}(T_b^x < T_a^x) = f(x).$$

Since

$$\mathbb{P}(T_a^x < T_b^x) + \mathbb{P}(T_b^x < T_a^x) = 1,$$

we deduce that

$$\mathbb{P}(T_a^x < T_b^x) = \begin{cases} \frac{\ln b - \ln \|x\|}{\ln b - \ln a}, & n = 2, \\ \frac{\|x\|^{2-n} - b^{2-n}}{a^{2-n} - b^{2-n}}, & n \geq 3. \end{cases}$$

□

By letting $b \to \infty$, we get

Corollary 5.48. *For $0 < a < \|x\|$,*

$$\mathbb{P}(T_a^x < +\infty) = \begin{cases} 1, & n = 2, \\ \frac{\|x\|^{2-n}}{a^{2-n}}, & n \geq 3. \end{cases}$$

As a consequence, for $n = 2$ the Brownian motion is recurrent, that is, for every non-empty set $\mathcal{O} \subset \mathbb{R}^2$,

$$\mathbb{P}(\exists t \geq 0, B_t \in \mathcal{O}) = 1.$$

Though the two-dimensional Brownian motion is recurrent, points are always polar.

Proposition 5.49. *For every $x \in \mathbb{R}^n$, $\mathbb{P}(\exists t \geq 0, B_t = x) = 0$.*

Proof. It suffices to prove that for every $x \in \mathbb{R}^n$, $x \neq 0$, $\mathbb{P}(T_0^x < +\infty) = 0$. We have

$$\{T_0^x < +\infty\} = \bigcup_{n \geq 0} \bigcap_{m \geq \frac{1}{\|x\|}} \{T_{1/m}^x \leq T_n^x\}.$$

Since $\mathbb{P}(\bigcap_{m \geq \frac{1}{\|x\|}} \{T_{1/m}^x \leq T_n^x\}) = \lim_{m \to \infty} \mathbb{P}(T_{1/m}^x \leq T_n^x) = 0$, we get

$$\mathbb{P}(T_0^x < +\infty) = 0.$$

□

As we have just seen, the two-dimensional Brownian motion will hit every non-empty open set with probability one. The situation is different in dimension higher than or equal to 3: Brownian motion paths will eventually leave any bounded set with probability one.

Proposition 5.50. *Let $(B_t)_{t\geq 0} = (B_t^1, \dots, B_t^n)_{t\geq 0}$ be an n-dimensional Brownian motion. If $n \geq 3$ then almost surely*

$$\lim_{t\to\infty} \|B_t\| = +\infty.$$

Proof. Let us assume $n \geq 3$. Let $\Phi(x) = \frac{1}{\|x+a\|^{n-2}}$ where $a \in \mathbb{R}^n$, $a \neq 0$. Since $(B_t)_{t\geq 0}$ will never hit the point $-a$, we can consider the process $(\Phi(B_t))_{t\geq 0}$ which is seen to be a positive local martingale from Itô's formula. A positive local martingale is always a supermartingale. Therefore from the Doob convergence theorem (see Exercise 1.42), the process $(\Phi(B_t))_{t\geq 0}$ converges almost surely when $t \to \infty$ to an integrable and non-negative random variable Z. From Fatou's lemma, we have $\mathbb{E}(Z) \leq \liminf_{t\to+\infty} \mathbb{E}(\Phi(B_t))$. By the scaling property of the Brownian motion, it is clear that $\liminf_{t\to+\infty} \mathbb{E}(\Phi(B_t)) = 0$. We conclude that $Z = 0$. □

Exercise 5.51 (Probabilistic proof of Liouville's theorem). By using martingale methods, prove that if $f\colon \mathbb{R}^n \to \mathbb{R}$ is a bounded harmonic function, then f is constant.

Exercise 5.52. Let $(B_t)_{t\geq 0} = (B_t^1, \dots, B_t^n)_{t\geq 0}$ be an n-dimensional Brownian motion. Show that for $n \geq 3$, the process $\left(\frac{1}{\|B_t+a\|^{n-2}}\right)_{t\geq 0}$ is a local martingale which is not a martingale.

5.7 Itô representation theorem

In this section we show that, remarkably, any square integrable random variable which is measurable with respect to a Brownian motion can be expressed as a stochastic integral with respect to this Brownian motion. A striking consequence of this result, which is known as Itô's representation theorem, is that any square martingale of the filtration has a continuous version.

Let $(B_t)_{t\geq 0}$ be a Brownian motion. In the sequel, we consider the filtration $(\mathcal{F}_t)_{t\geq 0}$ which is the usual completion of the natural filtration of $(B_t)_{t\geq 0}$ (such a filtration is called a Brownian filtration).

The following lemma is a straightforward consequence of the Döblin–Itô formula.

Lemma 5.53. *Let $f\colon \mathbb{R}_{\geq 0} \to \mathbb{R}$ be a locally square integrable function. The process*

$$\left(\exp\left(\int_0^t f(s)dB_s - \frac{1}{2}\int_0^t f(s)^2 ds\right)\right)_{t\geq 0}$$

is a square integrable martingale.

Proof. From the Döblin–Itô formula we have

$$\exp\Big(\int_0^t f(s)dB_s - \frac{1}{2}\int_0^t f(s)^2 ds\Big) = 1 + \int_0^t f(s)\exp\Big(\int_0^s f(u)dB_u - \frac{1}{2}\int_0^s f(u)^2 du\Big)dB_s.$$

The random variable $\int_0^s f(u)dB_u$ is a Gaussian random variable with mean 0 and variance $\int_0^s f(u)^2 du$. As a consequence

$$\mathbb{E}\Big(\int_0^t f(s)^2 \exp\Big(2\int_0^s f(u)dB_u\Big)ds\Big) < +\infty,$$

and the process

$$\int_0^t f(s)\exp\Big(\int_0^s f(u)dB_u - \frac{1}{2}\int_0^s f(u)^2 du\Big)dB_s$$

is a martingale. □

Lemma 5.54. *Let $\mathcal{D}$ be the set of compactly supported and piecewise constant functions $\mathbb{R}_{\geq 0} \to \mathbb{R}$, i.e. the set of functions f that can be written as*

$$f = \sum_{i=1}^n a_i \mathbf{1}_{(t_{i-1},t_i]},$$

for some $0 \leq t_1 \leq \cdots \leq t_n$ and $a_1, \ldots, a_n \in \mathbb{R}$. Then the linear span of the family

$$\Big\{\exp\Big(\int_0^{+\infty} f(s)dB_s - \frac{1}{2}\int_0^{+\infty} f(s)^2 ds\Big), f \in \mathcal{D}\Big\}$$

is dense in $L^2(\mathcal{F}_\infty, \mathbb{P})$.

Proof. Let $F \in L^2(\mathcal{F}_\infty, \mathbb{P})$ such that for every $f \in \mathcal{D}$,

$$\mathbb{E}\Big(F\exp\Big(\int_0^{+\infty} f(s)dB_s - \frac{1}{2}\int_0^{+\infty} f(s)^2 ds\Big)\Big) = 0.$$

Let $t_1, \ldots, t_n \geq 0$. We have for every $\lambda_1, \ldots, \lambda_n \in \mathbb{R}$,

$$\mathbb{E}\Big(F\exp\Big(\sum_{i=1}^n \lambda_i (B_{t_i} - B_{t_{i-1}})\Big)\Big) = 0.$$

By analytic continuation we see that

$$\mathbb{E}\Big(F \exp\Big(\sum_{i=1}^{n} \lambda_i (B_{t_i} - B_{t_{i-1}})\Big)\Big) = 0$$

actually also holds for every $\lambda_1, \dots, \lambda_n \in \mathbb{C}$. By using the Fourier transform, it implies that

$$\mathbb{E}(F \mid B_{t_1}, \dots, B_{t_n}) = 0.$$

Since $t_1, \dots, t_n$ were arbitrary, we conclude that $\mathbb{E}(F \mid \mathcal{F}_\infty) = 0$. As a conclusion $F = 0$. □

We are now in a position to state the representation theorem.

Theorem 5.55 (Itô representation theorem). *For every $F \in L^2(\mathcal{F}_\infty, \mathbb{P})$, there is a unique progressively measurable process $(u_t)_{t \ge 0}$ such that $\mathbb{E}\left(\int_0^\infty u_s^2 ds\right) < +\infty$ and*

$$F = \mathbb{E}(F) + \int_0^{+\infty} u_s dB_s.$$

Proof. The uniqueness is immediate as a consequence of Itô's isometry for stochastic integrals. Let $\mathcal{A}$ be the set of random variables $F \in L^2(\mathcal{F}_\infty, \mathbb{P})$ such that there exists a progressively measurable process $(u_t)_{t \ge 0}$ such that $\mathbb{E}(\int_0^\infty u_s^2 ds) < +\infty$ and

$$F = \mathbb{E}(F) + \int_0^{+\infty} u_s dB_s.$$

From the above lemma, it is clear that $\mathcal{A}$ contains the set of set of random variables

$$\Big\{\exp\Big(\int_0^{+\infty} f(s) dB_s - \frac{1}{2}\int_0^{+\infty} f(s)^2 ds\Big),\ f \in \mathcal{D}\Big\}.$$

Since this set is in $L^2(\mathcal{F}_\infty, \mathbb{P})$, we just need to prove that $\mathcal{A}$ is closed in $L^2(\mathcal{F}_\infty, \mathbb{P})$. So, let $(F_n)_{n \in \mathbb{N}}$ be a sequence of random variables such that $F_n \in \mathcal{A}$ and $F_n \xrightarrow[n \to \infty]{} F$ in $L^2(\mathcal{F}_\infty, \mathbb{P})$. There is a progressively measurable process $(u_t^n)_{t \ge 0}$ such that $\mathbb{E}(\int_0^\infty (u_s^n)^2 ds) < +\infty$ and

$$F_n = \mathbb{E}(F_n) + \int_0^{+\infty} u_s^n dB_s.$$

By using Itô's isometry, it is seen that the sequence u^n is a Cauchy sequence and therefore converges to a process u which satisfies

$$F = \mathbb{E}(F_n) + \int_0^{+\infty} u_s^n dB_s.$$

□

As a consequence of the representation theorem, we obtain the following description of the square integrable martingales of the filtration $(\mathcal{F}_t)_{t\geq 0}$.

Corollary 5.56. *Let $(M_t)_{t\geq 0}$ be a square integrable martingale of the filtration $(\mathcal{F}_t)_{t\geq 0}$. There is a unique progressively measurable process $(u_t)_{t\geq 0}$ such that for every $t \geq 0$ we have $\mathbb{E}\big(\int_0^t u_s^2 ds\big) < +\infty$ and*

$$M_t = \mathbb{E}(M_0) + \int_0^t u_s dB_s.$$

In particular, $(M_t)_{t\geq 0}$ admits a continuous version.

Exercise 5.57. Show that if $(M_t)_{t\geq 0}$ is a local martingale of the filtration $(\mathcal{F}_t)_{t\geq 0}$, then there is a unique progressively measurable process $(u_t)_{t\geq 0}$ such that for every $t \geq 0$, $\mathbb{P}(\int_0^t u_s^2 ds < +\infty) = 1$ and

$$M_t = M_0 + \int_0^t u_s dB_s.$$

5.8 Time changed martingales and planar Brownian motion

In the previous section, we proved that any martingale which is adapted to a Brownian filtration can be written as a stochastic integral. In this section, we prove that any martingale can also be represented as a time changed Brownian motion. To prove this fact, we first give a characterization of the Brownian motion which is interesting in itself. In this section we denote by $(\mathcal{F}_t)_{t\geq 0}$ a filtration that satisfies the usual conditions.

Proposition 5.58 (Lévy characterization theorem). *Let $(M_t)_{t\geq 0}$ be a continuous local martingale such that $M_0 = 0$ and such that for every $t \geq 0$, $\langle M \rangle_t = t$. The process $(M_t)_{t\geq 0}$ is a standard Brownian motion.*

Proof. Let $N_t = e^{i\lambda M_t + \frac{1}{2}\lambda^2 t}$. By using the Döblin–Itô formula, we obtain that for $s \leq t$,

$$N_t = N_s + \int_s^t N_u dM_u.$$

As a consequence, the process $(N_t)_{t\geq 0}$ is a martingale and from the above equality we get

$$\mathbb{E}(e^{i\lambda(M_t - M_s)} \mid \mathcal{F}_s) = e^{-\frac{1}{2}\lambda^2(t-s)}.$$

The process $(M_t)_{t\geq 0}$ is therefore a continuous process with stationary and independent increments such that M_t is normally distributed with mean 0 and variance t. It is thus a Brownian motion. □

Exercise 5.59 (Bessel process). Let $(B_t)_{t\geq 0}$ be a d-dimensional Brownian motion $d \geq 2$. We write

$$\rho_t = \|x + B_t\|,$$

where $x \neq 0$. Show that

$$\rho_t - \int_0^t \frac{d-1}{2\rho_s} ds$$

is a standard Brownian motion.

The next proposition shows that continuous martingales behave in a nice way with respect to time changes.

Proposition 5.60. *Let $(C_t)_{t\geq 0}$ be a continuous and increasing process such that for every $t \geq 0$, C_t is a finite stopping time of the filtration $(\mathcal{F}_t)_{t\geq 0}$. Let $(M_t)_{t\geq 0}$ be a continuous martingale with respect to $(\mathcal{F}_t)_{t\geq 0}$. The process $(M_{C_t})_{t\geq 0}$ is a local martingale with respect to the filtration $(\mathcal{F}_{C_t})_{t\geq 0}$. Moreover, $\langle M_C \rangle = \langle M \rangle_C$.*

Proof. By using localization, if necessary, we can assume that C is bounded. According to the Doob stopping theorem (see Theorem 1.34), we need to prove that for every bounded stopping time T of the filtration $(\mathcal{F}_{C_t})_{t\geq 0}$, we have $\mathbb{E}(M_{C_T}) = 0$. But C_T is obviously a bounded stopping time of the filtration $(\mathcal{F}_t)_{t\geq 0}$ and thus from Doob's stopping theorem we have $\mathbb{E}(M_{C_T}) = 0$. The same argument shows that $M_C^2 - \langle M \rangle_C$. □

Exercise 5.61. Let $(C_t)_{t\geq 0}$ be an increasing and right continuous process such that for every $t \geq 0$, C_t is a finite stopping time of the filtration $(\mathcal{F}_t)_{t\geq 0}$. Let $(M_t)_{t\geq 0}$ be a continuous martingale with respect to $(\mathcal{F}_t)_{t\geq 0}$ such that M is constant on each interval $[C_{t-}, C_t]$. Show that the process $(M_{C_t})_{t\geq 0}$ is a continuous local martingale with respect to the filtration $(\mathcal{F}_{C_t})_{t\geq 0}$ and that $\langle M_C \rangle = \langle M \rangle_C$.

We can now prove the following nice representation result for martingales.

Theorem 5.62 (Dambis, Dubins–Schwarz theorem). *Let $(M_t)_{t\geq 0}$ be a continuous martingale such that $M_0 = 0$ and $\langle M \rangle_\infty = +\infty$. There exists a Brownian motion $(B_t)_{t\geq 0}$ such that for every $t \geq 0$,*

$$M_t = B_{\langle M \rangle_t}.$$

Proof. Let $C_t = \inf\{s \geq 0, \langle M \rangle_s > t\}$. $(C_t)_{t\geq 0}$ is an increasing and right continuous process such that for every $t \geq 0$, C_t is a finite stopping time of the filtration $(\mathcal{F}_t)_{t\geq 0}$ and M is obviously constant on each interval $[C_{t-}, C_t]$. From the previous exercise the process $(M_{C_t})_{t\geq 0}$ is a continuous local martingale whose quadratic variation is equal to t. From Lévy's characterization theorem, it is thus a Brownian motion. □

Exercise 5.63. Show that if $(M_t)_{t\geq 0}$ is a continuous local martingale such that $M_0 = 0$ and $\langle M\rangle_\infty = +\infty$, there exists a Brownian motion $(B_t)_{t\geq 0}$, such that for every $t \geq 0$,

$$M_t = B_{\langle M\rangle_t}.$$

Exercise 5.64. Let $(u_t)_{t\geq 0}$ be a continuous adapted process and let $(B_t)_{t\geq 0}$ be a Brownian motion. Show that for every $T \geq 0$, the process $(\int_0^t u_s dB_s)_{0\leq t\leq T}$ has $\frac{1}{2} - \varepsilon$ Hölder paths, where $0 < \varepsilon \leq \frac{1}{2}$.

Exercise 5.65 (Construction of one-dimensional diffusions). Let $b\colon \mathbb{R} \to \mathbb{R}$ and let $\sigma\colon \mathbb{R} \to (0, +\infty)$ be two continuous functions. Let

$$s(x) = \int_0^x \exp\Big(- 2\int_0^y \frac{|b(z)|}{\sigma(z)^2} dz\Big) dy, \quad x \in \mathbb{R},$$

and let $(B_t)_{t\geq 0}$ be a Brownian motion. For $u \geq 0$, we write

$$A_u = \int_0^u \frac{ds}{s'(B_s)\sigma^2(B_s)} = \int_0^u \frac{\exp\big(2\int_0^{B_s} \frac{|b(z)|}{\sigma(z)^2} dz\big)}{\sigma^2(B_s)} ds.$$

Show that the process

$$(s^{-1}(B_{\inf\{u, A_u > t\}}))_{t\geq 0}$$

is a Markov process whose semigroup is generated by the operator

$$Lf = b(x) f'(x) + \frac{1}{2}\sigma(x)^2 f''(x), \quad f \in \mathscr{C}_c(\mathbb{R}, \mathbb{R}).$$

Therefore, all one-dimensional diffusions may be constructed from the Brownian motion.

The study of the planar Brownian is deeply connected to the theory of analytic functions. The fundamental property of the Brownian curve is that it is a conformal invariant. By definition, a complex Brownian motion is a process $(B_t)_{t\geq 0}$ in the complex plane that can be decomposed as $B_t = B_t^1 + iB_t^2$ where B^1 and B^2 are independent Brownian motions. The following proposition is easily proved as a consequence of the Döblin–Itô formula and of the Dambins, Dubins–Schwarz theorem.

Proposition 5.66 (Conformal invariance). *Let $(B_t)_{t\geq 0}$ be a complex Brownian motion and let $f : \mathbb{C} \to \mathbb{C}$ be an analytic function. Then we have for every $t \geq 0$,*

$$f(B_t) = f(0) + \int_0^t f'(B_s) dB_s.$$

As a consequence, there exists a complex Brownian motion $(\beta_t)_{t\geq 0}$ such that for every $t \geq 0$,

$$f(B_t) = f(0) + \beta_{\int_0^t |f'(B_s)|^2 ds}.$$

To study the complex Brownian motion, it is useful to look at it in polar coordinates. This leads to the so-called skew-product decomposition of the complex Brownian motion that we study in the following exercise.

Exercise 5.67 (Skew-product decomposition). Let $(B_t)_{t\geq 0}$ be a complex Brownian motion started at $z \neq 0$.

(1) Show that for $t \geq 0$,
$$B_t = z \exp\left(\int_0^t \frac{dB_s}{B_s}\right).$$

(2) Show that there exists a complex Brownian motion $(\beta_t)_{t\geq 0}$ such that
$$B_t = z \exp(\beta_{\int_0^t \frac{ds}{\rho_s^2}}),$$
where $\rho_t = |B_t|$.

(3) Show that the process $(\rho_t)_{t\geq 0}$ is independent from the Brownian motion $(\gamma_t)_{t\geq 0} = (\mathrm{Im}(\beta_t))_{t\geq 0}$.

(4) We write $\theta_t = \mathrm{Im}(\int_0^t \frac{dB_s}{B_s})$ which can be interpreted as a winding number around 0 of the complex Brownian motion paths. For $r > |z|$, we consider the stopping time
$$T_r = \inf\{t \geq 0, |B_t| = r\}.$$
Compute for every $r > |z|$, the distribution of the random variable
$$\frac{1}{\ln(r/|z|)}\theta_{T_r}.$$

(5) Prove Spitzer theorem: In distribution, we have the following convergence
$$\frac{2\theta_t}{\ln t} \to_{+\infty} C,$$
where C is a Cauchy random variable with parameter 1 that is a random variable with density $\frac{1}{\pi(1+x^2)}$.

Exercise 5.68. Let $(B_t)_{t\geq 0}$ be a complex Brownian motion started at z such that $\mathrm{Im}(z) > 0$. We consider the stopping time
$$T = \inf\{t \geq 0, \mathrm{Im}(B_t) = 0\}.$$
Compute the distribution of the random variable $\mathrm{Re}(B_T)$.

5.9 Burkholder–Davis–Gundy inequalities

In this section, we study some of the most important martingale inequalities: The Burkholder–Davis–Gundy inequalities. Interestingly, the range of application of these inequalities is very large and they play an important role in harmonic analysis and the study of singular integrals. These inequalities admit several proofs. We present here a proof using the Döblin–Itô formula and an interesting domination inequality which is due to Lenglart.

Proposition 5.69 (Lenglart inequality). *Let $(N_t)_{t\geq 0}$ be a positive adapted continuous process and let $(A_t)_{t\geq 0}$ be an increasing process such that $A_0 = 0$. Assume that for every bounded stopping time τ,*

$$\mathbb{E}(N_\tau) \leq \mathbb{E}(A_\tau).$$

Then, for every $k \in (0,1)$, and $T \geq 0$,

$$\mathbb{E}((\sup_{0\leq t\leq T} N_t)^k) \leq \frac{2-k}{1-k}\mathbb{E}(A_T^k).$$

Proof. Let $x, y > 0$. We write $R = \inf\{0 \leq t \leq T, A_t \geq y\}$ and $S = \inf\{0 \leq t \leq T, X_t \geq x\}$ with the usual convention that the infimum of the empty set is ∞. We first observe that

$$\mathbb{P}(\sup_{0\leq t\leq T} X_t \geq x, A_T \leq y) \leq \mathbb{P}(X_{T\wedge S\wedge R} \geq x) \leq \frac{1}{x}\mathbb{E}(A_{T\wedge S\wedge R}) \leq \frac{1}{x}\mathbb{E}(A_T \wedge y).$$

Let now F be a continuous increasing function from $\mathbb{R}_{>0}$ into $\mathbb{R}_{>0}$ with $F(0) = 0$. By using the previous inequality and Fubini's theorem, we have

$$\begin{aligned}
\mathbb{E}(F(\sup_{0\leq t\leq T} X_t)) &= \mathbb{E}\Big(\int_0^{+\infty} \mathbf{1}_{\sup_{0\leq t\leq T} X_t > x} dF(x)\Big)\\
&\leq \int_0^{+\infty} [\mathbb{P}(\sup_{0\leq t\leq T} X_t \geq x, A_T \leq x) + \mathbb{P}(A_T \leq x)] dF(x)\\
&\leq \int_0^{+\infty} \Big[\frac{1}{x}\mathbb{E}(A_T \wedge x) + \mathbb{P}(A_T \leq x)\Big] dF(x)\\
&\leq \int_0^{+\infty} \Big[\frac{1}{x}\mathbb{E}(A_T 1_{A_T\leq x}) + 2\mathbb{P}(A_T \leq x)\Big] dF(x)\\
&\leq 2\mathbb{E}(F(A_T)) + \mathbb{E}\Big(A_T \int_{A_T}^{+\infty} \frac{dF(x)}{x}\Big).
\end{aligned}$$

Taking $f(x) = x^k$, we obtain the desired result. □

We now turn to the Burkholder–Davis–Gundy inequalities.

Theorem 5.70 (Burkholder–Davis–Gundy inequalities). *Let $T > 0$ and $(M_t)_{0\le t\le T}$ be a continuous local martingale such that $M_0 = 0$. For every $0 < p < \infty$, there exist universal constants c_p and C_p, independent of T and $(M_t)_{0\le t\le T}$ such that*

$$c_p \mathbb{E}(\langle M\rangle_T^{\frac{p}{2}}) \le \mathbb{E}((\sup_{0\le t\le T} |M_t|)^p) \le C_p \mathbb{E}(\langle M\rangle_T^{\frac{p}{2}}).$$

Proof. By stopping it is enough to prove the result for bounded M. Let $q \ge 2$. From the Döblin–Itô formula we have

$$\begin{aligned} d|M_t|^q &= q|M_t|^{q-1}\operatorname{sgn}(M_t)dM_t + \frac{1}{2}q(q-1)|M_t|^{q-2}d\langle M\rangle_t \\ &= q\operatorname{sgn}(M_t)|M_t|^{q-1}dM_t + \frac{1}{2}q(q-1)|M_t|^{q-2}d\langle M\rangle_t. \end{aligned}$$

As a consequence of the Doob stopping theorem, we get that for every bounded stopping time τ,

$$\mathbb{E}(|M_\tau|^q) \le \frac{1}{2}q(q-1)\mathbb{E}\Big(\int_0^\tau |M_t|^{q-2}d\langle M\rangle_t\Big).$$

From Lenglart's domination inequality, we then deduce that for every $k \in (0,1)$,

$$\mathbb{E}((\sup_{0\le t\le T}|M_t|^q)^k) \le \frac{2-k}{1-k}\Big(\frac{1}{2}q(q-1)\Big)^k \mathbb{E}\Big(\Big(\int_0^T |M_t|^{q-2}d\langle M\rangle_t\Big)^k\Big).$$

We now bound

$$\begin{aligned} \mathbb{E}\Big(\Big(\int_0^T |M_t|^{q-2}d\langle M\rangle_t\Big)^k\Big) &\le \mathbb{E}\Big((\sup_{0\le t\le T}|M_t|)^{k(q-2)}\Big(\int_0^T d\langle M\rangle_t\Big)^k\Big) \\ &\le \mathbb{E}((\sup_{0\le t\le T}|M_t|)^{kq})^{1-\frac{2}{q}}\mathbb{E}(\langle M\rangle_T^{\frac{kq}{2}})^{\frac{2}{q}}. \end{aligned}$$

As a consequence we obtain

$$\mathbb{E}((\sup_{0\le t\le T}|M_t|^q)^k) \le \frac{2-k}{1-k}\Big(\frac{1}{2}q(q-1)\Big)^k \mathbb{E}((\sup_{0\le t\le T}|M_t|)^{kq})^{1-\frac{2}{q}}\mathbb{E}(\langle M\rangle_T^{\frac{kq}{2}})^{\frac{2}{q}}.$$

Letting $p = qk$ yields the claimed result, that is,

$$\mathbb{E}((\sup_{0\le t\le T}|M_t|)^p) \le C_p\mathbb{E}(\langle M\rangle_T^{\frac{p}{2}}).$$

We proceed now to the proof of the left-hand side inequality. We have

$$M_t^2 = \langle M \rangle_t + 2 \int_0^t M_s dM_s.$$

Therefore, we get

$$\mathbb{E}(\langle M \rangle_T^{\frac{p}{2}}) \leq A_p \Big(\mathbb{E}((\sup_{0 \leq t \leq T} |M_t|)^p) + \mathbb{E}\Big(\sup_{0 \leq t \leq T} \Big| \int_0^t M_s dM_s \Big|^{p/2} \Big) \Big).$$

By using the previous argument, we now have

$$\begin{aligned} \mathbb{E}\Big(\sup_{0 \leq t \leq T} \Big| \int_0^t M_s dM_s \Big|^{p/2} \Big) &\leq B_p \mathbb{E}\Big(\Big(\int_0^T M_s^2 d \langle M \rangle_s \Big)^{p/4} \Big) \\ &\leq B_p \mathbb{E}((\sup_{0 \leq t \leq T} |M_t|)^{p/2} \langle M \rangle_T^{p/4}) \\ &\leq B_p \mathbb{E}((\sup_{0 \leq t \leq T} |M_t|)^p)^{1/2} \mathbb{E}(\langle M \rangle_T^{p/2})^{1/2}. \end{aligned}$$

As a conclusion, we obtain

$$\mathbb{E}(\langle M \rangle_T^{\frac{p}{2}}) \leq A_p (\mathbb{E}((\sup_{0 \leq t \leq T} |M_t|)^p) + B_p \mathbb{E}((\sup_{0 \leq t \leq T} |M_t|)^p)^{1/2} \mathbb{E}(\langle M \rangle_T^{p/2})^{1/2}).$$

This is an inequality of the form

$$x^2 \leq A_p (y^2 + B_p x y),$$

which easily implies that

$$c_p x^2 \leq y^2,$$

due to the inequality $2xy \leq \frac{1}{\delta} x^2 + \delta y^2$ with a conveniently chosen δ. □

Exercise 5.71. Let $T > 0$ and $(M_t)_{0 \leq t \leq T}$ be a continuous local martingale. Consider the process

$$Z_t = e^{\int_0^t V_s ds} \int_0^t e^{-\int_0^s V_u du} dM_s,$$

where $(V_t)_{0 \leq t \leq T}$ is a non-positive adapted and continuous process. Show that for every $0 < p < \infty$, there is a universal constant C_p, independent of T, $(M_t)_{0 \leq t \leq T}$ and $(V_t)_{0 \leq t \leq T}$ such that

$$\mathbb{E}((\sup_{0 \leq t \leq T} |Z_t|)^p) \leq C_p \mathbb{E}(\langle M \rangle_T^{\frac{p}{2}}).$$

5.10 Girsanov theorem

In this section we describe a theorem which has far reaching consequences in mathematical finance: The Girsanov theorem. It describes the impact of a probability change on stochastic calculus.

Let $(\Omega, (\mathcal{F}_t)_{t\geq 0}, \mathbb{P})$ be a filtered probability space. We assume that $(\mathcal{F}_t)_{t\geq 0}$ is the usual completion of the filtration of a Brownian motion $(B_t)_{t\geq 0}$. Let $\mathbb{Q}$ be a probability measure on $\mathcal{F}_\infty$ which is equivalent to $\mathbb{P}$. We denote by D the density of $\mathbb{Q}$ with respect to $\mathbb{P}$.

Theorem 5.72 (Girsanov theorem). *There exists a progressively measurable process $(\Theta_t)_{t\geq 0}$ such that for every $t \geq 0$, $\mathbb{P}(\int_0^t \Theta_s^2 ds < +\infty) = 1$ and*

$$\mathbb{E}(D \mid \mathcal{F}_t) = \exp\left(\int_0^t \Theta_s dB_s - \frac{1}{2}\int_0^t \Theta_s^2 ds\right).$$

Moreover, the process

$$B_t - \int_0^t \Theta_s ds$$

is a Brownian motion on the filtered probability space $(\Omega, (\mathcal{F}_t)_{t\geq 0}, \mathbb{Q})$. As a consequence, a continuous and adapted process $(X_t)_{t\geq 0}$ is a $\mathbb{P}$-semimartingale if and only if it is a $\mathbb{Q}$-semimartingale.

Proof. Since $\mathbb{P}$ and $\mathbb{Q}$ are equivalent on $\mathcal{F}_\infty$, they are of course also equivalent on $\mathcal{F}_t$ for every $t \geq 0$. The density of $\mathbb{Q}_{/\mathcal{F}_t}$ with respect to $\mathbb{P}_{/\mathcal{F}_t}$ is given by $D_t = \mathbb{E}^{\mathbb{P}}(D \mid \mathcal{F}_t)$. As a consequence, the process D_t is a positive martingale. From Itô's representation theorem, we therefore deduce that there exists a progressively measurable process $(u_t)_{t\geq 0}$ such that

$$D_t = 1 + \int_0^t u_s dB_s.$$

Let now $\Theta_t = \frac{u_t}{D_t}$. We then have

$$D_t = 1 + \int_0^t \Theta_s D_s dB_s.$$

Applying the Döblin–Itô formula to the process $D_t \exp\left(-\int_0^t \Theta_s dB_s + \frac{1}{2}\int_0^t \Theta_s^2 ds\right)$, we see that it implies

$$D_t = \exp\left(\int_0^t \Theta_s dB_s - \frac{1}{2}\int_0^t \Theta_s^2 ds\right).$$

We now want to prove that the process $B_t - \int_0^t \Theta_s ds$ is a $\mathbb{Q}$-Brownian motion. It is clear that the $\mathbb{Q}$-quadratic variation of this process is t. From Lévy's characterization result, we therefore just need to prove that it is a $\mathbb{Q}$ local martingale. For this, we are going to prove that the process

$$N_t = (B_t - \int_0^t \Theta_s ds) \exp\left(\int_0^t \Theta_s dB_s - \frac{1}{2}\int_0^t \Theta_s^2 ds\right)$$

is a $\mathbb{P}$-local martingale. Indeed, from the integration by parts formula, it is immediate that

$$dN_t = D_t dB_t + \left(B_t - \int_0^t \Theta_s ds\right) dD_t.$$

Since D_t is the density of $\mathbb{Q}_{\mathcal{F}_t}$ with respect to $\mathbb{P}_{\mathcal{F}_t}$, it is then easy to deduce that N_t is a $\mathbb{P}$-local martingale and thus a $\mathbb{P}$-Brownian motion. □

Exercise 5.73. Let $(\Omega, (\mathcal{F}_t)_{t\geq 0}, \mathbb{P})$ be a filtered probability space that satisfies the usual conditions. As before, let $\mathbb{Q}$ be a probability measure on $\mathcal{F}_\infty$ that is equivalent to $\mathbb{P}$. We denote by D the density of $\mathbb{Q}$ with respect to $\mathbb{P}$ and $D_t = \mathbb{E}^{\mathbb{P}}(D \mid \mathcal{F}_t)$. Let $(M_t)_{t\geq 0}$ be a $\mathbb{P}$ local martingale. Show that the process

$$N_t = M_t - \int_0^t \frac{d\langle M, D\rangle_s}{D_s}$$

is a $\mathbb{Q}$ local martingale. As a consequence, a continuous and adapted process $(X_t)_{t\geq 0}$ is a $\mathbb{P}$-semimartingale if and only if it is a $\mathbb{Q}$-semimartingale.

Exercise 5.74. Let $(B_t)_{t\geq 0}$ be a Brownian motion. We denote by $\mathbb{P}$ the Wiener measure, by $(\pi_t)_{t\geq 0}$ the coordinate process and by $(\mathcal{F}_t)_{t\geq 0}$ its natural filtration.

(1) Let $\mu \in \mathbb{R}$ and $\mathbb{P}^\mu$ be the distribution of the process $(B_t + \mu t)_{t\geq 0}$. Show that for every $t \geq 0$,

$$\mathbb{P}^\mu_{/\mathcal{F}_t} \ll \mathbb{P}_{/\mathcal{F}_t},$$

and that

$$\frac{d\mathbb{P}^\mu_{/\mathcal{F}_t}}{d\mathbb{P}_{/\mathcal{F}_t}} = e^{\mu\pi_t - \frac{\mu^2}{2}t}.$$

(2) Is it true that

$$\mathbb{P}^\mu_{/\mathcal{F}_\infty} \ll \mathbb{P}_{/\mathcal{F}_\infty} ?$$

(3) For $a \in \mathbb{R}_{\geq 0}$, we write

$$T_a = \inf\{t \geq 0, B_t + \mu t = a\}.$$

Compute the density function of T_a. (You may use the previous question.)

(4) More generally, let $f : \mathbb{R}_{\geq 0} \to \mathbb{R}$ be a measurable function such that for every $t \geq 0$, $\int_0^t f^2(s)ds < +\infty$. We denote by $\mathbb{P}^f$ the distribution of the process $(B_t + \int_0^t f(s)ds)_{t \geq 0}$. Show that for every $t \geq 0$,

$$\mathbb{P}^f_{/\mathcal{F}_t} \ll \mathbb{P}_{/\mathcal{F}_t},$$

and that

$$\frac{d\,\mathbb{P}^f_{/\mathcal{F}_t}}{d\,\mathbb{P}_{/\mathcal{F}_t}} = e^{\int_0^t f(s)d\pi_s - \frac{1}{2}\int_0^t f^2(s)ds}.$$

Let $(\Omega, (\mathcal{F}_t)_{t \geq 0}, \mathcal{F}, \mathbb{P})$ be a filtered probability space that satisfies the usual conditions and let $(B_t)_{t \geq 0}$ be a Brownian motion on it. Let now $(\Theta_t)_{t \geq 0}$ be a progressively measurable process such that for every $t \geq 0$, $\mathbb{P}(\int_0^t \Theta_s^2 ds < +\infty) = 1$. We write

$$Z_t = \exp\left(\int_0^t \Theta_s dB_s - \frac{1}{2}\int_0^t \Theta_s^2 ds\right), \quad t \geq 0.$$

As a consequence of the Döblin–Itô formula, it is clear that $(Z_t)_{t \geq 0}$ is a local martingale. If we assume that $(Z_t)_{t \geq 0}$ is a uniformly integrable martingale, then it is easy to see that on the σ-field $\mathcal{F}_\infty$, there is a unique probability measure $\mathbb{Q}$ equivalent to $\mathbb{P}$ such that for every $t \geq 0$,

$$\frac{d\,\mathbb{Q}_{/\mathcal{F}_t}}{d\,\mathbb{P}_{/\mathcal{F}_t}} = Z_t, \quad \mathbb{P}\text{-p.s.}$$

The same argument as before shows then that with respect to $\mathbb{Q}$, the process

$$B_t - \int_0^t \Theta_s ds$$

is a Brownian motion.

It is thus important to decide whether or not $(Z_t)_{t \geq 0}$ is a uniformly martingale. The following two lemmas provide sufficient conditions that it is.

Lemma 5.75. *If for every $t \geq 0$,*

$$\mathbb{E}(Z_t) = 1,$$

then $(Z_t)_{t \geq 0}$ is a uniformly integrable martingale.

Proof. The process Z is a non-negative local martingale and thus a supermartingale. □

The second condition is known as the Novikov condition. It is often easier to check in practice than the previous one.

Lemma 5.76 (Novikov condition). *If*

$$\mathbb{E}\Big(\exp\Big(\frac{1}{2}\int_0^{+\infty} \Theta_s^2 ds\Big)\Big) < +\infty,$$

then $(Z_t)_{t\geq 0}$ is a uniformly integrable martingale.

Proof. We write $M_t = \int_0^t \Theta_s dB_s$. As a consequence of the integrability condition

$$\mathbb{E}\Big(\exp\Big(\frac{1}{2}\langle M\rangle_\infty\Big)\Big) < +\infty,$$

the random variable $\langle M\rangle_\infty$ has moments of every order. So from the Burkholder–Davis–Gundy inequalities, $\sup_{t\geq 0}|M_t|$ has moments of all orders, which implies that M is a uniformly integrable martingale. We then have

$$\exp\Big(\frac{1}{2}M_\infty\Big) = \exp\Big(\frac{1}{2}M_\infty - \frac{1}{4}\langle M\rangle_\infty\Big)\exp\Big(\frac{1}{4}\langle M\rangle_\infty\Big).$$

The Cauchy–Schwarz inequality then implies that $\mathbb{E}(\exp(\frac{1}{2}M_\infty)) < +\infty$. We deduce from the Doob convergence theorem that the process $\exp(\frac{1}{2}M)$ is a uniformly integrable submartingale. Let now $\eta < 1$. We have

$$\exp\Big(\eta M_t - \frac{\eta^2}{2}\langle M\rangle_t\Big) = \Big(\exp\Big(M_t - \frac{1}{2}\langle M\rangle_t\Big)\Big)^{\eta^2}\exp\Big(\frac{\eta M_t}{1+\eta}\Big)^{1-\eta^2}.$$

Then Hölder's inequality shows that

$$\begin{aligned}
\mathbb{E}\Big(\exp\Big(\eta M_t - \frac{\eta^2}{2}\langle M\rangle_t\Big)\Big) &\leq \mathbb{E}\Big(\exp\Big(M_t - \frac{1}{2}\langle M\rangle_t\Big)\Big)^{\eta^2}\mathbb{E}\Big(\exp\Big(\frac{\eta M_t}{1+\eta}\Big)\Big)^{1-\eta^2}\\
&\leq \mathbb{E}\Big(\exp\Big(M_t - \frac{1}{2}\langle M\rangle_t\Big)\Big)^{\eta^2}\mathbb{E}\Big(\exp\Big(\frac{M_t}{2}\Big)\Big)^{2\eta(1-\eta)}\\
&\leq \mathbb{E}\Big(\exp\Big(M_t - \frac{1}{2}\langle M\rangle_t\Big)\Big)^{\eta^2}\mathbb{E}\Big(\exp\Big(\frac{M_\infty}{2}\Big)\Big)^{2\eta(1-\eta)}.
\end{aligned}$$

If we can prove that $\mathbb{E}(\exp(\eta M_t - \frac{\eta^2}{2}\langle M\rangle_t)) = 1$, then by letting $\eta \to 1$ in the above inequality, we would get

$$\mathbb{E}\Big(\exp\Big(M_t - \frac{1}{2}\langle M\rangle_t\Big)\Big) \geq 1$$

and thus $\mathbb{E}(\exp(M_t - \frac{1}{2}\langle M\rangle_t)) = 1$.

Let $p > 1$ such that $\frac{\eta\sqrt{p}}{\sqrt{p}-1} \leq 1$. Consider $r = \frac{\sqrt{p}+1}{\sqrt{p}-1}$ and $s = \frac{\sqrt{p}+1}{2}$ so that $1/r + 1/s = 1$. Using

$$\exp\Big(\eta M_t - \frac{\eta^2}{2}\langle M\rangle_t\Big)^p = \exp\Big(\sqrt{\frac{p}{r}}\eta M_t - \frac{p}{2}\eta^2\langle M\rangle_t\Big)\exp\Big(\Big(p\eta - \sqrt{\frac{p}{r}}\Big)M_t\Big)$$

and then Hölder's inequality shows that there is a constant C (depending only on M) such that for any stopping time T,

$$\mathbb{E}\Big(\exp\Big(\eta M_T - \frac{\eta^2}{2}\langle M\rangle_T\Big)^p\Big) \leq C.$$

By the Doob maximal inequality, this implies that the local martingale

$$\exp\Big(\eta M_t - \frac{\eta^2}{2}\langle M\rangle_t\Big)$$

is actually a true martingale. This implies that

$$\mathbb{E}\Big(\exp\Big(\eta M_t - \frac{\eta^2}{2}\langle M\rangle_t\Big)\Big) = 1,$$

and the desired conclusion follows. □

Exercise 5.77 (Lévy area formula). Let $(B_t^1, B_t^2)_{t\geq 0}$ be a two-dimensional Brownian motion. The random variable

$$S_t = \int_0^t B_s^1 dB_s^2 - B_s^2 dB_s^1$$

is two times the algebraic area swept out by the path $s \to (B_s^1, B_s^2)$ in the time interval $[0, t]$.

(1) Show that
$$(B_t^1, B_t^2, S_t)_{t\geq 0} = (B_t^1, B_t^2, \beta_{\int_0^t \rho_s^2 ds})_{t\geq 0},$$
where $(\beta_t)_{t\geq 0}$ is a Brownian motion independent from $(\rho_t)_{t\geq 0}$ with $\rho_t^2 = (B_t^1)^2 + (B_t^2)^2$.
Hint. You may write $S_t = \int_0^t \rho_s d\gamma_s$ where $(\gamma_t)_{t\geq 0}$ is a Brownian motion independent from $\int_0^t \frac{B_s^1 dB_s^1 + B_s^2 dB_s^2}{\rho_s}$.

(2) Show that for $\lambda \in \mathbb{R}$,
$$\mathbb{E}(e^{i\lambda S_t} \mid B_t^1 = x, B_t^2 = y) = \mathbb{E}(e^{-\frac{\lambda^2}{2}\int_0^t \rho_s^2 ds} \mid \rho_t = \sqrt{x^2+y^2}).$$

(3) Show that the process
$$D_t = \exp\Big(\frac{\lambda}{2}(\rho_t^2 - 2t) - \frac{\lambda^2}{2}\int_0^t \rho_s^2 ds\Big)$$
is a martingale.

(4) Consider the probability measure $\mathbb{Q}_{\mathcal{F}_t} = D_t \mathbb{P}_{\mathcal{F}_t}$. Show that under $\mathbb{Q}$, $(B_t^1, B_t^2)_{t\geq 0}$ is a Gaussian process, whose covariance is to be computed.

(5) Conclude that

$$\mathbb{E}(e^{i\lambda S_t} \mid B_t^1 = x, B_t^2 = y) = \frac{\lambda}{\sinh \lambda} \exp\Big(-\frac{1}{2}(x^2 + y^2)(\lambda \coth \lambda - 1)\Big).$$

Exercise 5.78. Let $(B_t)_{t\geq 0}$ be a standard Brownian motion and Y be an independent random variable such that $\mathbb{P}(Y = -1) = \frac{1}{2}$ and $\mathbb{P}(Y = 1) = \frac{1}{2}$. We consider the process $X_t = B_t + \mu t Y$, where $\mu \in \mathbb{R}$. By using Girsanov theorem provide the semimartingale decomposition of $(X_t)_{t\geq 0}$ in its own filtration.

Notes and comments

Section 5.1. It can be shown (see for instance [29]) that almost surely

$$\sup_{\Pi \in \Delta[0,T]} \sum_{k=0}^{n-1} |B_{t_{k+1}} - B_{t_k}|^2 = +\infty,$$

that is, the 2-variation of Brownian motion paths is almost surely $+\infty$.

Sections 5.2, 5.3, 5.4. The study of integrals with respect to Brownian motion has a long history going back at least to Wiener, but the breakthrough of understanding the role of adaptedness of integrands is due to Itô in his famous 1944 paper [39]. The general theory of stochastic integration with respect to martingales, local martingales and semimartingales was later developed and popularized in particular by Watanabe (see [38]) and Meyer (see [17]). The hypothesis of the continuity of the integrator process is not strictly necessary and an integral with respect to a possibly discontinuous local martingale can be defined (see the books by Bichteler [11] or Protter [61]).

Section 5.5. The work of Döblin has long been unknown to mathematicians. Before being killed during the second world war, in a letter sent to the French academy of sciences in 1940 and only opened in 2000, Döblin presented a version of the Döblin–Itô formula. His formula for a Brownian motion $(B_t)_{t\geq 0}$ writes as

$$f(B_t) = f(0) + \beta_{\int_0^t f'(B_s)^2 ds} + \frac{1}{2}\int_0^t f''(B_s)ds,$$

where β is another Brownian motion. Observe that this formula is of course consistent with the formula we stated in the text since according to the Dambis, Dubins–Schwarz theorem

$$\int_0^t f'(B_s)dB_s = \beta_{\int_0^t f'(B_s)^2 ds}$$

for some Brownian motion β. What is remarkable is that Döblin obtained his formula even before stochastic integration was conceived! We refer to [12] and [18] for more details about the fascinating story of Döblin.

Section 5.6. Results of this section may be extended to more general diffusion processes and hint at the deep and rich connection between Markov processes and potential theory. There are several books devoted to the interactions between potential analysis and Markov processes, we mention in particular the classical books by Chung [14] and Doob [20] or to the more recent book by Durett [21].

Section 5.7. The representation theorem plays a fundamental role in mathematical finance, where the integrand appears as a hedging strategy, see the book by Karatzas and Shreve [45]. A formula for the integrand can be obtained through Malliavin calculus (see Exercise 6.36 in Chapter 6)

Section 5.8. There is a deep interaction between planar Brownian motion and complex analysis, due to the conformal invariance of the planar Brownian motion paths. Conformally invariant processes in the plane play an important in statistical mechanics, see the book by Lawler [51] for a detailed account.

Section 5.9. Martingale techniques in general and the Burkholder–Davis–Gundy inequalities in particular have been found to play an important role in the analysis of singular integrals where sharp, dimension free estimates can be obtained, see the survey [4] by Bañuelos.

Section 5.10. Girsanov's theorem is another pillar of mathematical finance. As illustrated in the Lévy area formula exercise (Exercise 5.77) it may be used to compute distributions of Brownian functionals. This formula for the Fourier transform of the area swept out by the planar Brownian motion is due to Lévy [53]. An independent and analytic proof can be found in Gaveau [31]. For many other computations related to the stochastic area formula we refer to the book by Yor [80].

Chapter 6
Stochastic differential equations and Malliavin calculus

In this chapter we study the stochastic differential equations. These are the equations associated to the Itô integral. Stochastic differential equations are a great device to study the existence of a Markov process with a given generator, a problem studied in Chapter 4. In the first part of the chapter, we study existence and uniqueness questions and then prove that solutions of stochastic differential equations are Markov processes. The second part of the chapter is an introduction to Malliavin calculus which is the Sobolev regularity theory of functionals defined on the Wiener space. Malliavin calculus is applied to show the existence of a smooth density for solutions of stochastic differential equations under an ellipticity assumption.

6.1 Existence and uniqueness of solutions

As usual, we consider a filtered probability space $(\Omega, (\mathcal{F}_t)_{t\geq 0}, \mathcal{F}, \mathbb{P})$ which satisfies the usual conditions and on which there is defined an n-dimensional Brownian motion $(B_t)_{t\geq 0}$. Let $b\colon \mathbb{R}^n \to \mathbb{R}^n$ and $\sigma\colon \mathbb{R}^n \to \mathbb{R}^{n\times n}$ be functions.

Theorem 6.1. *Let us assume that there exists $C > 0$ such that*

$$\|b(x) - b(y)\| + \|\sigma(x) - \sigma(y)\| \leq C\|x - y\|, \quad x, y \in \mathbb{R}^n.$$

Then, for every $x_0 \in \mathbb{R}^n$, there exists a unique continuous and adapted process $(X_t^{x_0})_{t\geq 0}$ such that for $t \geq 0$,

$$X_t^{x_0} = x_0 + \int_0^t b(X_s^{x_0})ds + \int_0^t \sigma(X_s^{x_0})dB_s. \tag{6.1}$$

Moreover, for every $T \geq 0$, we have

$$\mathbb{E}(\sup_{0\leq s\leq T} \|X_s\|^2) < +\infty.$$

Proof. Let us first observe that from our assumptions there exists $K > 0$ such that

(1) $\|b(x) - b(y)\| + \|\sigma(x) - \sigma(y)\| \leq K\|x - y\|$, $x, y \in \mathbb{R}^n$;

(2) $\|b(x)\| + \|\sigma(x)\| \leq K(1 + \|x\|)$, $x \in \mathbb{R}^n$.

The idea is to apply a fixed point theorem in a convenient Banach space. For $T > 0$, let us consider the space $\mathcal{E}_T$ of continuous and adapted processes such that

$$\mathbb{E}(\sup_{0\le s\le T} \|X_s\|^2) < +\infty$$

endowed with the norm

$$\| X \|^2 = \mathbb{E}(\sup_{0\le s\le T} \|X_s\|^2).$$

It is easily seen that $(\mathcal{E}_T, \|\cdot\|)$ is a Banach space.

Step 1. We first prove that if a continuous and adapted process $(X_t^{x_0})_{t\ge 0}$ is a solution of the equation (6.1) then, for every $T > 0$, $(X_t^{x_0})_{0\le t\le T} \in \mathcal{E}_T$.

Let us fix $T > 0$ and consider for $n \in \mathbb{N}$ the stopping times

$$T_n = \inf\{t \ge 0, \|X_t^{x_0}\| > n\},$$

For $t \le T$,

$$X_{t\wedge T_n}^{x_0} = x_0 + \int_0^{t\wedge T_n} b(X_s^{x_0})ds + \int_0^{t\wedge T_n} \sigma(X_s^{x_0})dB_s.$$

Therefore, by using the inequality

$$\|a+b+c\|^2 \le 3(\|a\|^2 + \|b\|^2 + \|c\|^2),$$

we get

$$\|X_{t\wedge T_n}^{x_0}\|^2 \le 3\Big(\|x_0\|^2 + \Big\|\int_0^{t\wedge T_n} b(X_s^{x_0})ds\Big\|^2 + \Big\|\int_0^{t\wedge T_n} \sigma(X_s^{x_0})dB_s\Big\|^2\Big).$$

Thus, we have

$$\begin{aligned}\mathbb{E}(\sup_{0\le u\le t\wedge T_n} \|X_u^{x_0}\|^2) \le 3\Big(\|x_0\|^2 &+ \mathbb{E}\Big(\sup_{0\le u\le t\wedge T_n} \Big\|\int_0^{u\wedge T_n} b(X_s^{x_0})ds\Big\|^2\Big) \\ &+ \mathbb{E}\Big(\sup_{0\le u\le t\wedge T_n} \Big\|\int_0^{u\wedge T_n} \sigma(X_s^{x_0})dB_s\Big\|^2\Big)\Big)\end{aligned}$$

By using our assumptions, we first estimate

$$\mathbb{E}\Big(\sup_{0\le u\le t\wedge T_n} \Big\|\int_0^{u\wedge T_n} b(X_s^{x_0})ds\Big\|^2\Big) \le K^2\mathbb{E}\Big(\Big(\int_0^{t\wedge T_n} (1+\|X_s^{x_0}\|)ds\Big)^2\Big).$$

By using our assumptions and Doob's inequality, we now estimate

$$\mathbb{E}\Big(\sup_{0\le u\le t\wedge T_n} \Big\|\int_0^{u\wedge T_n} \sigma(X_s^{x_0})dB_s\Big\|^2\Big) \le 4K^2\mathbb{E}\Big(\int_0^{t\wedge T_n} (1+\|X_s\|)^2 ds\Big).$$

Therefore, from the inequality $\|a+b\|^2 \leq 2(\|a\|^2+\|b\|^2)$ we get

$$\mathbb{E}\big(\sup_{0\leq u\leq t\wedge T_n} \|X_u^{x_0}\|^2\big)$$
$$\leq 3\Big(\|x_0\|^2 + 2(K^2T+4K^2)\int_0^t \big(1+\mathbb{E}\big(\sup_{0\leq u\leq s\wedge T_n} \|X_{u\wedge T_n}^{x_0}\|^2\big)ds\big)\Big)$$

We may now apply Gronwall's lemma to the function

$$t \to \mathbb{E}\big(\sup_{0\leq u\leq t\wedge T_n} \|X_u^{x_0}\|^2\big)$$

and deduce

$$\mathbb{E}\big(\sup_{0\leq u\leq T\wedge T_n} \|X_u^{x_0}\|^2\big) \leq C,$$

where C is a constant that does not depend on n. Fatou's lemma implies by passing to the limit when $n \to +\infty$ that

$$\mathbb{E}\big(\sup_{0\leq u\leq T} \|X_u^{x_0}\|^2\big) \leq C.$$

We conclude, as expected, that

$$(X_t^{x_0})_{0\leq t\leq T} \in \mathcal{E}_T.$$

More generally, by using the same arguments we can observe that if a continuous and adapted process satisfies

$$X_t = X_0 + \int_0^t b(X_s)ds + \int_0^t \sigma(X_s)dB_s$$

with $\mathbb{E}(X_0^2) < +\infty$, then $(X_t)_{0\leq t\leq T} \in \mathcal{E}_T$.

Step 2. We now show existence and uniqueness of solutions for equation (6.1) on a time interval $[0,T]$ where T is small enough.

Let us consider the mapping Φ that sends a continuous and adapted process $(X_t)_{0\leq t\leq T}$ to the process

$$\Phi(X)_t = x_0 + \int_0^t b(X_s)ds + \int_0^t \sigma(X_s)dB_s.$$

By using successively the inequalities $(a+b)^2 \leq 2(a^2+b^2)$, the Cauchy–Schwarz inequality and Doob's inequality, we get

$$\| \Phi(X) - \Phi(Y) \|^2 \leq 2(K^2T^2 + 4K^2T) \| X - Y \|^2 .$$

Moreover, arguing in the same way as above, we can prove

$$\| \Phi(0) \|^2 \leq 3(x_0^2 + K^2 T^2 + 4K^2 T).$$

Therefore, if T is small enough, Φ is a Lipschitz map $\mathcal{E}_T \to \mathcal{E}_T$ whose Lipschitz constant is strictly less than 1. Consequently, it has a unique fixed point. This fixed point is, of course the unique solution of (6.1) on the time interval $[0, T]$. Here again, we can observe that the same reasoning applies if x_0 is replaced by a random variable X_0 that satisfies $\mathbb{E}(X_0^2) < +\infty$.

Step 3. In order to get a solution of (6.1) on $[0, +\infty)$, we may apply the previous step to get a solution on intervals $[T_n, T_{n+1}]$, where $T_{n+1} - T_n$ is small enough and $T_n \to +\infty$. This will provide a solution of (6.1) on $[0, +\infty)$. This solution is unique, from the uniqueness on each interval $[T_n, T_{n+1}]$. □

Definition 6.2. An equation like (6.1) is called a *stochastic differential equation.*

Remark 6.3. As it appears from the very definition, solutions of stochastic differential equations are constructed by using stochastic integrals and thus probabilistic techniques. A theory developed in recent years, the rough paths theory, allows us to give a pathwise interpretation of solutions of differential equations driven by rough signals, including the Brownian motion. We refer the reader to Chapter 7 for an overview of this theory.

Exercise 6.4. Let $(X_t^{x_0})_{t \geq 0}$ denote the solution of (6.1). Show that under the same assumptions as in the previous theorem, we actually have for every $T \geq 0$ and $p \geq 1$,

$$\mathbb{E}(\sup_{0 \leq s \leq T} \|X_s\|^p) < +\infty.$$

Exercise 6.5 (Ornstein–Uhlenbeck process). Let $\theta \in \mathbb{R}$. We consider the following stochastic differential equation:

$$dX_t = \theta X_t dt + dB_t, \quad X_0 = x.$$

(1) Show that it admits a unique solution that is given by

$$X_t = e^{\theta t} x + \int_0^t e^{\theta(t-s)} dB_s.$$

(2) Show that $(X_t)_{t \geq 0}$ is Gaussian process. Compute its mean and covariance function.

(3) Show that if $\theta < 0$ then, when $t \to +\infty$, X_t converges in distribution toward a Gaussian distribution.

Exercise 6.6 (Brownian bridge). We consider for $0 \le t < 1$ the following stochastic differential equation:

$$dX_t = -\frac{X_t}{1-t} dt + dB_t, \quad X_0 = 0.$$

(1) Show that

$$X_t = (1-t) \int_0^t \frac{dB_s}{1-s}.$$

(2) Deduce that $(X_t)_{t \ge 0}$ is Gaussian process. Compute its mean and covariance function.

(3) Show that in L^2, when $t \to 1$, $X_t \to 0$.

Exercise 6.7 (Black–Scholes process). Let $\mu \in \mathbb{R}$ and $\sigma > 0$. We consider the following stochastic differential equation:

$$dX_t = \mu X_t dt + \sigma X_t dB_t, \quad X_0 = x > 0.$$

Show that for every $t \ge 0$,

$$X_t = x e^{\sigma B_t + (\mu - \frac{\sigma^2}{2})t}.$$

Exercise 6.8 (Explosion time). Let $b\colon \mathbb{R}^n \to \mathbb{R}^n$ and $\sigma\colon \mathbb{R}^n \to \mathbb{R}^{n \times n}$ be locally Lipschitz functions, that is, for every $N \ge 0$, there exists $C_N > 0$ such that

$$\|b(x) - b(y)\| + \|\sigma(x) - \sigma(y)\| \le C_N \|x - y\|, \quad x, y \in [-N, N]^n.$$

Show that for every $x \in \mathbb{R}^n$ we can find a stopping time $\boldsymbol{e}(x)$, almost surely positive, and a stochastic process $(X_t^x)_{t < \boldsymbol{e}(x)}$ such that

$$X_t^x = x + \int_0^t b(X_s^x) ds + \int_0^t \sigma(X_s^x) dB_s, \quad t < \boldsymbol{e}(x). \tag{6.2}$$

The process $(X_t^x)_{t < \boldsymbol{e}(x)}$ is unique in the sense that if $\boldsymbol{f}(x)$ is an almost surely positive stopping time and if $(Y_t^x)_{t < \boldsymbol{f}(x)}$ is a stochastic process such that

$$Y_t^x = x + \int_0^t b(Y_s^x) ds + \int_0^t \sigma(Y_s^x) dB_s, \quad t < \boldsymbol{f}(x),$$

then $\boldsymbol{f}(x) \le \boldsymbol{e}(x)$ and for every $t \ge 0$, $Y_t^x \mathbf{1}_{t < \boldsymbol{f}(x)} = X_t^x \mathbf{1}_{t < \boldsymbol{f}(x)}$ almost surely.

The process $(X_t^x)_{t < \boldsymbol{e}(x)}$ is called the solution of the stochastic differential equation (6.2) up to the explosion time $\boldsymbol{e}(x)$.

The next proposition shows that solutions of stochastic differential equations are intrinsically related to a second order differential operator, which will turn out to be the generator of the Markov process $(X_t^x)_{t\geq 0}$.

Proposition 6.9. *Let $(X_t^x)_{t\geq 0}$ be the solution of a stochastic differential equation*

$$X_t^{x_0} = x_0 + \int_0^t b(X_s^{x_0})ds + \int_0^t \sigma(X_s^{x_0})dB_s,$$

where $b\colon \mathbb{R}^n \to \mathbb{R}^n$ and $\sigma\colon \mathbb{R}^n \to \mathbb{R}^{n\times n}$ are Borel functions. Let now $f : \mathbb{R}^n \to \mathbb{R}^n$ be a C^2 function. The process

$$M_t^f = f(X_t^x) - \int_0^t Lf(X_s^x)ds$$

is a local martingale, where L is the second order differential operator

$$L = \sum_{i=1}^n b_i(x)\frac{\partial}{\partial x_i} + \frac{1}{2}\sum_{i,j=1}^n a_{ij}(x)\frac{\partial^2}{\partial x_i \partial x_j}$$

and $a_{ij}(x) = (\sigma(x)\sigma^*(x))_{ij}$.

Proof. The proof is as an easy application of the Döblin–Itô formula applied to $f(X_t^x)$. □

6.2 Continuity and differentiability of stochastic flows

In this section, we study the regularity of the solution of a stochastic differential equation with respect to its initial condition. The key tool is a multi-dimensional parameter extension of the Kolmogorov continuity theorem whose proof is almost identical to the one-dimensional case and left as an exercise to the reader.

Theorem 6.10. *Let $(\Theta_x)_{x\in[0,1]^d}$ be an n-dimensional stochastic process such that there exist positive constants γ, c, ε such that, for every $x, y \in [0,1]^d$,*

$$\mathbb{E}(\|\Theta_x - \Theta_y\|^\gamma) \leq C\|x-y\|^{d+\varepsilon}.$$

There exists a modification $(\widetilde{\Theta}_x)_{x\in[0,1]^d}$ of the process $(\Theta_x)_{x\in[0,1]^d}$ such that for every $\alpha \in [0, \varepsilon/\gamma)$ there exists a finite random variable K_α such that for every $x, y \in [0,1]^d$,

$$\|\widetilde{\Theta}_x - \widetilde{\Theta}_y\| \leq K_\alpha \|x-y\|^\alpha.$$

As above, we consider two functions $b\colon \mathbb{R}^n \to \mathbb{R}^n$ and $\sigma : \mathbb{R}^{n\times n}$ and we assume that there exists $C > 0$ such that

$$\|b(x) - b(y)\| + \|\sigma(x) - \sigma(y)\| \leq C\|x-y\|, \quad x, y \in \mathbb{R}^n.$$

As we already know, for every $x \in \mathbb{R}^n$, there exists a continuous and adapted process $(X_t^x)_{t \geq 0}$ such that for $t \geq 0$,

$$X_t^x = x + \int_0^t b(X_s^x) ds + \int_0^t \sigma(X_s^x) dB_s. \tag{6.3}$$

Proposition 6.11. *Let $T > 0$. For every $p \geq 2$, there exists a constant $C_{p,T} > 0$ such that for every $0 \leq s \leq t \leq T$ and $x, y \in \mathbb{R}^n$,*

$$\mathbb{E}(\|X_t^x - X_s^y\|^p) \leq C_{p,T}(\|x - y\|^p + |t - s|^{p/2}).$$

As a consequence, there exists a modification $(\widetilde{X}_t^x)_{t \geq 0, x \in \mathbb{R}^n}$ of the process $(X_t^x)_{t \geq 0, x \in \mathbb{R}^n}$ such that for $t \geq 0$, $x \in \mathbb{R}^n$,

$$\widetilde{X}_t^x = x + \int_0^t b(\widetilde{X}_s^x) ds + \int_0^t \sigma(\widetilde{X}_s^x) dB_s$$

and such that $(t, x) \to X_t^x(\omega)$ is continuous for almost every ω.

Proof. As before, we can find $K > 0$ such that

(1) $\|b(x) - b(y)\| + \|\sigma(x) - \sigma(y)\| \leq K\|x - y\|$, $x, y \in \mathbb{R}^n$;

(2) $\|b(x)\| + \|\sigma(x)\| \leq K(1 + \|x\|)$, $x \in \mathbb{R}^n$.

We fix $x, y \in \mathbb{R}^n$ and $p \geq 2$. Let

$$h(t) = \mathbb{E}(\|X_t^x - X_t^y\|^p).$$

By using the inequality $\|a + b + c\|^p \leq 3^{p-1}(\|a\|^p + \|b\|^p + \|c\|^p)$, we obtain

$$\begin{aligned}\|X_t^x - X_t^y\|^p \leq 3^{p-1}\Big(&\|x - y\|^p + \Big(\int_0^t \|b(X_s^x) - b(X_s^y)\| ds\Big)^p \\ &+ \Big\|\int_0^t (\sigma(X_s^x) - \sigma(X_s^y)) dB_s\Big\|^p\Big).\end{aligned}$$

We now have

$$\begin{aligned}\Big(\int_0^t \|b(X_s^x) - b(X_s^y)\| ds\Big)^p &\leq t^{p-1} \int_0^t \|b(X_s^x) - b(X_s^y)\|^p ds \\ &\leq K^p t^{p-1} \int_0^t \|X_s^x - X_s^y\|^p ds,\end{aligned}$$

and from the Burkholder–Davis–Gundy inequality

$$\begin{aligned}\mathbb{E}\Big(\Big\|\int_0^t (\sigma(X_s^x) - \sigma(X_s^y)) dB_s\Big\|^p\Big) &\leq C_p \mathbb{E}\Big(\Big(\int_0^t \|\sigma(X_s^x) - \sigma(X_s^y)\|^2 ds\Big)^{p/2}\Big) \\ &\leq C_p K^2 \mathbb{E}\Big(\Big(\int_0^t \|X_s^x - X_s^y\|^2 ds\Big)^{p/2}\Big) \\ &\leq C_p K^2 t^{p/2-1} \mathbb{E}\Big(\int_0^t \|X_s^x - X_s^y\|^p ds\Big).\end{aligned}$$

As a conclusion we obtain

$$h(t) \leq 3^{p-1}\Big(\|x-y\|^p + (K^p t^{p-1} + C_p K^2 t^{p/2-1}) \int_0^t h(s)ds\Big).$$

Gronwall's inequality then yields

$$h(t) \leq \phi(t)\|x-y\|^p,$$

where ϕ is a continuous function.

On the other hand, for $0 \leq s \leq t \leq T$ we have

$$\|X_t^x - X_s^x\|^p \leq 2^{p-1}\Big(\Big\|\int_s^t b(X_u^x)ds\Big\|^p + \Big\|\int_s^t \sigma(X_u^x)dB_u\Big\|^p\Big)$$

and

$$\Big\|\int_s^t b(X_u^x)ds\Big\|^p \leq K^p(t-s)^p(1+\sup_{0\leq s\leq T}\|X_s\|)^p,$$

$$\begin{aligned}\mathbb{E}\Big(\Big\|\int_s^t \sigma(X_u^x)dB_u\Big\|^p\Big) &\leq C_p\mathbb{E}\Big(\Big(\int_s^t \|\sigma(X_u^x)\|^2du\Big)^{p/2}\Big)\\ &\leq C_p K^p (t-s)^{p/2}\mathbb{E}((1+\sup_{0\leq s\leq T}\|X_s\|)^p).\end{aligned}$$

The conclusion then easily follows by combining the two previous estimates. □

In the sequel, of course, we shall always work with this bicontinuous version of the solution.

Definition 6.12. The continuous process of continuous maps $\Psi_t : x \to X_t^x$ is called the *stochastic flow* associated to the equation (6.3).

If the maps b and σ are moreover C^1, then the stochastic flow is itself differentiable and the equation for the derivative can be obtained by formally differentiating the equation with respect to its initial condition. We state this result without giving a proof (for a proof see [47]):

Theorem 6.13. *Let us assume that b and σ are C^1 Lipschitz functions. Then for every $t \geq 0$, the flow Ψ_t associated to the equation* (6.3) *is a flow of differentiable maps. Moreover, the first variation process $\boldsymbol{J}_t(x)$, which is defined by the Jacobian matrix $\frac{\partial \Psi_t}{\partial x}(x)$, is the unique solution of the matrix stochastic differential equation:*

$$\boldsymbol{J}_t(x) = \mathrm{Id} + \int_0^t \frac{\partial b}{\partial x}(X_s^x)\boldsymbol{J}_s(x)ds + \sum_{i=1}^n \int_0^t \frac{\partial \sigma_i}{\partial x}(X_s^x)\boldsymbol{J}_s(x)dB_s^i.$$

Exercise 6.14. For $x \in \mathbb{R}^n$, let $(\boldsymbol{K}_t(x))_{t\geq 0}$ be the unique solution of the matrix linear stochastic differential equation

$$\boldsymbol{K}_t(x) = \mathrm{Id} + \sum_{i=1}^n \int_0^t \boldsymbol{K}_s(x)\left(\frac{\partial \sigma_i}{\partial x}\right)^2 (X_s^x)ds - \int_0^t \boldsymbol{K}_s(x)\frac{\partial b}{\partial x}(X_s^x)ds$$
$$- \sum_{i=1}^n \int_0^t \boldsymbol{K}_s(x)\frac{\partial \sigma_i}{\partial x}(X_s^x)dB_s^i.$$

Show that the process $(\boldsymbol{K}_t(x)\boldsymbol{J}_t(x))_{t\geq 0}$ is constant and equal to Id. As a consequence, the Jacobian matrix $\boldsymbol{J}_t(x)$ is always almost surely invertible.

6.3 The Feynman–Kac formula

It is now time to give some applications of the theory of stochastic differential equations to parabolic second order partial differential equations. In particular we are going to prove that solutions of such equations can represented by using solutions of stochastic differential equations. This representation formula is called the Feynman–Kac formula. As usual, we consider a filtered probability space $(\Omega, (\mathcal{F}_t)_{t\geq 0}, \mathcal{F}, \mathbb{P})$ which satisfies the usual conditions and on which there is defined an n-dimensional Brownian motion $(B_t)_{t\geq 0}$. Again, we consider two functions $b\colon \mathbb{R}^n \to \mathbb{R}^n$ and $\sigma\colon \mathbb{R}^n \to \mathbb{R}^{n\times n}$ and we assume that there exists $C > 0$ such that

$$\|b(x) - b(y)\| + \|\sigma(x) - \sigma(y)\| \leq C\|x - y\|, \quad x, y \in \mathbb{R}^n.$$

Let L be the diffusion operator

$$L = \sum_{i=1}^n b_i(x)\frac{\partial}{\partial x_i} + \frac{1}{2}\sum_{i,j=1}^n a_{ij}(x)\frac{\partial^2}{\partial x_i \partial x_j},$$

where $a_{ij}(x) = (\sigma(x)\sigma^*(x))_{ij}$.

As we know, there exists a bicontinuous process $(X_t^x)_{t\geq 0, x\in\mathbb{R}^d}$ such that for $t \geq 0$,

$$X_t^x = x + \int_0^t b(X_s^x)ds + \int_0^t \sigma(X_s^x)dB_s.$$

Moreover, as it has been stressed before, for every $p \geq 1$ and $T \geq 0$,

$$\mathbb{E}(\sup_{0\leq t\leq T} \|X_t^x\|^p) < +\infty.$$

As a consequence, if $f\colon \mathbb{R}^n \to \mathbb{R}$ is a Borel function with polynomial growth, we can consider the function

$$\boldsymbol{P}_t f(x) = \mathbb{E}(f(X_t^x)).$$

Theorem 6.15. *For every $x \in \mathbb{R}^n$, $(X_t^x)_{t\geq 0, x\in\mathbb{R}^d}$ is a Markov process with semigroup $(\boldsymbol{P}_t)_{t\geq 0}$. More precisely, for every Borel function $f : \mathbb{R}^n \to \mathbb{R}$ with polynomial growth and every $t \geq s$,*

$$\mathbb{E}(f(X_t^x) \mid \mathcal{F}_s) = (\boldsymbol{P}_{t-s} f)(X_s^x).$$

Proof. The key point here is to observe that solutions are actually adapted to the natural filtration of the Brownian motion $(B_t)_{t\geq 0}$. More precisely, on the space of continuous functions $[0, +\infty) \to \mathbb{R}^n$ there exists a predictable functional such that for $t \geq 0$,

$$X_t^{x_0} = F(x_0, (B_u)_{0\leq u\leq t}).$$

Indeed, let us first work on $[0, T]$ where T is small enough. In this case, as seen previously, the process $(X_t^{x_0})_{0\leq t\leq T}$ is the unique fixed point of the mapping Φ defined by

$$\Phi(X)_t = x_0 + \int_0^t b(X_s^{x_0})ds + \int_0^t \sigma(X_s^{x_0})dB_s.$$

Alternatively, one can interpret this by observing that $(X_t^{x_0})_{0\leq t\leq T}$ is the limit of the sequence of processes $(X_t^n)_{0\leq t\leq T}$ inductively defined by

$$X^{n+1} = \Phi(X^n), \quad X^0 = x_0.$$

It is easily checked that for each X^n there is a predictable functional F_n such that

$$X_t^n = F_n(x_0, (B_u)_{0\leq u\leq t}),$$

which proves the above claim when T is small enough. To get the existence of F for any T, we can proceed analogously on the intervals $[T, 2T]$, $[2T, 3T]$ and so on.

With this hands, we can now prove the Markov property. Let $s \geq 0$. For $t \geq 0$, we have

$$\begin{aligned} X_{s+t}^{x_0} &= X_s + \int_s^{s+t} b(X_u^{x_0})du + \int_s^{s+t} \sigma(X_u^{x_0})dB_u \\ &= X_s + \int_0^t b(X_{u+s}^{x_0})du + \int_0^t \sigma(X_s^{x_0})d(B_{u+s} - B_s). \end{aligned}$$

Consequently, from the uniqueness of solutions,

$$X_{s+t}^{x_0} = F(X_s^{x_0}, (B_{u+s} - B_s)_{0\leq u\leq t}).$$

We deduce that for a Borel function $f : \mathbb{R}^n \to \mathbb{R}$ with polynomial growth,

$$\mathbb{E}(f(X_{s+t}^{x_0}) \mid \mathcal{F}_s) = \mathbb{E}(f(F(X_s^{x_0}, (B_{u+s} - B_s)_{0\leq u\leq t})) \mid \mathcal{F}_s) = \boldsymbol{P}_t f(X_s^{x_0})$$

because $(B_{u+s} - B_s)_{0\leq u\leq t}$ is a Brownian motion independent of $\mathcal{F}_s$. □

If the coefficients b and σ are bounded, then solutions of stochastic differential equations are more than Markov processes, they are Feller–Dynkin diffusion processes in the sense of Definition 3.33, Chapter 3.

Proposition 6.16. *If the functions b and σ are bounded, then, for every $x_0 \in \mathbb{R}^n$, $(X_t^{x_0})_{t\geq 0}$ is a Feller–Dynkin diffusion process with infinitesimal generator*

$$Lf = \frac{1}{2}\sum_{i,j=1}^{n} a_{ij}(x)\frac{\partial^2 f}{\partial x_i \partial x_j} + \sum_{i=1}^{n} b_i(x)\frac{\partial f}{\partial x_i}, \quad f \in \mathcal{C}_c(\mathbb{R}^n, \mathbb{R}).$$

Proof. We first need to prove that $(\boldsymbol{P}_t)_{t\geq 0}$ is a Feller–Dynkin semigroup. Let $f \in \mathcal{C}_0(\mathbb{R}^n, \mathbb{R})$. First, it is clear that the function $\boldsymbol{P}_t$ is continuous. We now have for $x \in \mathbb{R}^n$ and $r > 0$,

$$|\boldsymbol{P}_t f(x)| \leq \sup_{y\in \mathrm{B}(x,r)} |f(y)| + \|f\|_\infty \mathbb{P}(\|X_t^x - x\| > r)$$

and

$$\begin{aligned}
\mathbb{P}(\|X_t^x - x\| > r) &\leq \frac{1}{r^2}\mathbb{E}(\|X_t^x - x\|^2) \\
&\leq \frac{2}{r^2}\Big(\mathbb{E}\Big(\Big\|\int_0^t b(X_s^x)ds\Big\|^2\Big) + \mathbb{E}\Big(\int_0^t \|\sigma(X_s^x)\|^2 ds\Big)\Big) \\
&\leq \frac{2M}{r^2}(t + t^2),
\end{aligned}$$

where M is a common upper bound for b and σ. By letting x and then r to ∞, we conclude that $P_t f \in \mathcal{C}_0(\mathbb{R}^n, \mathbb{R})$. Thus $\boldsymbol{P}_t$ is a Feller–Dynkin semigroup. Let us compute its generator on $\mathcal{C}_c(\mathbb{R}^n, \mathbb{R})$. Let $f \in \mathcal{C}_c(\mathbb{R}^n, \mathbb{R})$. From Itô's formula we have

$$f(X_t^x) = f(x) + \int_0^t Lf(X_s^x)ds + M_t,$$

where M_t is a martingale. As a consequence, by taking expectations, we obtain

$$\boldsymbol{P}_t f(x) = f(x) + \int_0^t \boldsymbol{P}_s Lf(x)ds.$$

This implies that

$$\lim_{t\to 0}\left\|\frac{\boldsymbol{P}_t f - f}{t} - Lf\right\|_\infty = 0.$$

Thus the domain of the generator of $\boldsymbol{P}_t$ contains $\mathcal{C}_c(\mathbb{R}^n, \mathbb{R})$ and on that space is equal to L. □

Theorem 6.17. *Let $f : \mathbb{R}^n \to \mathbb{R}$ be a Borel function with polynomial growth and assume that the function*

$$u(t,x) = (\boldsymbol{P}_t f)(x)$$

is $C^{1,2}$, that is, once differentiable with respect to t and twice differentiable with respect to x. Then u solves the Cauchy problem

$$\frac{\partial u}{\partial t}(t,x) = Lu(t,x)$$

in $[0,+\infty) \times \mathbb{R}^n$, with the initial condition

$$u(0,x) = f(x).$$

Proof. Let $T > 0$ and consider the function $v(t,x) = u(T-t,x)$. According the previous theorem, we have

$$\mathbb{E}(f(X_T^x) \mid \mathcal{F}_t) = v(t, X_t^x).$$

As a consequence, the process $v(t, X_t^x)$ is a martingale. But from Döblin–Itô's formula the bounded variation part of $v(t, X_t^x)$ is $\int_0^t \big(\frac{\partial v}{\partial t}(s, X_s^x) + Lv(s, X_s^x)\big)ds$ which is therefore 0. We conclude that

$$\frac{\partial v}{\partial t}(0,x) + Lv(0,x) = \lim_{t \to 0} \frac{1}{t} \int_0^t \Big(\frac{\partial v}{\partial t}(s, X_s^x) + Lv(s, X_s^x)\Big)ds = 0. \qquad \square$$

Exercise 6.18. Show that if f is a C^2 function such that ∇f and $\nabla^2 f$ have polynomial growth, then the function $\boldsymbol{P}_t f(x)$ is $C^{1,2}$. Here, we denote by $\nabla^2 f$ the Hessian matrix of f.

Theorem 6.19. *Let $f : \mathbb{R}^n \to \mathbb{R}$ be a Borel function with polynomial growth. Let $u : [0,+\infty) \times \mathbb{R}^n \to \mathbb{R}$ be a solution of the Cauchy problem*

$$\frac{\partial u}{\partial t}(t,x) = Lu(t,x)$$

with the initial condition

$$u(0,x) = f(x).$$

If there exists a locally integrable function C and $p \geq 0$ such that for every $t \geq 0$ and $x \in \mathbb{R}^n$,

$$\|\nabla u(t,x)\| \leq C(t)(1 + \|x\|^p),$$

then $u(t,x) = \boldsymbol{P}_t f(x)$.

Proof. Let $T > 0$ and, as before, consider the function $v(t,x) = u(T-t,x)$. As a consequence of Itô's formula, we have

$$v(t, X_t^x) = u(T,x) + M_t,$$

where M_t is a local martingale with quadratic variation

$$\sum_{i,j}^{n} \int_0^t a_{ij}(X_s^x) \frac{\partial u}{\partial x_i}(X_s^x) \frac{\partial u}{\partial x_j}(X_s^x) ds.$$

The conditions on σ and u imply that this quadratic variation is integrable. As a consequence, $v(t, X_t^x)$ is a martingale and thus $\mathbb{E}(v(T, X_t^x)) = u(T,x)$. □

The previous results may be extended to study parabolic equations with potential as well. More precisely, let $V : \mathbb{R}^n \to \mathbb{R}$ be a bounded function. If $f : \mathbb{R}^n \to \mathbb{R}$ is a Borel function with polynomial growth, we define

$$\boldsymbol{P}_t^V f(x) = \mathbb{E}(e^{\int_0^t V(X_s^x) ds} f(X_t^x)).$$

The same proofs as before will give the following theorems.

Theorem 6.20. *For every $x \in \mathbb{R}^n$ and every Borel function $f : \mathbb{R}^n \to \mathbb{R}$ with polynomial growth and every $t \geq s$,*

$$\mathbb{E}(e^{\int_0^t V(X_u^x) du} f(X_t^x) \mid \mathcal{F}_s) = e^{\int_0^s V(X_u^x) du} (\boldsymbol{P}_{t-s}^V f)(X_s^x).$$

Theorem 6.21 (Feynman–Kac formula). *Let $f : \mathbb{R}^n \to \mathbb{R}$ be a Borel function with polynomial growth and assume that the function*

$$u(t,x) = (\boldsymbol{P}_t^V f)(x)$$

is $C^{1,2}$, that is, once differentiable with respect to t and twice differentiable with respect to x. Then u solves the Cauchy problem

$$\frac{\partial u}{\partial t}(t,x) = Lu(t,x) + V(x)u(t,x)$$

in $[0,+\infty) \times \mathbb{R}^n$, with the initial condition

$$u(0,x) = f(x).$$

Theorem 6.22. *Let $f : \mathbb{R}^n \to \mathbb{R}$ be a Borel function with polynomial growth. Let $u : [0,+\infty) \times \mathbb{R}^n \to \mathbb{R}$ be a solution of the Cauchy problem*

$$\frac{\partial u}{\partial t}(t,x) = Lu(t,x) + V(x)u(t,x)$$

with the initial condition

$$u(0,x) = f(x).$$

If there exists a locally integrable function C and $p \geq 0$ such that for every $t \geq 0$ and $x \in \mathbb{R}^n$,

$$\|\nabla u(t,x)\| \leq C(t)(1 + \|x\|^p),$$

then $u(t,x) = \boldsymbol{P}_t^V f(x)$.

Exercise 6.23 (Arcsine law). Let $(B_t)_{t \geq 0}$ be a standard Brownian motion and let $\lambda, \alpha \in \mathbb{R}$. We write

$$\nu(x) = \mathbb{E}\Big(\int_0^{+\infty} \exp\Big(-\lambda t - \alpha \int_0^t \mathbf{1}_{[0,+\infty)}(B_s + x) ds \Big) dt \Big).$$

(1) Show that ν is the unique solution of the differential equation

$$y'' - (\alpha \mathbf{1}_{[0,+\infty)} + \lambda) y = 1$$

that satisfies

$$\lim_{x \to +\infty} \nu(x) = \frac{1}{\alpha + \lambda}, \quad \lim_{x \to -\infty} \nu(x) = \frac{1}{\lambda}.$$

(2) Deduce that

$$\int_0^{+\infty} e^{-\lambda t} \mathbb{E}(e^{-\alpha t A_t}) dt = \frac{1}{\lambda} \frac{\lambda + \sqrt{\lambda(\alpha + \lambda)}}{\alpha + \lambda + \sqrt{\lambda(\alpha + \lambda)}},$$

where $A_t = \frac{1}{t} \int_0^t \mathbf{1}_{[0,+\infty)}(B_s) ds$.

(3) Conclude that the density of A_t is given by

$$s \to \frac{1}{\pi \sqrt{s(1-s)}}, \quad s \in (0,1).$$

This is the arc-sine law for Brownian motion paths that was already proved in Theorem 2.51, Chapter 2, by using random walks.

Exercise 6.24 (Exponential functional). Let $(B_t)_{t \geq 0}$ be a standard Brownian motion and $\mu > 0$. We write

$$A_\infty = \int_0^{+\infty} e^{-2(B_t + \mu t)} dt.$$

(1) Let $h(x) = e^{\mu x}\mathbb{E}(\exp(-\frac{1}{2}e^{-2x}A_\infty))$. Show that h is the unique solution of the differential equation

$$\left(\frac{d^2}{dx^2} - e^{-2x}\right)h = \mu^2 h$$

such that $h(x) \sim_{x\to+\infty} e^{\mu x}$.

(2) Deduce that

$$h(x) = \frac{2^{1-\mu}}{\Gamma(\mu)}K_\mu(e^{-x}),$$

where K_μ is the McDonald function

$$K_\mu(x) = \frac{1}{2}\left(\frac{x}{2}\right)^\mu \int_0^{+\infty} \frac{e^{-\frac{x^2}{4t}-t}}{t^{1+\mu}}dt.$$

(3) Compute the density of A_∞.

(4) Consider the process $(\rho_t)_{t\geq 0}$ such that for every $t \geq 0$,

$$e^{-B_t-\mu t} = \rho_{\int_0^t e^{-2(B_s+\mu s)}ds}.$$

Show that the process

$$\rho_t - \int_0^t \frac{-\mu+\frac{1}{2}}{\rho_s}ds$$

is a Brownian motion. The process $(\rho_t)_{t\geq 0}$ is called a Bessel process (see Exercises 3.8 and 5.59).

(5) By using A_∞, compute the density of the stopping time

$$T_0 = \inf\{t \geq 0,\ \rho_t = 0\}.$$

6.4 The strong Markov property of solutions

In the previous section, we have seen that if $(X_t^x)_{t\geq 0}$ is the solution of a stochastic differential equation

$$X_t^x = x + \int_0^t b(X_s^x)ds + \int_0^t \sigma(X_s^x)dB_s,$$

then $(X_t^x)_{t\geq 0}$ is a Markov process, that is, for every $t, T \geq 0$,

$$\mathbb{E}(f(X_{t+T}^x) \mid \mathcal{F}_T) = (\boldsymbol{P}_t f)(X_T^x),$$

where $\boldsymbol{P}_t f(x) = \mathbb{E}(f(X_t^x))$. It is remarkable that this property still holds when T is now any finite stopping time; That is solutions of stochastic differential equations enjoy the strong Markov property. The key lemma is Proposition 3.20 in Chapter 3.

Theorem 6.25. *For every $x \in \mathbb{R}^n$, $(X_t^x)_{t\geq 0, x\in\mathbb{R}^d}$ is a strong Markov process with semigroup $(P_t)_{t\geq 0}$: For every Borel function $f : \mathbb{R}^n \to \mathbb{R}$ with polynomial growth, every $t \geq 0$, and every finite stopping time T,*

$$\mathbb{E}(f(X_{t+T}^x) \mid \mathcal{F}_T) = (\boldsymbol{P}_t f)(X_T^x).$$

Proof. The proof is identical to the proof of Theorem 6.15 with the additional ingredient given by Proposition 3.20 in Chapter 2. □

Exercise 6.26. Let $b\colon \mathbb{R}^n \to \mathbb{R}^n$ and $\sigma\colon \mathbb{R}^n \to \mathbb{R}^{n\times n}$ be locally Lipschitz functions and consider the solution of the stochastic differential equation

$$X_t^x = x + \int_0^t b(X_s^x)ds + \int_0^t \sigma(X_s^x)dB_s,$$

up to the explosion time $\boldsymbol{e}(x)$ (see Exercise 6.8). Show that $(X_t)_{t<\boldsymbol{e}(x)}$ is a sub-Markov process with semigroup

$$\boldsymbol{P}_t f(x) = \mathbb{E}(f(X_t^x)\mathbf{1}_{t<\boldsymbol{e}(x)}).$$

The strong Markov property for solutions of stochastic differential equations is useful to solve boundary value problems in the theory of partial differential equations. Let K be a bounded closed set in $\mathbb{R}^n$. For $x \in \Omega$, we write $T_x = \inf\{t \geq 0, X_t \in \partial K\}$. If f is a bounded Borel function such that $f_{\partial K} = 0$, we define

$$\boldsymbol{P}_t^K f(x) = \mathbb{E}(f(X_t^x)\mathbf{1}_{t\leq T_x}).$$

The proof of the following theorem is left to the reader.

Theorem 6.27. *Let $f : K \to \mathbb{R}$ be a bounded Borel function and assume that the function*

$$u(t,x) = (\boldsymbol{P}_t^K f)(x)$$

is $C^{1,2}$. Then u is the unique solution of the Dirichlet boundary value problem

$$\frac{\partial u}{\partial t}(t,x) = Lu(t,x)$$

in $[0,+\infty) \times \mathbb{R}^n$, with the initial condition

$$u(0,x) = f(x),$$

and the boundary condition

$$u(t,x) = 0, \quad x \in \partial K.$$

6.5 Stratonovitch stochastic differential equations and the language of vector fields

Again let $(\Omega, (\mathcal{F}_t)_{t\geq 0}, \mathbb{P})$ be a filtered probability space which satisfies the usual conditions. It is often useful to use the language of Stratonovitch's integration to study stochastic differential equations because Itô's formula takes a much nicer form. If $(N_t)_{0\leq t\leq T}$, $T > 0$, is an $\mathcal{F}$-adapted real-valued local martingale and if $(\Theta_t)_{0\leq t\leq T}$ is an $\mathcal{F}$-adapted continuous semimartingale then the Stratonovitch integral of $(\Theta_t)_{0\leq t\leq T}$ with respect to $(N_t)_{t\geq 0}$ is given as

$$\int_0^T \Theta_t \circ dN_t = \int_0^T \Theta_t dN_t + \frac{1}{2}\langle \Theta, N\rangle_T,$$

where

(1) $\int_0^T \Theta_t dN_t$ is the Itô integral of $(\Theta_t)_{0\leq t\leq T}$ against $(N_t)_{0\leq t\leq T}$;

(2) $\langle \Theta, N\rangle_T$ is the quadratic covariation at time T between $(\Theta_t)_{0\leq t\leq T}$ and $(N_t)_{0\leq t\leq T}$.

By using the Stratonovitch integral instead of Itô's, we can see that the Döblin–Itô formula reduces to the classical change of variable formula.

Theorem 6.28. *Let $(X_t)_{t\geq 0} = (X_t^1, \dots, X_t^n)_{t\geq 0}$ be an n-dimensional continuous semimartingale. Let now $f : \mathbb{R}^n \to \mathbb{R}$ be a C^2 function. We have*

$$f(X_t) = f(X_0) + \sum_{i=1}^n \int_0^t \frac{\partial f}{\partial x_i}(X_s) \circ dX_s^i, \quad t \geq 0.$$

Let $\mathcal{O} \subset \mathbb{R}^n$ be a non-empty open set. A smooth vector field V on $\mathcal{O}$ is simply a smooth map

$$V : \mathcal{O} \to \mathbb{R}^n, \quad x \mapsto (v_1(x), \dots, v_n(x)).$$

The vector field V defines a differential operator acting on smooth functions $f : \mathcal{O} \to \mathbb{R}$ as follows:

$$Vf(x) = \sum_{i=1}^n v_i(x) \frac{\partial f}{\partial x_i}.$$

We note that V is a derivation, that is, a map on $\mathcal{C}^\infty(\mathcal{O}, \mathbb{R})$, linear over $\mathbb{R}$, satisfying for $f, g \in \mathcal{C}^\infty(\mathcal{O}, \mathbb{R})$,

$$V(fg) = (Vf)g + f(Vg).$$

Interestingly, conversely, any derivation on $\mathcal{C}^\infty(\mathcal{O}, \mathbb{R})$ is a vector field.

Let now

$$(B_t)_{t\geq 0} = (B_t^1, \dots, B_t^d)_{t\geq 0}$$

be a d-dimensional Brownian motion and consider $d+1$ C^1 vector fields $V_i : \mathbb{R}^n \to \mathbb{R}^n$, $n \geq 1$, $i = 0, \dots, d$. By using the language of vector fields and Stratonovitch integrals, the fundamental theorem for the existence and the uniqueness of solutions for stochastic differential equations is the following:

Theorem 6.29. *Assume that $V_0, V_1, \dots, V_d$ are C^2 vector fields with bounded derivatives up to order 2. Let $x_0 \in \mathbb{R}^n$. On $(\Omega, (\mathcal{F}_t)_{t\geq 0}, \mathbb{P})$, there exists a unique continuous and adapted process $(X_t^{x_0})_{t\geq 0}$ such that for $t \geq 0$,*

$$X_t^{x_0} = x_0 + \sum_{i=0}^{d} \int_0^t V_i(X_s^{x_0}) \circ dB_s^i, \tag{6.4}$$

with the convention that $B_t^0 = t$.

Due to the Döblin–Itô formula, Itô's formulation of a Stratonovitch equation is

$$X_t^{x_0} = x_0 + \frac{1}{2} \sum_{i=1}^{d} \int_0^t \nabla_{V_i} V_i(X_s^{x_0}) ds + \sum_{i=0}^{d} \int_0^t V_i(X_s^{x_0}) dB_s^i,$$

where for $1 \leq i \leq d$, $\nabla_{V_i} V_i$ is the vector field given by

$$\nabla_{V_i} V_i(x) = \sum_{j=1}^{n} \Big(\sum_{k=1}^{n} v_i^k(x) \frac{\partial v_i^j}{\partial x_k}(x) \Big) \frac{\partial}{\partial x_j}, \qquad x \in \mathbb{R}^n.$$

If $f : \mathbb{R}^n \to \mathbb{R}$ is a C^2 function, from Itô's formula we have for $t \geq 0$,

$$f(X_t^{x_0}) = f(x_0) + \sum_{i=0}^{d} \int_0^t (V_i f)(X_s^{x_0}) \circ dB_s^i,$$

and the process

$$\Big(f(X_t^{x_0}) - \int_0^t (Lf)(X_s^{x_0}) ds \Big)_{t \geq 0}$$

is a local martingale where L is the second order differential operator

$$L = V_0 + \frac{1}{2} \sum_{i=1}^{d} V_i^2.$$

Exercise 6.30. Let $(B_t^1, B_t^2)_{t\geq 0}$ be a two-dimensional Brownian motion. Show that the process

$$X_t = \Big(B_t^1, B_t^2, \frac{1}{2} \int_0^t B_s^1 dB_s^2 - B_s^2 dB_s^1 \Big)_{t \geq 0}$$

solves a stochastic differential equation that may be written

$$dX_t = V_1(X_t) \circ dB_t^1 + V_2(X_t) \circ dB_t^2,$$

where V_1, V_2 are two vector fields to be computed.

Exercise 6.31. Show that the diffusion operator

$$L = V_0 + \frac{1}{2} \sum_{i=1}^{n} V_i^2$$

is elliptic in $\mathbb{R}^n$ if and only if for every $x \in \mathbb{R}^n$, the vectors $V_1(x), \dots, V_n(x)$ form a basis of $\mathbb{R}^n$.

A general theorem by Phillips and Sarason states that if L is a diffusion operator

$$L = \frac{1}{2} \sum_{i,j=1}^{n} a_{ij}(x) \frac{\partial^2}{\partial x_i \partial x_j} + \sum_{i=1}^{n} b_i(x) \frac{\partial}{\partial x_i},$$

where the coefficients a_{ij}'s and b_i's are two times continuously differentiable, then L can be written as

$$L = V_0 + \frac{1}{2} \sum_{i=1}^{n} V_i^2,$$

where $V_1, \dots, V_n$' are locally Lipschitz vector fields. In this book we are mostly interested in elliptic operators. In this case we have the following stronger result:

Proposition 6.32. *Let*

$$L = \frac{1}{2} \sum_{i,j=1}^{n} a_{ij}(x) \frac{\partial^2}{\partial x_i \partial x_j} + \sum_{i=1}^{n} b_i(x) \frac{\partial}{\partial x_i}$$

be a diffusion operator on $\mathbb{R}^n$ such that the a_{ij}'s and the b_i's are smooth functions. Let us assume that for every $x \in \mathbb{R}^n$, the rank of the matrix $(a_{ij}(x))_{1 \le i,j \le n}$ is equal to n. Then there exist smooth vector fields $V_0, V_1, \dots, V_n$ on $\mathbb{R}^n$ such that $V_1, \dots, V_n$ are linearly independent and

$$L = V_0 + \sum_{i=1}^{n} V_i^2.$$

Proof. Since the matrix $(a_{ij}(x))_{1 \le i,j \le n}$ is symmetric and positive it admits a unique symmetric and positive square root $v(x) = (v_{ij}(x))_{1 \le i,j \le n}$. Let us assume for a moment that the v_{ij}'s are smooth functions. In this case by writing

$$V_i = \sum_{j=1}^{n} v_{ij} \frac{\partial}{\partial x_j},$$

the vector fields $V_1, \dots, V_n$ are linearly independent and it is readily seen that the differential operator

$$L - \frac{1}{2}\sum_{i=1}^{n} V_i^2$$

is actually a first order differential operator and thus a vector field. We therefore are led to prove that the v_{ij}'s are smooth functions. Let $\mathcal{O}$ be a bounded non-empty subset of $\mathbb{R}^n$ and Γ be any contour in the half plane $\mathrm{Re}(z) > 0$ that contains all the eigenvalues of $\sigma(x)$, $x \in \mathcal{O}$. We claim that

$$v(x) = \frac{1}{2\pi}\int_\Gamma \sqrt{z}(\sigma(x) - z\mathrm{I}_n)^{-1} dz, \quad x \in \mathcal{O}.$$

Indeed, if Γ' is another contour in the half plane $\mathrm{Re}(z) > 0$ whose interior contains Γ, as a straightforward application of Fubini's theorem and Cauchy's formula we have

$$\begin{aligned}
&\frac{1}{2\pi}\int_\Gamma \sqrt{z}(\sigma(x) - z\mathrm{I}_n)^{-1} dz \times \frac{1}{2\pi}\int_{\Gamma'} \sqrt{z}(\sigma(x) - z\mathrm{I}_n)^{-1} dz \\
&= \frac{1}{4\pi^2}\int_\Gamma\int_{\Gamma'} \sqrt{zz'}(\sigma(x) - z\mathrm{I}_n)^{-1}(\sigma(x) - z'\mathrm{I}_n)^{-1} dz dz' \\
&= \frac{1}{4\pi^2}\int_\Gamma\int_{\Gamma'} \sqrt{zz'}\frac{(\sigma(x) - z\mathrm{I}_n)^{-1} - (\sigma(x) - z'\mathrm{I}_n)^{-1}}{z - z'} dz dz' \\
&= -\frac{1}{4\pi^2}\int_\Gamma\int_{\Gamma'} \sqrt{zz'}\frac{(\sigma(x) - z\mathrm{I}_n)^{-1}}{z' - z} dz' dz \\
&= -\frac{1}{4\pi^2}\int_\Gamma (\sigma(x) - z\mathrm{I}_n)^{-1}\Big(\int_{\Gamma'} \frac{\sqrt{zz'}}{z' - z} dz'\Big) dz \\
&= \frac{1}{2i\pi}\int_\Gamma (\sigma(x) - z\mathrm{I}_n)^{-1} z dz \\
&= \frac{1}{2i\pi}\int_\Gamma (\sigma(x) - z\mathrm{I}_n)^{-1} \sigma(x) dz.
\end{aligned}$$

In the last expression above, we may modify Γ into a circle $\Gamma_R = \{z, |z| = R\}$. Then by choosing R big enough ($R > \sup_{x\in\mathcal{O}} \sqrt{\|\sigma(x)\|}$), and expanding $\int_\Gamma (\sigma(x) - z\mathrm{I}_n)^{-1} dz$ in powers of z, we see that

$$\frac{1}{2i\pi}\int_{\Gamma_R} (\sigma(x) - z\mathrm{I}_n)^{-1} dz = \sigma(x).$$

As a conclusion,

$$\Big(\frac{1}{2\pi}\int_\Gamma \sqrt{z}(\sigma(x) - z\mathrm{I}_n)^{-1} dz\Big)^2 = \sigma(x),$$

so that, as we claimed,

$$v(x) = \frac{1}{2\pi} \int_\Gamma \sqrt{z}(\sigma(x) - z\mathrm{I}_n)^{-1} dz, \quad x \in \mathcal{O}.$$

This expression of the square root of σ clearly shows that the v_{ij}'s are smooth functions. □

Exercise 6.33. Prove the following extension of the above theorem. Let

$$L = \frac{1}{2} \sum_{i,j=1}^{n} a_{ij}(x) \frac{\partial^2}{\partial x_i \partial x_j} + \sum_{i=1}^{n} b_i(x) \frac{\partial}{\partial x_i},$$

be a diffusion operator on $\mathbb{R}^n$ such that the a_{ij}'s and the b_i's are smooth functions. Let us assume that for every $x \in \mathbb{R}^n$, the rank of the matrix $(a_{ij}(x))_{1 \le i,j \le n}$ is equal to a constant $d \le n$. Then there exist smooth vector fields $V_0, V_1, \dots, V_d$ on $\mathbb{R}^n$ such that $V_1, \dots, V_d$ are linearly independent and

$$L = V_0 + \frac{1}{2} \sum_{i=1}^{d} V_i^2.$$

6.6 Malliavin calculus

This section is an introduction to the techniques of the so-called Malliavin calculus. Our eventual goal will be to prove that if $(X_t^x)_{t \ge 0}$ is the solution of a stochastic differential equation

$$X_t^x = x + \int_0^t b(X_s^x) ds + \int_0^t \sigma(X_s^x) dB_s,$$

where $b \colon \mathbb{R}^n \to \mathbb{R}^n$ and $\sigma \colon \mathbb{R}^n \to \mathbb{R}^{n \times n}$ are smooth functions with bounded derivatives, and if the matrix $\sigma(x)$ is uniformly elliptic, then for every $t > 0$ the random variable X_t^x has a smooth density with respect to the Lebesgue measure (see Theorem 6.51).

This Theorem 6.51 actually provides a probabilistic proof of the hypoellipticity of elliptic operators (see Appendix B for the relevant definitions). Indeed, let L be the diffusion operator

$$L = \sum_{i=1}^{n} b_i(x) \frac{\partial}{\partial x_i} + \frac{1}{2} \sum_{i,j=1}^{n} a_{ij}(x) \frac{\partial^2}{\partial x_i \partial x_j},$$

where $a_{ij}(x) = (\sigma(x)\sigma^*(x))_{ij}$. If L is elliptic, then Theorem 6.51 implies that for every $x \in \mathbb{R}^n$ and $t > 0$ the random variable X_t^x has a smooth density with respect to the Lebesgue measure. In other words, there exists a smooth kernel $p(t,x,y)$ such that for every bounded Borel function f,

$$\mathbb{E}(f(X_t^x)) = \boldsymbol{P}_t f(x) = \int_{\mathbb{R}^n} p(t,x,y) f(y).$$

Combining this fact with the interpretation of $\boldsymbol{P}_t f$ as the solution of a Cauchy problem, we see that the operator $L - \frac{\partial}{\partial t}$ needs to be hypoelliptic which implies the hypoellipticity of L. The fact that elliptic operators are hypoelliptic is a special case of a celebrated theorem by Hörmander and the motivation of Malliavin to introduce his calculus was actually precisely to re-prove Hörmander's result by using probabilistic techniques (see [58]).

We also remark that Theorem 6.51 implies Theorem 4.23 that we proved in Chapter 4 in the case where L is essentially self-adjoint. It should be observed that the advantage of the probabilistic method is that L is not required to be symmetric with respect to any measure, however strong assumptions are required on the coefficients b and σ.

6.6.1 The Malliavin derivative

In this section, we introduce the basic tools of Malliavin calculus, which is a set of tools devoted to the study of the Sobolev regularity of Brownian functionals. This is just an introduction to this theory. The interested reader is referred to the classical book [60] by Nualart for a detailed presentation of the theory.

Let us consider a filtered probability space $(\Omega, (\mathcal{F}_t)_{0\le t\le 1}, \mathbb{P})$ on which there is defined an n-dimensional Brownian motion $(B_t)_{0\le t\le 1}$. We assume that $(\mathcal{F}_t)_{0\le t\le 1}$ is the usual completion of the natural filtration of $(B_t)_{0\le t\le 1}$.

An $\mathcal{F}_1$ measurable real-valued random variable F is said to be cylindric if it can be written

$$F = f\Big(\int_0^1 h^1(s)dB_s, \dots, \int_0^1 h^m(s)dB_s\Big),$$

where $h^i \in L^2([0,1], \mathbb{R}^n)$ and $f\colon \mathbb{R}^m \to \mathbb{R}$ is a C^∞ function such that f and all its partial derivatives have polynomial growth. The set of cylindric random variables is denoted by $\mathcal{S}$. It is easy to see that $\mathcal{S}$ is dense in $L^p(\mathcal{F}_1, \mathbb{P})$ for every $p \ge 1$.

The Malliavin derivative of $F \in \mathcal{S}$ is the $\mathbb{R}^n$-valued stochastic process $(\boldsymbol{D}_t F)_{0\le t\le 1}$ given by

$$\boldsymbol{D}_t F = \sum_{i=1}^m h^i(t) \frac{\partial f}{\partial x_i}\Big(\int_0^1 h^1(s)dB_s, \dots, \int_0^1 h^m(s)dB_s\Big).$$

We can regard $\boldsymbol{D}$ as an (unbounded) operator from the space $\mathcal{S} \subset L^p(\mathcal{F}_1, \mathbb{P})$ into the Banach space

$$\mathcal{L}^p = \left\{(X_t)_{0\le t\le 1}, \mathbb{E}\left(\left(\int_0^1 \|X_t\|^2 dt\right)^p\right) < +\infty\right\}.$$

Our first task will be to prove that $\boldsymbol{D}$ is closable (see Appendix A for the definition of closability). This will be a consequence of the following fundamental integration by parts formula which is interesting in itself.

Proposition 6.34 (Integration by parts formula). *Let $F \in \mathcal{S}$ and let $(h(s))_{0\le s\le 1}$ be a progressively measurable process such that $\mathbb{E}\big(\int_0^1 \|h(s)\|^2 ds\big) < +\infty$. We have*

$$\mathbb{E}\left(\int_0^1 (\boldsymbol{D}_s F) h(s) ds\right) = \mathbb{E}\left(F \int_0^1 h(s) dB_s\right).$$

Proof. Let

$$F = f\left(\int_0^1 h^1(s) dB_s, \dots, \int_0^1 h^m(s) dB_s\right) \in \mathcal{S}.$$

Let us now fix $\varepsilon \ge 0$ and write

$$F_\varepsilon = f\left(\int_0^1 h^1(s) d\left(B_s + \varepsilon \int_0^s h(u) du\right), \dots, \int_0^1 h^m(s) d\left(B_s + \varepsilon \int_0^s h(u) du\right)\right).$$

From Girsanov's theorem (Theorem 5.72 in Chapter 3), we have

$$\mathbb{E}(F_\varepsilon) = \mathbb{E}\left(\exp\left(\varepsilon \int_0^1 h(u) dB_u - \frac{\varepsilon^2}{2} \int_0^1 \|h(u)\|^2 du\right) F\right).$$

Now on one hand we compute

$$\begin{aligned} &\lim_{\varepsilon \to 0} \frac{1}{\varepsilon}(\mathbb{E}(F_\varepsilon) - \mathbb{E}(F)) \\ &\quad = \mathbb{E}\left(\int_0^1 \sum_{i=1}^m \frac{\partial f}{\partial x_i}\left(\int_0^1 h^1(s) dB_s, \dots, \int_0^1 h^m(s) dB_s\right) h^i(s) h(s) dt\right) \\ &\quad = \mathbb{E}\left(\int_0^1 (\boldsymbol{D}_s F) h(s) dt\right), \end{aligned}$$

and on the other hand, we obtain

$$\lim_{\varepsilon \to 0} \frac{1}{\varepsilon}(\mathbb{E}(F_\varepsilon) - \mathbb{E}(F)) = \mathbb{E}\left(F \int_0^1 h(s) dB_s\right). \qquad \square$$

Proposition 6.35. *Let $p \ge 1$. As a densely defined operator from $L^p(\mathcal{F}_1, \mathbb{P})$ into $\mathcal{L}^p$, $\boldsymbol{D}$ is closable.*

Proof. Let $(F_n)_{n\in\mathbb{N}}$ be a sequence in $\mathcal{S}$ that converges in $L^p(\mathcal{F}_1, \mathbb{P})$ to 0 and such that $\boldsymbol{D} F_n$ converges in $\mathcal{L}^p$ to X. We want to prove that $X = 0$. Let $(h(s))_{0\le s\le 1}$ be a function in $L^2([0,1])$. Let us first assume $p > 1$. We have

$$\lim_{n\to\infty} \mathbb{E}\Big(\int_0^1 (\boldsymbol{D}_s F_n) h(s) ds\Big) = \mathbb{E}\Big(\int_0^1 X_s h(s) ds\Big)$$

and

$$\lim_{n\to\infty} \mathbb{E}\Big(F_n \int_0^1 h(s) dB_s\Big) = 0.$$

As a consequence we obtain

$$\mathbb{E}\Big(\int_0^1 X_s h(s) ds\Big) = 0.$$

Since h is arbitrary, we conclude $X = 0$. Let us now assume $p = 1$. Let η be a smooth and compactly supported function and let $\Theta = \eta(\int_0^1 h(s) dB_s) \in \mathcal{S}$. We have

$$\boldsymbol{D}(F_n \Theta) = F_n(\boldsymbol{D}\Theta) + (\boldsymbol{D} F_n)\Theta.$$

As a consequence we get

$$\begin{aligned}\mathbb{E}\Big(\int_0^1 &\boldsymbol{D}_s(F_n\Theta) h(s) ds\Big) \\ &= \mathbb{E}\Big(F_n \int_0^1 (\boldsymbol{D}_s\Theta) h(s) ds\Big) + \mathbb{E}\Big(\Theta \int_0^1 (\boldsymbol{D}_s F_n) h(s) ds\Big),\end{aligned}$$

and thus

$$\lim_{n\to\infty} \mathbb{E}\Big(\int_0^1 \boldsymbol{D}_s(F_n\Theta) h(s) ds\Big) = \mathbb{E}\Big(\Theta \int_0^1 X_s h(s) ds\Big).$$

On the other hand, we have

$$\mathbb{E}\Big(\int_0^1 \boldsymbol{D}_s(F_n\Theta) h(s) ds\Big) = \mathbb{E}\Big(F_n\Theta \int_0^1 h(s) dB_s\Big) \xrightarrow[n\to\infty]{} 0.$$

We conclude that

$$\mathbb{E}\Big(\Theta \int_0^1 X_s h(s) ds\Big) = 0. \qquad \square$$

The closure of $\boldsymbol{D}$ in $L^p(\mathcal{F}_1, \mathbb{P})$ shall still be denoted by $\boldsymbol{D}$. Its domain $\mathbb{D}^{1,p}$ is the closure of $\mathcal{S}$ with respect to the norm

$$\|F\|_{1,p} = (\mathbb{E}(F^p) + \mathbb{E}(\|\boldsymbol{D} F\|^p_{L^2([0,1],\mathbb{R}^n)}))^{\frac{1}{p}}.$$

For $p > 1$, we can consider the adjoint operator δ of $\boldsymbol{D}$. This is a densely defined operator $\mathcal{L}^q \to L^p(\mathcal{F}_1, \mathbb{P})$ with $1/p + 1/q = 1$ which is characterized by the duality formula

$$\mathbb{E}(F\delta u) = \mathbb{E}\Big(\int_0^1 (\boldsymbol{D}_s F) u_s ds\Big), \quad F \in \mathbb{D}^{1,p}.$$

From Proposition 6.34 and Burkholder–Davis–Gundy inequalities, it is clear that the domain of δ in $\mathcal{L}^q$ contains the set of progressively measurable processes $(u_t)_{0 \le t \le 1}$ such that $\mathbb{E}((\int_0^1 \|u_s\|^2 ds)^{q/2}) < +\infty$, and that in this case

$$\delta u = \int_0^1 u_s dB_s.$$

The operator δ can thus be thought of as an extension of the Itô integral. It is often called the Skohorod integral or the divergence operator.

Exercise 6.36 (Clark–Ocone formula). Show that for $F \in \mathbb{D}^{1,2}$,

$$F = \mathbb{E}(F) + \int_0^1 \mathbb{E}(\boldsymbol{D}_1 F \mid \mathcal{F}_t) dB_t.$$

More generally, we can introduce iterated derivatives. If $F \in \mathcal{S}$, we set

$$\boldsymbol{D}^k_{t_1,\dots,t_k} F = \boldsymbol{D}_{t_1} \dots \boldsymbol{D}_{t_k} F.$$

We may consider $\boldsymbol{D}^k F$ as a square integrable random process indexed by $[0,1]^k$ and valued in $\mathbb{R}^n$. For any $p \ge 1$, the operator $\boldsymbol{D}^k$ is closable on $\mathcal{S}$. We denote by $\mathbb{D}^{k,p}$ the closure of the class of cylindric random variables with respect to the norm

$$\|F\|_{k,p} = \Big(\mathbb{E}(F^p) + \sum_{j=1}^k \mathbb{E}(\|\boldsymbol{D}^j F\|^p_{L^2([0,1]^j,\mathbb{R}^n)})\Big)^{\frac{1}{p}},$$

and

$$\mathbb{D}^\infty = \bigcap_{p \ge 1} \bigcap_{k \ge 1} \mathbb{D}^{k,p}.$$

6.6.2 The Wiener chaos expansion

As in the previous section, we consider a filtered probability space $(\Omega, (\mathcal{F}_t)_{0\le t\le 1}, \mathbb{P})$ on which there is defined a Brownian motion $(B_t)_{0\le t\le 1}$, and we assume that $(\mathcal{F}_t)_{0\le t\le 1}$ is the usual completion of the natural filtration of $(B_t)_{0\le t\le 1}$. Our goal is here to write an orthogonal decomposition of the space $L^2(\mathcal{F}_1, \mathbb{P})$ that is particularly suited to the study of the space $\mathbb{D}^{1,2}$. For simplicity of the exposition, we restrict ourselves to the case where the Brownian motion $(B_t)_{0\le t\le 1}$ is one-dimensional but similar results obviously hold in higher dimensions provided the suitable changes are made.

In the sequel, for $n \ge 1$, we denote by Δ_n the simplex

$$\Delta_n = \{0 \le t_1 \le \cdots \le t_n \le 1\}$$

and if $f_n \in L^2(\Delta_n)$,

$$\begin{aligned} I_n(f_n) &= \int_0^1 \int_0^{t_n} \cdots \int_0^{t_2} f_n(t_1,\dots,t_n)dB_{t_1}\dots dB_{t_n} \\ &= \int_{\Delta_n} f_n(t_1,\dots,t_n)dB_{t_1}\dots dB_{t_n}. \end{aligned}$$

The set

$$\boldsymbol{K}_n = \left\{ \int_{\Delta_n} f_n(t_1,\dots,t_n)dB_{t_1}\dots dB_{t_n},\ f_n \in L^2(\Delta_n) \right\}$$

is called the space of chaos of order n. By convention the set of constant random variables shall be denoted by $\boldsymbol{K}_0$.

By using Itô's isometry, we readily compute that

$$\mathbb{E}(I_n(f_n)I_p(f_p)) = \begin{cases} 0 & \text{if } p \ne n, \\ \|f_n\|^2_{L^2(\Delta_n)} & \text{if } p = n. \end{cases}$$

As a consequence, the spaces $\boldsymbol{K}_n$ are orthogonal in $L^2(\mathcal{F}_1, \mathbb{P})$. It is easily seen that $\boldsymbol{K}_n$ is the closure of the linear span of the family

$$\{I_n(f^{\otimes n}),\ f \in L^2([0,1])\},$$

where for $f \in L^2([0,1])$, we denoted by $f^{\otimes n}$ the map $\Delta_n \to \mathbb{R}$ such that $f^{\otimes n}(t_1,\dots,t_n) = f(t_1)\dots f(t_n)$. It turns out that $I_n(f^{\otimes n})$ can be computed by using Hermite polynomials. The Hermite polynomial of order n is defined by

$$H_n(x) = (-1)^n \frac{1}{n!} e^{\frac{x^2}{2}} \frac{d^n}{dx^n} e^{-\frac{x^2}{2}}.$$

By the very definition of H_n, we see that for every $t, x \in \mathbb{R}$,

$$\exp\left(tx - \frac{t^2}{2}\right) = \sum_{k=0}^{+\infty} t^k H_k(x).$$

Lemma 6.37. *If $f \in L^2([0,1])$ then*

$$I_n(f^{\otimes n}) = \|f\|^n_{L^2([0,1])} H_n\left(\frac{\int_0^1 f(s)dB_s}{\|f\|_{L^2([0,1])}}\right).$$

Proof. On one hand, we have for $\lambda \in \mathbb{R}$,

$$\exp\Big(\lambda \int_0^1 f(s)dB_s - \frac{\lambda^2}{2}\int_0^1 f(s)^2 ds\Big) = \sum_{n=0}^{+\infty} \lambda^n \|f\|^n_{L^2([0,1])} H_n\Bigg(\frac{\int_0^1 f(s)dB_s}{\|f\|_{L^2([0,1])}}\Bigg).$$

On the other hand, for $0 \le t \le 1$, let us consider

$$M_t(\lambda) = \exp\Big(\lambda \int_0^t f(s)dB_s - \frac{\lambda^2}{2}\int_0^t f(s)^2 ds\Big).$$

From Itô's formula, we have

$$M_t(\lambda) = 1 + \lambda \int_0^t M_s f(s) dB_s.$$

By iterating the previous linear relation, we easily obtain that for every $n \ge 1$,

$$\begin{aligned} M_1(\lambda) = 1 &+ \sum_{k=1}^{n} \lambda^k I_k(f^{\otimes k}) \\ &+ \lambda^{n+1} \int_0^1 M_t f(t) \Big(\int_{\Delta_n([0,t])} f(t_1)\dots f(t_n) dB_{t_1}\dots dB_{t_n}\Big) dB_t. \end{aligned}$$

We conclude that

$$I_n(f^{\otimes n}) = \frac{1}{n!}\frac{d^k M_1}{d\lambda^n}(0) = \|f\|^n_{L^2([0,1])} H_n\Bigg(\frac{\int_0^1 f(s)dB_s}{\|f\|_{L^2([0,1])}}\Bigg). \qquad \square$$

As we pointed out, for $p \ne n$, the spaces $\boldsymbol{K}_n$ and $\boldsymbol{K}_p$ are orthogonal. We have the following orthogonal decomposition of L^2:

Theorem 6.38 (Wiener chaos expansion).

$$L^2(\mathcal{F}_1, \mathbb{P}) = \bigoplus_{n\ge 0} \boldsymbol{K}_n.$$

Proof. As a by-product of the previous proof, we easily obtain that for $f \in L^2([0,1])$,

$$\exp\Big(\lambda \int_0^1 f(s)dB_s - \frac{\lambda^2}{2}\int_0^1 f(s)^2 ds\Big) = \sum_{n=1}^{+\infty} I_n(f^{\otimes n}),$$

where the convergence of the series is almost sure and also in $L^2(\mathcal{F}_1, \mathbb{P})$. Therefore, if $F \in L^2(\mathcal{F}_1, \mathbb{P})$ is orthogonal to $\bigoplus_{n\ge 1} \boldsymbol{K}_n$, then F is orthogonal to every $\exp\big(\lambda \int_0^1 f(s)dB_s - \frac{\lambda^2}{2}\int_0^1 f(s)^2 ds\big)$, $f \in L^2([0,1])$. This implies that $F = 0$. $\square$

As we are going to see, the space $\mathbb{D}^{1,2}$ or, more generally, $\mathbb{D}^{k,2}$ is easy to describe by using the Wiener chaos expansion. The keypoint is the following proposition:

Proposition 6.39. *Let $F = I_n(f_n) \in \boldsymbol{K}_n$. Then $F \in \mathbb{D}^{1,2}$ and*

$$\boldsymbol{D}_t F = I_{n-1}(\tilde{f}_n(\cdot, t)),$$

where for $0 \le t_1 \le \dots \le t_{n-1} \le 1$,

$$\tilde{f}_n(t_1, \dots, t_{n-1}, t) = f_n(t_1, \dots, t_k, t, t_{k+1}, \dots, t_{n-1}) \quad \text{if } t_k \le t \le t_{k+1}.$$

Proof. Let $f \in L^2([0,1])$. We have

$$I_n(f^{\otimes n}) = \|f\|^n_{L^2([0,1])} H_n\left(\frac{\int_0^1 f(s)dB_s}{\|f\|_{L^2([0,1])}}\right).$$

Thus $F = I_n(f^{\otimes n})$ is a smooth cylindric functional and

$$\boldsymbol{D}_t F = \|f\|^{n-1}_{L^2([0,1])} f(t) H_n'\left(\frac{\int_0^1 f(s)dB_s}{\|f\|_{L^2([0,1])}}\right).$$

It is easy to see that $H_n' = H_{n-1}$, therefore we have

$$\boldsymbol{D}_t F = \|f\|^{n-1}_{L^2([0,1])} f(t) H_{n-1}\left(\frac{\int_0^1 f(s)dB_s}{\|f\|_{L^2([0,1])}}\right) = f(t) I_{n-1}(f^{\otimes(n-1)}).$$

As a consequence we compute that

$$\mathbb{E}\Big(\int_0^1 (\boldsymbol{D}_t F)^2 dt\Big) = n\mathbb{E}(F^2).$$

We now observe that $\boldsymbol{K}_n$ is the closure in $L^2(\mathcal{F}_1, \mathbb{P})$ of the linear span of the family

$$\{I_n(f^{\otimes n}), f \in L^2([0,1])\}$$

to conclude the proof of the proposition. □

We can finally turn to the description of $\mathbb{D}^{1,2}$ using the chaos decomposition:

Theorem 6.40. *Let $F \in L^2(\mathcal{F}_1, \mathbb{P})$ and let*

$$F = \mathbb{E}(F) + \sum_{m \ge 1} I_m(f_m)$$

be the chaotic decomposition of F. Then $F \in \mathbb{D}^{1,2}$ if and only if

$$\sum_{m \ge 1} m\mathbb{E}(I_m(f_m)^2) < +\infty,$$

and in this case,

$$\boldsymbol{D}_t F = \mathbb{E}(\boldsymbol{D}_t F) + \sum_{m \geq 2} I_{m-1}(\tilde{f}_m(\cdot, t)).$$

Proof. It is a consequence of the fact that for $F \in \boldsymbol{K}_n$,

$$\mathbb{E}\Big(\int_0^1 (\boldsymbol{D}_t F)^2 dt\Big) = n\mathbb{E}(F^2). \qquad \square$$

An immediate but useful corollary of the previous theorem is the following result:

Corollary 6.41. *Let $(F_n)_{n \geq 0}$ be a sequence in $\mathbb{D}^{1,2}$ that converges to F in $L^2(\mathcal{F}_1, \mathbb{P})$ and such that*

$$\sup_{n \geq 0} \mathbb{E}\Big(\int_0^1 (\boldsymbol{D}_t F_n)^2 dt\Big) < +\infty.$$

Then $F \in \mathbb{D}^{1,2}$.

This corollary can actually be generalized in the following way (see [60] for a proof).

Proposition 6.42. *Let $(F_n)_{n \geq 0}$ be a sequence in $\mathbb{D}^{k,p}$, $k \geq 1$, $p > 1$, that converges to F in $L^p(\mathcal{F}_1, \mathbb{P})$ and such that*

$$\sup_{n \geq 0} \| F_n \|_{k,p} < +\infty.$$

Then $F \in \mathbb{D}^{k,p}$.

Exercise 6.43. Let $F \in L^2(\mathcal{F}_1, \mathbb{P})$ and let

$$F = \mathbb{E}(F) + \sum_{m \geq 1} I_m(f_m)$$

be the chaotic decomposition of F. Show that $F \in \mathbb{D}^{k,2}$ with $k \geq 1$ if and only if

$$\sum_{m \geq 1} m^k \mathbb{E}(I_m(f_m)^2) < +\infty.$$

Exercise 6.44. Let $L = \delta \boldsymbol{D}$. Show that for $F \in \boldsymbol{K}_n$, $LF = nF$. Deduce that the domain of L in $L^2(\mathcal{F}_1, \mathbb{P})$ is $\mathbb{D}^{2,2}$.

6.6.3 Regularity of probability distributions using Malliavin calculus

We have the following key result which makes Malliavin calculus so useful when one wants to study the existence of densities for random variables.

Theorem 6.45. *Let $F = (F_1, \dots, F_m)$ be an $\mathcal{F}_1$ measurable random vector such that*

(1) *for every $i = 1, \dots, m$, $F_i \in \mathbb{D}^\infty$;*

(2) *the matrix*

$$\Gamma = \left(\int_0^1 \langle \boldsymbol{D}_s F^i, \boldsymbol{D}_s F^j \rangle_{\mathbb{R}^n} ds \right)_{1 \le i,j \le m}$$

is invertible.

Then F has a density with respect to the Lebesgue measure. If moreover, for every $p > 1$,

$$\mathbb{E}\left(\frac{1}{|\det \Gamma|^p} \right) < +\infty,$$

then this density is smooth.

Remark 6.46. The matrix Γ is often called the Malliavin matrix of the random vector F.

This theorem relies on the following lemma from Fourier analysis for which we shall use the following notation: If $\phi \colon \mathbb{R}^n \to \mathbb{R}$ is a smooth function then for $\alpha = (i_1, \dots, i_k) \in \{1, \dots, n\}^k$, we write

$$\partial_\alpha \phi = \frac{\partial^k}{\partial x_{i_1} \dots \partial x_{i_k}} \phi.$$

Lemma 6.47. *Let μ be a probability measure on $\mathbb{R}^n$ such that for every smooth and compactly supported function $\phi \colon \mathbb{R}^n \to \mathbb{R}$,*

$$\left| \int_{\mathbb{R}^n} \partial_\alpha \phi d\mu \right| \le C_\alpha \|\phi\|_\infty,$$

where $\alpha \in \{1, \dots, n\}^k$, $k \ge 1$, $C_\alpha > 0$. Then μ is absolutely continuous with respect to the Lebesgue measure with a smooth density.

Proof. The idea is to show that we may assume that μ is compactly supported and then use Fourier transform techniques. Let $x_0 \in \mathbb{R}^n$, $R > 0$ and $R' > R$. Let Ψ be a smooth function on $\mathbb{R}^n$ such that $\Psi = 1$ on the ball $\mathrm{B}(x_0, R)$ and $\Psi = 0$ outside the ball $\mathrm{B}(x_0, R')$. Let ν be the measure on $\mathbb{R}^n$ that has a density Ψ with respect to

μ. It is easily seen by induction and integrating by parts that for every smooth and compactly supported function $\phi\colon \mathbb{R}^n \to \mathbb{R}$,

$$\left| \int_{\mathbb{R}^n} \partial_\alpha \phi d\nu \right| \leq C'_\alpha \|\phi\|_\infty,$$

where $\alpha \in \{1,\ldots,n\}^k$, $k \geq 1$, $C'_\alpha > 0$. Now, if we can prove that under the above assumption ν has a smooth density, then we will able to conclude that ϕ has a smooth density because $x_0 \in \mathbb{R}^n$ and R, R' are arbitrary. Let

$$\hat{\nu}(y) = \int_{\mathbb{R}^n} e^{i\langle y,x\rangle} \nu(dx)$$

be the Fourier transform of the measure μ. The assumption implies that $\hat{\nu}$ is rapidly decreasing (apply the inequality with $\phi(x) = e^{i\langle y,x\rangle}$). We conclude that ν has a smooth density with respect to the Lebesgue measure and that this density f is given by the inverse Fourier transform formula:

$$f(x) = \frac{1}{(2\pi)^n} \int_{\mathbb{R}^n} e^{-i\langle y,x\rangle} \hat{\nu}(y) dy. \qquad \square$$

We may now turn to the proof of Theorem 6.45.

Proof. The proof relies on an integration by parts formula which is interesting by itself. Let ϕ be a smooth and compactly supported function on $\mathbb{R}^n$. Since $F_i \in \mathbb{D}^\infty$, we easily deduce that $\phi(F) \in \mathbb{D}^\infty$ and that

$$\boldsymbol{D}\phi(F) = \sum_{i=1}^n \partial_i \phi(F) \boldsymbol{D} F_i.$$

Therefore we obtain

$$\int_0^1 \langle \boldsymbol{D}_t \phi(F), \boldsymbol{D}_t F_j \rangle dt = \sum_{i=1}^n \partial_i \phi(F) \int_0^1 \langle \boldsymbol{D}_t F_i, \boldsymbol{D}_t F_j \rangle dt.$$

We conclude that

$$\partial_i \phi(F) = \sum_{j=1}^n (\Gamma^{-1})_{i,j} \int_0^1 \langle \boldsymbol{D}_t \phi(F), \boldsymbol{D}_t F_j \rangle dt.$$

As a consequence we obtain

$$\begin{aligned}
\mathbb{E}(\partial_i \phi(F)) &= \mathbb{E}\Big(\sum_{j=1}^{n} (\Gamma^{-1})_{i,j} \int_0^1 \langle \boldsymbol{D}_t \phi(F), \boldsymbol{D}_t F_j \rangle dt\Big) \\
&= \sum_{j=1}^{n} \mathbb{E}\Big(\int_0^1 \langle \boldsymbol{D}_t \phi(F), (\Gamma^{-1})_{i,j} \boldsymbol{D}_t F_j \rangle dt\Big) \\
&= \sum_{j=1}^{n} \mathbb{E}(\phi(F) \delta((\Gamma^{-1})_{i,j} \boldsymbol{D} F_j)) \\
&= \mathbb{E}\Big(\phi(F) \delta\Big(\sum_{j=1}^{n} (\Gamma^{-1})_{i,j} \boldsymbol{D} F_j\Big)\Big).
\end{aligned}$$

Remark that the integration by parts is licit because $(\Gamma^{-1})_{i,j} \boldsymbol{D} F_j$ belongs to the domain of δ. Indeed, from our assumptions, $\boldsymbol{D} F_j$ is in $\mathbb{D}^{2,2}$ and thus in the domain of δ from Exercise 6.44. Also $(\Gamma^{-1})_{i,j} \in \mathbb{D}^{1,2}$ and $(\Gamma^{-1})_{i,j} \boldsymbol{D} F_j \in L^2(\mathcal{F}_1, \mathbb{P})$ which easily implies that $(\Gamma^{-1})_{i,j} \boldsymbol{D} F_j$ belongs to the domain of δ.

By using inductively this integration by parts formula, it is seen that for every $\alpha \in \{1, \dots, n\}^k$, $k \geq 1$, there exists an integrable random variable Z_α such that

$$\mathbb{E}(\partial_\alpha \phi(F)) = \mathbb{E}(\phi(F) Z_\alpha).$$

This finishes the proof. □

6.7 Existence of a smooth density

As usual, we consider a filtered probability space $(\Omega, (\mathcal{F}_t)_{t \geq 0}, \mathcal{F}, \mathbb{P})$ which satisfies the usual conditions and on which there is defined an n-dimensional Brownian motion $(B_t)_{t \geq 0}$. Our first purpose here is to prove that solutions of stochastic differential equations are differentiable in the sense of Malliavin. The following lemma is easy to prove by using the Wiener chaos expansion.

Lemma 6.48. *Let $(u_s)_{0 \leq s \leq 1}$ be a progressively measurable process such that for every $0 \leq s \leq 1$, $u_s^i \in \mathbb{D}^{1,2}$ and*

$$\mathbb{E}\Big(\int_0^1 \|u_s\|^2 ds\Big) < +\infty, \quad \mathbb{E}\Big(\int_0^1 \int_0^1 \|\boldsymbol{D}_s u_t\|^2 ds dt\Big) < +\infty.$$

Then $\int_0^1 u_s dB_s \in \mathbb{D}^{1,2}$ and

$$\boldsymbol{D}_t\Big(\int_0^1 u_s dB_s\Big) = u_t + \sum_{i=1}^{n} \int_t^1 (\boldsymbol{D}_t u_s^i) dB_s^i.$$

Proof. We make the proof when $n = 1$ and use the notations introduced in the Wiener chaos expansion section. For $f \in L^2([0,1])$, we have

$$\mathbf{D}_t I_n(f^{\otimes n}) = f(t) I_{n-1}(f^{\otimes(n-1)}).$$

But we can write,

$$I_n(f^{\otimes n}) = \int_0^1 f(t) \Big(\int_{\Delta_{n-1}[0,t]} f^{\otimes(n-1)} dB_{t_1} \dots dB_{t_{n-1}} \Big) dB_t,$$

and thus

$$I_n(f^{\otimes n}) = \int_0^1 u_s dB_s$$

with $u_t = f(t) \int_{\Delta_{n-1}[0,t]} f^{\otimes(n-1)} dB_{t_1} \dots dB_{t_{n-1}}$. Since

$$\begin{aligned} f(t) I_{n-1}(f^{\otimes(n-1)}) &= f(t) \Big(\int_{\Delta_{n-1}[0,t]} f^{\otimes(n-1)} dB_{t_1} \dots dB_{t_{n-1}} \Big) \\ &\quad + f(t) \int_t^1 f(s) \Big(\int_{\Delta_{n-2}[0,s]} f^{\otimes(n-1)} dB_{t_1} \dots dB_{t_{n-2}} \Big) dB_s, \end{aligned}$$

we get the result when $\int_0^1 u_s dB_s$ can be written as $I_n(f^{\otimes n})$. By continuity of the Malliavin derivative on the space of chaos of order n, we conclude that the formula is true if $\int_0^1 u_s dB_s$ is a chaos of order n. The result finally holds in all generality by using the Wiener chaos expansion. □

We consider two functions $b\colon \mathbb{R}^n \to \mathbb{R}^n$ and $\sigma\colon \mathbb{R}^n \to \mathbb{R}^{n \times n}$ and we assume that b and σ are C^∞ with derivatives at any order (greater than 1) bounded.

As we know, there exists a bicontinuous process $(X_t^x)_{t \ge 0, x \in \mathbb{R}^d}$ such that for $t \ge 0$,

$$X_t^x = x + \int_0^t b(X_s^x) ds + \sum_{k=1}^n \int_0^t \sigma_k(X_s^x) dB_s^k.$$

Moreover, for every $p \ge 1$, and $T \ge 0$,

$$\mathbb{E}\big(\sup_{0 \le t \le T} \|X_t^x\|^p \big) < +\infty.$$

Theorem 6.49. *For every $i = 1, \dots, n$, $0 \le t \le 1$, $X_t^{x,i} \in \mathbb{D}^\infty$ and for $r \le t$,*

$$\begin{aligned} \mathbf{D}_r^j X_t^{x,i} &= \sigma_{i,j}(X_r^x) + \sum_{l=1}^n \int_r^t \partial_l b_i(X_s^x) \mathbf{D}_r^j X_s^{x,l} ds \\ &\quad + \sum_{k,l=1}^n \int_r^t \partial_l \sigma_{i,k}(X_s^x) \mathbf{D}_r^j X_s^{x,l} dB_s^k, \end{aligned}$$

where $\mathbf{D}_r^j X_t^i$ is the j-th component of $\mathbf{D}_r X_t^i$. If $r > t$, then $\mathbf{D}_r^j X_t^{x,i} = 0$.

Proof. We first prove that $X_1^{x,i} \in \mathbb{D}^{1,p}$ for every $p \geq 1$. We consider the Picard approximations given by $X_0(t) = x$ and

$$X_{n+1}(t) = x + \int_0^t b(X_n(s))ds + \sum_{k=1}^n \int_0^t \sigma_k(X_n(s))dB_s^k.$$

By induction, it is easy to see that $X_n(t) \in \mathbb{D}^{1,p}$ and that for every $p \geq 1$, we have

$$\Psi_n(t) = \sup_{0 \leq r \leq t} \mathbb{E}(\sup_{s \in [r,t]} \| \boldsymbol{D}_r X_n(s) \|^p) < +\infty$$

and

$$\Psi_{n+1}(t) \leq \alpha + \beta \int_0^t \Psi_n(s) ds.$$

Then we observe that $X_n(t)$ converges to X_t^x in L^p and that the sequence $\| X_n(t) \|_{1,p}$ is bounded. As a consequence $X_1^{x,i} \in \mathbb{D}^{1,p}$ for every $p \geq 1$. The equation for the Malliavin derivative is obtained by differentiating the equation satisfied by X_t^x. Higher order derivatives may be treated in a similar way with a few additional work, the details are left to the reader. □

Combining this theorem with Theorem 6.13 and using the uniqueness property for solutions of linear stochastic differential equations, we obtain the following representation for the Malliavin derivative of a solution of a stochastic differential equation:

Corollary 6.50. *We have*

$$\boldsymbol{D}_r^j X_t^x = \boldsymbol{J}_{0 \to t}(x) \boldsymbol{J}_{0 \to r}^{-1}(x) \sigma_j(X_r^x), \quad j = 1, \dots, n, \ 0 \leq r \leq t,$$

where $(\boldsymbol{J}_{0 \to t}(x))_{t \geq 0}$ is the first variation process defined by

$$\boldsymbol{J}_{0 \to t}(x) = \frac{\partial X_t^x}{\partial x}(x).$$

We now fix $x \in \mathbb{R}^n$ as the initial condition for our equation and denote by

$$\Gamma_t = \Big(\sum_{j=1}^n \int_0^1 \boldsymbol{D}_r^j X_t^{i,x} \boldsymbol{D}_r^j X_t^{i',x} dr \Big)_{1 \leq i,i' \leq n}$$

the Malliavin matrix of X_t^x. From the above corollary we deduce that

$$\Gamma_t(x) = \boldsymbol{J}_{0 \to t}(x) \int_0^t \boldsymbol{J}_{0 \to r}^{-1}(x) \sigma(X_r^x) \sigma(X_r^x)^* \boldsymbol{J}_{0 \to r}^{-1}(x)^* dr \, \boldsymbol{J}_{0 \to t}(x)^*. \tag{6.5}$$

We are now finally in a position to state the main theorem of the section:

Theorem 6.51. *Assume that there exists $\lambda > 0$ such that for every $x \in \mathbb{R}^n$,*

$$\|\sigma(x)\|^2 \geq \lambda \|x\|^2.$$

Then for every $t > 0$ and $x \in \mathbb{R}^n$, the random variable X_t^x has a smooth density with respect to the Lebesgue measure.

Proof. We use Theorem 6.45. We therefore want to prove that $\Gamma_t(x)$ is invertible with inverse in L^p for $p \geq 1$. Since $\boldsymbol{J}_{0\to t}(x)$ is invertible and its inverse solves a linear equation (see Exercise 6.14), we deduce that for every $p \geq 1$,

$$\mathbb{E}\left(\|\boldsymbol{J}_{0\to t}^{-1}(x)\|^p\right) < +\infty.$$

From formula (6.5), we conclude that it is enough to prove that $C_t(x)$ is invertible with inverse in L^p where

$$C_t(x) = \int_0^t \boldsymbol{J}_{0\to r}^{-1}(x)\sigma(X_r^x)\sigma(X_r^x)^* \boldsymbol{J}_{0\to r}^{-1}(x)^* dr.$$

By the uniform ellipticity assumption, we have

$$C_t(x) \geq \lambda \int_0^t \boldsymbol{J}_{0\to r}^{-1}(x)\boldsymbol{J}_{0\to r}^{-1}(x)^* dr,$$

where the inequality is understood in the sense that the difference of the two symmetric matrices is non-negative. This implies that $C_t(x)$ is invertible. Moreover, it is an easy exercise to prove that if M_t is a continuous map taking its values in the set of positive definite matrices, then we have

$$\left(\int_0^t M_s ds\right)^{-1} \leq \frac{1}{t^2}\left(\int_0^t M_s^{-1} ds\right).$$

As a consequence we obtain

$$C_t^{-1}(x) \leq \frac{1}{t^2\lambda}\int_0^t \boldsymbol{J}_{0\to r}(x)^* \boldsymbol{J}_{0\to r}(x) dr.$$

Since $\boldsymbol{J}_{0\to r}(x)$ has moments in L^p for all $p \geq 1$, we conclude that $C_t(x)$ is invertible with inverse in L^p. □

Notes and comments

Sections 6.1, 6.2, 6.3, 6.4. Stochastic differential equations were first considered by Itô [40], with the purpose of providing a pathwise construction of a diffusion process associated to a given diffusion operator. We have to mention that another pathwise construction of diffusion processes was made by Döblin in 1940, and thus before Itô, in a paper that was rediscovered in 2000 (see [12]). Essentially, to construct the diffusion with generator

$$L = \frac{1}{2}\sigma(x)^2 \frac{d}{dx} + b(x)\frac{d}{dx},$$

Döblin considers the equation

$$X_t = x + \beta_{\int_0^t \sigma(X_s)^2 ds} + \int_0^t b(X_s)ds.$$

Observe that this equation bypasses the theory of stochastic integrals that was not available at that time but is of course equivalent to Itô's equation when keeping in mind the Dambins, Dubins–Schwarz theorem. Concerning the pathwise construction of diffusion processes, we also have to mention the martingale problem approach by Stroock and Varadhan [77]. Stochastic differential equations with respect to general semimartingales may also be considered and we refer to the book by Protter [61] for an account of the theory. The theory of regularity of stochastic flows is difficult and the classical reference for these problems is the book by Kunita [47]. The rough paths theory presented in Chapter 7 simplifies several proofs of the regularity of flows (see the book by Friz and Victoir [29]).

Sections 6.5, 6.6. Malliavin calculus was first introduced by Malliavin in [58] and later developed by Shigekawa [70]. The motivation was to give a probabilistic proof of Hörmander's theorem. The nowadays classical reference to study Malliavin calculus is the book by Nualart [60]. For an alternative and more geometric approach we refer to the book by Bell [8], to the book by Malliavin himself [59] or the book by Shigekawa [71]. For a more specialized reading, we refer the interested reader to the lecture notes [76] and the very influential papers by Kusuoka and Stroock [48], [49] and [50] for further details and applications of Malliavin calculus to stochastic differential equations and partial differential equations.

Chapter 7

An introduction to Lyons' rough paths theory

This chapter is an introduction to the theory of rough paths that was conceived by T. Lyons in the 1990s. The theory is purely deterministic and offers a convenient alternative to Itô's calculus to define and study differential equations driven by Brownian motions. We focus on the main ideas and prove Lyon's continuity result in the linear case and then show that rough paths theory can be applied to study stochastic differential equations.

7.1 Continuous paths with finite p-variation

As before, if $s \le t$ we will denote by $\Delta[s,t]$ the set of subdivisions of the interval $[s,t]$, that is, $\Pi \in \Delta[s,t]$ can be written

$$\Pi = \{s = t_0 < t_1 < \cdots < t_n = t\}.$$

The following definition is basic.

Definition 7.1. A continuous path $x\colon [s,t] \to \mathbb{R}^d$ is said to have *bounded variation* on $[s,t]$ if the 1-variation of x on $[s,t]$, which is defined by

$$\|x\|_{1\text{-var};[s,t]} := \sup_{\Pi \in \Delta[s,t]} \sum_{k=0}^{n-1} \|x(t_{k+1}) - x(t_k)\|,$$

is finite. The space of continuous bounded variation paths $x\colon [s,t] \to \mathbb{R}^d$ will be denoted by $C^{1\text{-var}}([s,t],\mathbb{R}^d)$.

Of course $\|\cdot\|_{1\text{-var};[s,t]}$ is not a norm, because constant functions have a zero 1-variation, but it is obviously a semi-norm. If x is continuously differentiable on $[s,t]$, it is easily seen that

$$\|x\|_{1\text{-var},[s,t]} = \int_s^t \|x'(s)\| ds.$$

Proposition 7.2. *Let $x \in C^{1\text{-var}}([0,T],\mathbb{R}^d)$. The function $(s,t) \to \|x\|_{1\text{-var},[s,t]}$ is additive, i.e., for $0 \le s \le t \le u \le T$,*

$$\|x\|_{1\text{-var},[s,t]} + \|x\|_{1\text{-var},[t,u]} = \|x\|_{1\text{-var},[s,u]},$$

and controls x in the sense that for $0 \le s \le t \le T$,

$$\|x(s) - x(t)\| \le \|x\|_{1\text{-var},[s,t]}.$$

The function $s \to \|x\|_{1\text{-var},[0,s]}$ is moreover continuous and non-decreasing.

Proof. If $\Pi_1 \in \Delta[s,t]$ and $\Pi_2 \in \Delta[t,u]$, then $\Pi_1 \cup \Pi_2 \in \Delta[s,u]$. As a consequence we obtain

$$\begin{aligned}
\sup_{\Pi_1 \in \Delta[s,t]} \sum_{k=0}^{n-1} \|x(t_{k+1}) - x(t_k)\| &+ \sup_{\Pi_2 \in \Delta[t,u]} \sum_{k=0}^{n-1} \|x(t_{k+1}) - x(t_k)\| \\
&\le \sup_{\Pi \in \Delta[s,u]} \sum_{k=0}^{n-1} \|x(t_{k+1}) - x(t_k)\|,
\end{aligned}$$

thus

$$\|x\|_{1\text{-var},[s,t]} + \|x\|_{1\text{-var},[t,u]} \le \|x\|_{1\text{-var},[s,u]}.$$

Let now $\Pi \in \Delta[s,u]$:

$$\Pi = \{s = t_0 < t_1 < \cdots < t_n = t\}.$$

Let $k = \max\{j, t_j \le t\}$. By the triangle inequality, we have

$$\begin{aligned}
\sum_{j=0}^{n-1} \|x(t_{j+1}) - x(t_j)\| &\le \sum_{j=0}^{k-1} \|x(t_{j+1}) - x(t_j)\| + \sum_{j=k}^{n-1} \|x(t_{j+1}) - x(t_j)\| \\
&\le \|x\|_{1\text{-var},[s,t]} + \|x\|_{1\text{-var},[t,u]}.
\end{aligned}$$

Taking the supremum of $\Pi \in \Delta[s,u]$ gives

$$\|x\|_{1\text{-var},[s,t]} + \|x\|_{1\text{-var},[t,u]} \ge \|x\|_{1\text{-var},[s,u]},$$

which completes the proof. We leave the proof of the continuity and monotonicity of $s \to \|x\|_{1\text{-var},[0,s]}$ to the reader. □

This control of the path by the 1-variation norm is an illustration of the notion of controlled path which is central in rough paths theory.

Definition 7.3. A map $\omega\colon \{0 \le s \le t \le T\} \to [0,\infty)$ is called *superadditive* if

$$\omega(s,t) + \omega(t,u) \le \omega(s,u)$$

for all $s \le t \le u$. If, in addition, ω is continuous and $\omega(t,t) = 0$, we call ω a *control*. We say that a path $x\colon [0,T] \to \mathbb{R}$ is controlled by a control ω if there exists a constant $C > 0$ such that

$$\|x(t) - x(s)\| \le C\omega(s,t)$$

for every $0 \le s \le t \le T$.

Obviously, Lipschitz functions have a bounded variation. The converse is not true: $t \to \sqrt{t}$ has a bounded variation on $[0, 1]$ but is not Lipschitz. However, any continuous path with bounded variation is the reparametrization of a Lipschitz path in the following sense.

Proposition 7.4. *Let $x \in C^{1\text{-var}}([0,T], \mathbb{R}^d)$. There exist a Lipschitz function $y\colon [0,1] \to \mathbb{R}^d$ and a continuous and non-decreasing function $\phi\colon [0,T] \to [0,1]$ such that $x = y \circ \phi$.*

Proof. We assume that $\|x\|_{1\text{-var},[0,T]} \neq 0$ and consider

$$\phi(t) = \frac{\|x\|_{1\text{-var},[0,t]}}{\|x\|_{1\text{-var},[0,T]}}.$$

It is continuous and non-decreasing. There exists a function y such that $x = y \circ \phi$ because $\phi(t_1) = \phi(t_2)$ implies that $x(t_1) = x(t_2)$. We then have, for $s \le t$,

$$\|y(\phi(t)) - y(\phi(s))\| = \|x(t) - x(s)\| \le \|x\|_{1\text{-var},[s,t]} = \|x\|_{1\text{-var},[0,T]}(\phi(t) - \phi(s)).$$

□

The next result shows that the set of continuous paths with bounded variation is a Banach space.

Theorem 7.5. *The space $C^{1\text{-var}}([0,T], \mathbb{R}^d)$ endowed with the norm $\|x(0)\| + \|x\|_{1\text{-var},[0,T]}$ is a Banach space.*

Proof. Let $x^n \in C^{1\text{-var}}([0,T], \mathbb{R}^d)$ be a Cauchy sequence. It is clear that

$$\|x^n - x^m\|_\infty \le \|x^n(0) - x^m(0)\| + \|x^n - x^m\|_{1\text{-var},[0,T]}.$$

Thus x^n converges uniformly to a continuous path $x\colon [0,T] \to \mathbb{R}$. We need to prove that x has a bounded variation. Let

$$\Pi = \{0 = t_0 < t_1 < \cdots < t_n = T\}$$

be a subdivision of $[0,T]$. There is $m \ge 0$ such that $\|x - x^m\|_\infty \le \frac{1}{2n}$, hence

$$\begin{aligned}
\sum_{k=0}^{n-1} \|x(t_{k+1}) - x(t_k)\| &\le \sum_{k=0}^{n-1} \|x(t_{k+1}) - x^m(t_{k+1})\| \\
&\quad + \sum_{k=0}^{n-1} \|x^m(t_k) - x(t_k)\| + \|x^m\|_{1\text{-var},[0,T]} \\
&\le 1 + \sup_n \|x^n\|_{1\text{-var},[0,T]}.
\end{aligned}$$

Thus we have

$$\|x\|_{1\text{-var},[0,T]} \le 1 + \sup_n \|x^n\|_{1\text{-var},[0,T]} < \infty.$$

□

More generally, we shall be interested in paths that have finite p- variation, $p > 0$.

Definition 7.6. Let $s \le t$. A path $x\colon [s,t] \to \mathbb{R}^d$ is said to be of *finite p-variation*, $p > 0$, if the p-variation of x on $[s,t]$ defined by

$$\|x\|_{p\text{-var};[s,t]} := \Big(\sup_{\Pi \in \Delta[s,t]} \sum_{k=0}^{n-1} \|x(t_{k+1}) - x(t_k)\|^p \Big)^{1/p}$$

is finite. The space of continuous paths $x\colon [s,t] \to \mathbb{R}^d$ with a finite p-variation will be denoted by $C^{p\text{-var}}([s,t],\mathbb{R}^d)$.

Exercise 7.7. Show that if $x\colon [s,t] \to \mathbb{R}^d$ is an α-Hölder path, then it has finite $1/\alpha$-variation.

The notion of p-variation is only interesting when $p \ge 1$.

Proposition 7.8. *Let $x\colon [s,t] \to \mathbb{R}^d$ be a continuous path of finite p-variation with $p < 1$. Then x is constant.*

Proof. We have for $s \le u \le t$,

$$\begin{aligned} \|x(u) - x(s)\| &\le (\max \|x(t_{k+1}) - x(t_k)\|^{1-p}) \Big(\sum_{k=0}^{n-1} \|x(t_{k+1}) - x(t_k)\|^p \Big) \\ &\le (\max \|x(t_{k+1}) - x(t_k)\|^{1-p}) \|x\|^p_{p\text{-var};[s,t]}. \end{aligned}$$

Since x is continuous, it is also uniformly continuous on $[s,t]$. By taking a sequence of subdivisions whose mesh tends to 0, we deduce that

$$\|x(u) - x(s)\| = 0,$$

that is, x is constant. □

The next proposition is immediate and its proof is left to the reader:

Proposition 7.9. *Let $x\colon [s,t] \to \mathbb{R}^d$ be a continuous path. If $p \le p'$ then*

$$\|x\|_{p'\text{-var};[s,t]} \le \|x\|_{p\text{-var};[s,t]}.$$

As a consequence the following inclusion holds:

$$C^{p\text{-var}}([s,t],\mathbb{R}^d) \subset C^{p'\text{-var}}([s,t],\mathbb{R}^d)$$

The following proposition generalizes Proposition 7.2 to paths with finite p-variation.

Proposition 7.10. *Let $x \in C^{p\text{-var}}([0,T],\mathbb{R}^d)$. Then $\omega(s,t) = \|x\|^p_{p\text{-var};[s,t]}$ is a control such that for every $s \le t$,*

$$\|x(s) - x(t)\| \le \omega(s,t)^{1/p}.$$

Proof. It is immediate that

$$\|x(s) - x(t)\| \le \omega(s,t)^{1/p},$$

so we focus on the proof that ω is a control. If $\Pi_1 \in \Delta[s,t]$ and $\Pi_2 \in \Delta[t,u]$, then $\Pi_1 \cup \Pi_2 \in \Delta[s,u]$. As a consequence we obtain

$$\begin{aligned}
&\sup_{\Pi_1 \in \Delta[s,t]} \sum_{k=0}^{n-1} \|x(t_{k+1}) - x(t_k)\|^p + \sup_{\Pi_2 \in \Delta[t,u]} \sum_{k=0}^{n-1} \|x(t_{k+1}) - x(t_k)\|^p \\
&\quad \le \sup_{\Pi \in \Delta[s,u]} \sum_{k=0}^{n-1} \|x(t_{k+1}) - x(t_k)\|^p,
\end{aligned}$$

thus

$$\|x\|^p_{p\text{-var},[s,t]} + \|x\|^p_{p\text{-var},[t,u]} \le \|x\|^p_{p\text{-var},[s,u]}.$$

The proof of the continuity is left to the reader. □

Thanks to the next proposition, $\|x\|^p_{p\text{-var};[s,t]}$ is the minimal control of a path $x \in C^{p\text{-var}}([0,T],\mathbb{R}^d)$.

Proposition 7.11. *Let $x \in C^{p\text{-var}}([0,T],\mathbb{R}^d)$ and let $\omega\colon \{0 \le s \le t \le T\} \to [0,\infty)$ be a control such that for $0 \le s \le t \le T$,*

$$\|x(s) - x(t)\| \le C\omega(s,t)^{1/p}.$$

Then

$$\|x\|_{p\text{-var};[s,t]} \le C\omega(s,t)^{1/p}.$$

Proof. We have

$$\begin{aligned}
\|x\|_{p\text{-var};[s,t]} &= \Big(\sup_{\Pi \in \Delta[s,t]} \sum_{k=0}^{n-1} \|x(t_{k+1}) - x(t_k)\|^p \Big)^{1/p} \\
&\le \Big(\sup_{\Pi \in \Delta[s,t]} \sum_{k=0}^{n-1} C^p \omega(t_k, t_{k+1}) \Big)^{1/p} \le C\omega(s,t).
\end{aligned}$$

□

The next theorem states that the set of continuous paths with bounded p-variation is a Banach space. The proof is identical to the case $p = 1$ (see proof of Theorem 7.5).

Theorem 7.12. *Let $p \ge 1$. The space $C^{p\text{-var}}([0,T],\mathbb{R}^d)$ endowed with the norm $\|x(0)\| + \|x\|_{p\text{-var},[0,T]}$ is a Banach space.*

7.2 The signature of a bounded variation path

We now introduce the notion of the signature of a path $x \in C^{1\text{-var}}([0,T],\mathbb{R}^d)$ which is a convenient way to encode all the algebraic information on the path x which is relevant to study differential equations driven by x. The motivation for the definition of the signature comes from formal manipulations on Taylor series.

Let us consider the ordinary differential equation

$$y(t) = y(0) + \sum_{i=1}^{d} \int_0^t V_i(y(u))dx^i(u),$$

where the V_i's are smooth vector fields on $\mathbb{R}^n$ (see Section 6.5 in Chapter 6 for the definition of a vector field). If $f:\mathbb{R}^n \to \mathbb{R}$ is a C^∞ function, by the change of variable formula,

$$f(y(t)) = f(y(s)) + \sum_{i=1}^{d} \int_s^t V_i f(y(u))dx^i(u).$$

Now a new application of the change of variable formula to $V_i f(y(s))$ leads to

$$f(y(t)) = f(y(s)) + \sum_{i=1}^{d} V_i f(y(s)) \int_s^t dx^i(u)$$
$$+ \sum_{i,j=1}^{d} \int_s^t \int_0^u V_j V_i f(y(v))dx^j(v)dx^i(u).$$

We can continue this procedure to get after N steps

$$f(y(t)) = f(y(s)) + \sum_{k=1}^{N} \sum_{I=(i_1,\dots,i_k)} (V_{i_1}\dots V_{i_k} f)(y(s)) \int_{\Delta^k[s,t]} dx^I + R_N(s,t)$$

for some remainder term $R_N(s,t)$, where we used the notations

(1) $\Delta^k[s,t] = \{(t_1,\dots,t_k)\in[s,t]^k, s \le t_1 \le t_2 \cdots \le t_k \le t\}$;

(2) if $I=(i_1,\dots,i_k)\in\{1,\dots,d\}^k$ is a word of length k,

$$\int_{\Delta^k[s,t]} dx^I = \int_{s\le t_1\le t_2\le\cdots\le t_k\le t} dx^{i_1}(t_1)\dots dx^{i_k}(t_k).$$

If we let $N\to+\infty$, assuming $R_N(s,t)\to 0$ (which is by the way true for $t-s$ small enough if the V_i's are analytic), we are led to the formal expansion formula

$$f(y(t)) = f(y(s)) + \sum_{k=1}^{+\infty} \sum_{I=(i_1,\dots,i_k)} (V_{i_1}\dots V_{i_k} f)(y(s)) \int_{\Delta^k[s,t]} dx^I.$$

This shows, at least at the formal level, that all the information given by x about y is contained in the set of iterated integrals $\int_{\Delta^k[s,t]} dx^I$.

Let $\mathbb{R}[[X_1,\dots,X_d]]$ be the noncommutative algebra over $\mathbb{R}$ of the formal series with d indeterminates, that is, the set of series

$$Y = y_0 + \sum_{k=1}^{+\infty} \sum_{I\in\{1,\dots,d\}^k} a_{i_1,\dots,i_k} X_{i_1}\dots X_{i_k}.$$

Definition 7.13. Let $x \in C^{1\text{-var}}([0,T],\mathbb{R}^d)$. The *signature* of x (or Chen's series) is the formal series

$$\mathfrak{S}(x)_{s,t} = 1 + \sum_{k=1}^{+\infty} \sum_{I\in\{1,\dots,d\}^k} \Big(\int_{\Delta^k[s,t]} dx^I\Big) X_{i_1}\dots X_{i_k}, \quad 0\le s\le t\le T.$$

As we are going to see, the signature is a fascinating algebraic object. At the source of the numerous properties of the signature lie the following so-called Chen relations.

Lemma 7.14 (Chen relations). *Let $x \in C^{1\text{-var}}([0,T],\mathbb{R}^d)$. For any word $(i_1,\dots,i_n)\in\{1,\dots,d\}^n$ and any $0\le s\le t\le u\le T$,*

$$\int_{\Delta^n[s,u]} dx^{(i_1,\dots,i_n)} = \sum_{k=0}^{n} \int_{\Delta^k[s,t]} dx^{(i_1,\dots,i_k)} \int_{\Delta^{n-k}[t,u]} dx^{(i_{k+1},\dots,i_n)},$$

where we used the convention that if I is a word of length 0, then $\int_{\Delta^0[0,t]} dx^I = 1$.

Proof. This follows readily by induction on n by noticing that

$$\int_{\Delta^n[s,u]} dx^{(i_1,\dots,i_n)} = \int_s^u \Big(\int_{\Delta^{n-1}[s,t_n]} dx^{(i_1,\dots,i_{n-1})}\Big) dx^{i_n}(t_n). \qquad \square$$

To avoid heavy and cumbersome notations, it will be convenient to write

$$\int_{\Delta^k[s,t]} dx^{\otimes k} = \sum_{I\in\{1,\dots,d\}^k} \Big(\int_{\Delta^k[s,t]} dx^I\Big) X_{i_1}\dots X_{i_k}.$$

This notation reflects a natural algebra isomorphism between $\mathbb{R}[[X_1,\dots,X_d]]$ and $1\bigoplus_{k=1}^{+\infty}(\mathbb{R}^d)^{\otimes k}$. With this notation, observe that the signature then writes

$$\mathfrak{S}(x)_{s,t} = 1 + \sum_{k=1}^{+\infty} \int_{\Delta^k[s,t]} dx^{\otimes k},$$

and that Chen's relations become

$$\int_{\Delta^n[s,u]} dx^{\otimes n} = \sum_{k=0}^{n} \int_{\Delta^k[s,t]} dx^{\otimes k} \int_{\Delta^{n-k}[t,u]} dx^{\otimes (n-k)}.$$

The Chen relations imply the following flow property for the signature:

Lemma 7.15. *Let $x \in C^{1\text{-var}}([0,T],\mathbb{R}^d)$. For any $0 \le s \le t \le u \le T$,*

$$\mathfrak{S}(x)_{s,u} = \mathfrak{S}(x)_{s,t}\mathfrak{S}(x)_{t,u}.$$

Proof. Indeed, from the Chen relations, we have

$$\begin{aligned}
\mathfrak{S}(x)_{s,u} &= 1 + \sum_{k=1}^{+\infty} \int_{\Delta^k[s,u]} dx^{\otimes k} \\
&= 1 + \sum_{k=1}^{+\infty} \sum_{j=0}^{n} \int_{\Delta^j[s,t]} dx^{\otimes j} \int_{\Delta^{k-j}[t,u]} dx^{\otimes (k-j)} \\
&= \mathfrak{S}(x)_{s,t}\mathfrak{S}(x)_{t,u}.
\end{aligned}$$

□

7.3 Estimating iterated integrals

In the previous section we introduced the signature of a bounded variation path x as the formal series

$$\mathfrak{S}(x)_{s,t} = 1 + \sum_{k=1}^{+\infty} \int_{\Delta^k[s,t]} dx^{\otimes k}.$$

If now $x \in C^{p\text{-var}}([0,T],\mathbb{R}^d)$, $p \ge 1$, the iterated integrals $\int_{\Delta^k[s,t]} dx^{\otimes k}$ can only be defined by Riemann–Stieltjes integrals when $p = 1$ or by Young integrals when $p < 2$. As we are going to see, it is actually possible to define the signature of paths with a finite p variation even when $p \ge 2$. For $P \in \mathbb{R}[[X_1,\dots,X_d]]$, which can be written as

$$P = P_0 + \sum_{k=1}^{+\infty} \sum_{I \in \{1,\dots,d\}^k} a_{i_1,\dots,i_k} X_{i_1} \dots X_{i_k},$$

we define

$$\|P\| = |P_0| + \sum_{k=1}^{+\infty} \sum_{I \in \{1,\dots,d\}^k} |a_{i_1,\dots,i_k}| \in [0,\infty].$$

It is quite easy to check that for $P, Q \in \mathbb{R}[[X_1,\dots,X_d]]$,

$$\|PQ\| \le \|P\| \, \|Q\|.$$

Let $x \in C^{1\text{-var}}([0,T],\mathbb{R}^d)$. For $p \geq 1$, we write

$$\Big\| \int dx^{\otimes k} \Big\|_{p\text{-var},[s,t]} = \Big(\sup_{\Pi \in \mathcal{D}[s,t]} \sum_{i=0}^{n-1} \Big\| \int_{\Delta^k[t_i,t_{i+1}]} dx^{\otimes k} \Big\|^p \Big)^{1/p},$$

where $\mathcal{D}[s,t]$ is the set of subdivisions of the interval $[s,t]$. Observe that for $k \geq 2$ in general

$$\int_{\Delta^k[s,t]} dx^{\otimes k} + \int_{\Delta^k[t,u]} dx^{\otimes k} \neq \int_{\Delta^k[s,u]} dx^{\otimes k}.$$

Actually from the Chen relations we have

$$\begin{aligned}\int_{\Delta^n[s,u]} dx^{\otimes n} &= \int_{\Delta^n[s,t]} dx^{\otimes n} + \int_{\Delta^n[t,u]} dx^{\otimes n} \\ &\quad + \sum_{k=1}^{n-1} \int_{\Delta^k[s,t]} dx^{\otimes k} \int_{\Delta^{n-k}[t,u]} dx^{\otimes (n-k)}.\end{aligned}$$

It follows that $\| \int dx^{\otimes k} \|_{p\text{-var},[s,t]}$ does not need to be the p-variation of the path $t \to \int_{\Delta^k[s,t]} dx^{\otimes k}$. It is however easy to verify that

$$\Big\| \int_{\Delta^k[s,\cdot]} dx^{\otimes k} \Big\|_{p\text{-var},[s,t]} \leq \Big\| \int dx^{\otimes k} \Big\|_{p\text{-var},[s,t]}.$$

The first major result of rough paths theory is the following estimate:

Theorem 7.16. *Let $p \geq 1$. There exists a constant $C \geq 0$, depending only on p, such that for every $x \in C^{1\text{-var}}([0,T],\mathbb{R}^d)$ and $k \geq 0$,*

$$\Big\| \int_{\Delta^k[s,t]} dx^{\otimes k} \Big\| \leq \frac{C^k}{(\frac{k}{p})!} \Big(\sum_{j=1}^{[p]} \Big\| \int dx^{\otimes j} \Big\|^{1/j}_{\frac{p}{j}\text{-var},[s,t]} \Big)^k, \quad 0 \leq s \leq t \leq T.$$

By $\big(\frac{k}{p}\big)!$ of course we mean $\Gamma(\frac{k}{p}+1)$, where Γ is the Euler function. Some remarks are in order before we prove the result. If $p=1$, then the estimate becomes

$$\Big\| \int_{\Delta^k[s,t]} dx^{\otimes k} \Big\| \leq \frac{C^k}{k!} \|x\|^k_{1\text{-var},[s,t]},$$

which is immediately checked because

$$\begin{aligned}\Big\| \int_{\Delta^k[s,t]} dx^{\otimes k} \Big\| &\leq \sum_{I \in \{1,\dots,d\}^k} \Big\| \int_{\Delta^k[s,t]} dx^I \Big\| \\ &\leq \sum_{I \in \{1,\dots,d\}^k} \int_{s \leq t_1 \leq t_2 \leq \dots \leq t_k \leq t} d\|x^{i_1}\|_{1\text{-var},[s,t_1]} \dots d\|x^{i_k}\|_{1\text{-var},[s,t_k]} \\ &\leq \frac{1}{k!} \|x\|^k_{1\text{-var},[s,t]}.\end{aligned}$$

We also observe that for $k \le p$, the estimate is easy to obtain because

$$\left\| \int_{\Delta^k[s,t]} dx^{\otimes k} \right\| \le \left\| \int dx^{\otimes k} \right\|_{\frac{p}{k}\text{-var},[s,t]}.$$

So, all the work is to prove the estimate when $k > p$. The proof is split into two lemmas. The first one is the following binomial inequality which is proved as an exercise.

Lemma 7.17 (Binomial inequality). *For $x, y > 0$, $n \in \mathbb{N}$, $n \ge 0$, and $p \ge 1$,*

$$\sum_{j=0}^{n} \frac{x^{j/p}}{\left(\frac{j}{p}\right)!} \frac{y^{(n-j)/p}}{\left(\frac{n-j}{p}\right)!} \le p \frac{(x+y)^{n/p}}{\left(\frac{n}{p}\right)!}.$$

Proof. See Exercise 7.18. □

Exercise 7.18. Let D be the open unit disc in $\mathbb{C}$, $D = \{z \in \mathbb{C}, |z| < 1\}$. If f is a continuous on $\bar{D}$ and holomorphic on D, then we write for $\xi \in \mathbb{R}$,

$$\hat{f}(\xi) = \int_{-\frac{1}{2}}^{\frac{1}{2}} f(e^{2i\pi x}) e^{-2i\pi x \xi} dx.$$

(1) By writing $\hat{f}(\xi)$ as an integral over the unit circle and using the residue theorem, show that for $0 < \alpha < 2$ and $0 < \lambda < 1$ the following identities hold:

$$\alpha \sum_{j=0}^{\infty} \hat{f}(\alpha j) \lambda^{\alpha j} = f(\lambda) - \frac{\alpha \lambda^{\alpha} \sin(\alpha\pi)}{\pi} \int_0^1 \frac{t^{\alpha-1} f(-t)}{|t^{\alpha} - \lambda^{\alpha} e^{-i\alpha\pi}|^2} dt,$$

$$\alpha \sum_{j=-\infty}^{-1} \hat{f}(\alpha j) \lambda^{\alpha j} = \frac{\alpha \lambda^{\alpha} \sin(\alpha\pi)}{\pi} \int_0^1 \frac{t^{\alpha-1} f(-t)}{|e^{-i\alpha\pi} - \lambda^{\alpha} t^{\alpha}|^2} dt.$$

(2) Compute $\hat{f}$ when $f(z) = (1+z)^T$, $T > 0$.

(3) Show that for $\alpha \in (0,2)$, $n \in \mathbb{N}$ and $0 < \lambda \le 1$ we have

$$\begin{aligned} \alpha \sum_{j=0}^{n} \binom{\alpha n}{\alpha j} \lambda^{\alpha j} &= (1+\lambda)^{\alpha n} - \frac{\alpha \lambda^{\alpha} \sin(\alpha\pi)}{\pi} \\ &\quad \cdot \int_0^1 t^{\alpha-1} (1-t)^{\alpha n} \left(\frac{1}{|t^{\alpha} - \lambda^{\alpha} e^{-i\alpha\pi}|^2} + \frac{1}{|e^{-i\alpha\pi} - \lambda^{\alpha} t^{\alpha}|^2} \right) dt, \end{aligned}$$

where $\binom{\alpha n}{\alpha j} = \frac{(\alpha n)!}{(\alpha j)!((n-j)\alpha)!}$.

(4) Give the proof of Lemma 7.17.

The second lemma we need is the following.

Lemma 7.19. *Let $\Gamma \colon \{0 \le s \le t \le T\} \to \mathbb{R}^N$. Let us assume that*

(1) *there exists a control $\tilde{\omega}$ such that*

$$\lim_{r \to 0} \sup_{(s,t), \tilde{\omega}(s,t) \le r} \frac{\|\Gamma_{s,t}\|}{r} = 0;$$

(2) *there exists a control ω and $\theta > 1, \xi > 0$ such that for $0 \le s \le t \le u \le T$,*

$$\|\Gamma_{s,u}\| \le \|\Gamma_{s,t}\| + \|\Gamma_{t,u}\| + \xi \omega(s,u)^\theta.$$

Then, for all $0 \le s < t \le T$,

$$\|\Gamma_{s,t}\| \le \frac{\xi}{1 - 2^{1-\theta}} \omega(s,t)^\theta.$$

Proof. For $\varepsilon > 0$, let us consider the control

$$\omega_\varepsilon(s,t) = \omega(s,t) + \varepsilon \tilde{\omega}(s,t).$$

Define now

$$\Psi(r) = \sup_{s,u,\omega_\varepsilon(s,u) \le r} \|\Gamma_{s,u}\|.$$

If s, u is such that $\omega_\varepsilon(s,u) \le r$, we can find a t such that $\omega_\varepsilon(s,t) \le \frac{1}{2}\omega_\varepsilon(s,u)$, $\omega_\varepsilon(t,u) \le \frac{1}{2}\omega_\varepsilon(s,u)$. Indeed, the continuity of ω_ε forces the existence of a t such that $\omega_\varepsilon(s,t) = \omega_\varepsilon(t,u)$. We therefore obtain

$$\|\Gamma_{s,u}\| \le 2\Psi(r/2) + \xi r^\theta,$$

which implies by maximization that

$$\Psi(r) \le 2\Psi(r/2) + \xi r^\theta.$$

By iterating n times this inequality, we obtain

$$\Psi(r) \le 2^n \Psi\left(\frac{r}{2^n}\right) + \xi \sum_{k=0}^{n-1} 2^{k(1-\theta)} r^\theta \le 2^n \Psi\left(\frac{r}{2^n}\right) + \xi \frac{1}{1 - 2^{1-\theta}} r^\theta.$$

It is furthermore clear that

$$\lim_{n \to \infty} 2^n \Psi\left(\frac{r}{2^n}\right) = 0.$$

We conclude that

$$\Psi(r) \leq \frac{\xi}{1-2^{1-\theta}} r^{\theta}$$

and thus

$$\|\Gamma_{s,u}\| \leq \frac{\xi}{1-2^{1-\theta}} \omega_{\varepsilon}(s,u)^{\theta}.$$

Sending $\varepsilon \to 0$ finishes the proof. □

We can now turn to the proof of Theorem 7.16.

Proof. Let us write

$$\omega(s,t) = \Big(\sum_{j=1}^{[p]} \Big\| \int dx^{\otimes j} \Big\|_{\frac{p}{j}\text{-var},[s,t]}^{1/j} \Big)^{p}.$$

We claim that ω is a control. Indeed for $0 \leq s \leq t \leq u \leq T$, from the reverse Minkowski inequality $\|x+y\|_{1/p} \geq \|x\|_{1/p} + \|y\|_{1/p}$ we have

$$\begin{aligned}
\omega(s,t) + \omega(t,u) &= \Big(\sum_{j=1}^{[p]} \Big\| \int dx^{\otimes j} \Big\|_{\frac{p}{j}\text{-var},[s,t]}^{1/j} \Big)^{p} + \Big(\sum_{j=1}^{[p]} \Big\| \int dx^{\otimes j} \Big\|_{\frac{p}{j}\text{-var},[t,u]}^{1/j} \Big)^{p} \\
&\leq \Big(\sum_{j=1}^{[p]} \Big(\Big\| \int dx^{\otimes j} \Big\|_{\frac{p}{j}\text{-var},[s,t]}^{p/j} + \Big\| \int dx^{\otimes j} \Big\|_{\frac{p}{j}\text{-var},[t,u]}^{p/j} \Big)^{1/p} \Big)^{p} \\
&\leq \Big(\sum_{j=1}^{[p]} \Big\| \int dx^{\otimes j} \Big\|_{\frac{p}{j}\text{-var},[s,u]}^{1/j} \Big)^{p} = \omega(s,u).
\end{aligned}$$

It is clear that for some constant $\beta > 0$ which is small enough, we have for $k \leq p$,

$$\Big\| \int_{\Delta^k[s,t]} dx^{\otimes k} \Big\| \leq \frac{1}{\beta \big(\frac{k}{p}\big)!} \omega(s,t)^{k/p}.$$

Let us now consider

$$\Gamma_{s,t} = \int_{\Delta^{[p]+1}[s,t]} dx^{\otimes([p]+1)}.$$

From the Chen relations, for $0 \leq s \leq t \leq u \leq T$ we have

$$\Gamma_{s,u} = \Gamma_{s,t} + \Gamma_{t,u} + \sum_{j=1}^{[p]} \int_{\Delta^j[s,t]} dx^{\otimes j} \int_{\Delta^{[p]+1-j}[t,u]} dx^{\otimes([p]+1-j)}.$$

Therefore, we obtain

$$\begin{aligned}
&\|\Gamma_{s,u}\| \\
&\le \|\Gamma_{s,t}\| + \|\Gamma_{t,u}\| + \sum_{j=1}^{[p]} \left\| \int_{\Delta^j[s,t]} dx^{\otimes j} \right\| \left\| \int_{\Delta^{[p]+1-j}[t,u]} dx^{\otimes([p]+1-j)} \right\| \\
&\le \|\Gamma_{s,t}\| + \|\Gamma_{t,u}\| + \frac{1}{\beta^2} \sum_{j=1}^{[p]} \frac{1}{\left(\frac{j}{p}\right)!} \omega(s,t)^{j/p} \frac{1}{\left(\frac{[p]+1-j}{p}\right)!} \omega(t,u)^{([p]+1-j)/p} \\
&\le \|\Gamma_{s,t}\| + \|\Gamma_{t,u}\| + \frac{1}{\beta^2} \sum_{j=0}^{[p]+1} \frac{1}{\left(\frac{j}{p}\right)!} \omega(s,t)^{j/p} \frac{1}{\left(\frac{[p]+1-j}{p}\right)!} \omega(t,u)^{([p]+1-j)/p} \\
&\le \|\Gamma_{s,t}\| + \|\Gamma_{t,u}\| + \frac{1}{\beta^2} p \frac{(\omega(s,t)+\omega(t,u))^{([p]+1)/p}}{\left(\frac{[p]+1}{p}\right)!} \\
&\le \|\Gamma_{s,t}\| + \|\Gamma_{t,u}\| + \frac{1}{\beta^2} p \frac{\omega(s,u)^{([p]+1)/p}}{\left(\frac{[p]+1}{p}\right)!}.
\end{aligned}$$

On the other hand, for some constant A, we have

$$\|\Gamma_{s,t}\| \le A \|x\|^{[p]+1}_{1\text{-var},[s,t]}.$$

We deduce from the previous Lemma 7.19 that

$$\|\Gamma_{s,t}\| \le \frac{1}{\beta^2} \frac{p}{1-2^{1-\theta}} \frac{\omega(s,t)^{([p]+1)/p}}{\left(\frac{[p]+1}{p}\right)!},$$

with $\theta = \frac{[p]+1}{p}$. The general case $k \ge p$ is dealt by induction. The details are left to the reader. □

Let $x \in C^{1\text{-var}}([0,T], \mathbb{R}^d)$. Since

$$\omega(s,t) = \left(\sum_{j=1}^{[p]} \left\| \int dx^{\otimes j} \right\|^{1/j}_{\frac{p}{j}\text{-var},[s,t]} \right)^p$$

is a control, the estimate

$$\left\| \int_{\Delta^k[s,t]} dx^{\otimes k} \right\| \le \frac{C^k}{\left(\frac{k}{p}\right)!} \left(\sum_{j=1}^{[p]} \left\| \int dx^{\otimes j} \right\|^{1/j}_{\frac{p}{j}\text{-var},[s,t]} \right)^k, \quad 0 \le s \le t \le T,$$

easily implies that for $k > p$,

$$\left\| \int dx^{\otimes k} \right\|_{1\text{-var},[s,t]} \le \frac{C^k}{\left(\frac{k}{p}\right)!} \omega(s,t)^{k/p}.$$

We stress that it does not imply a bound on the 1-variation of the path $t \to \int_{\Delta^k[0,t]} dx^{\otimes k}$. What we can get for this path are bounds in p-variation:

Proposition 7.20. *Let $p \ge 1$. There exists a constant $C \ge 0$, depending only on p, such that for every $x \in C^{1\text{-var}}([0,T], \mathbb{R}^d)$ and $k \ge 0$,*

$$\left\| \int_{\Delta^k[0,\cdot]} dx^{\otimes k} \right\|_{p\text{-var},[s,t]} \le \frac{C^k}{\left(\frac{k}{p}\right)!} \omega(s,t)^{1/p} \omega(0,T)^{\frac{k-1}{p}},$$

where

$$\omega(s,t) = \left(\sum_{j=1}^{[p]} \left\| \int dx^{\otimes j} \right\|^{1/j}_{\frac{p}{j}\text{-var},[s,t]} \right)^p, \quad 0 \le s \le t \le T.$$

Proof. This is an easy consequence of the Chen relations and of the previous theorem. Indeed,

$$\begin{aligned}
&\left\| \int_{\Delta^k[0,t]} dx^{\otimes k} - \int_{\Delta^k[0,s]} dx^{\otimes k} \right\| \\
&\quad = \left\| \sum_{j=1}^{k} \int_{\Delta^j[s,t]} dx^{\otimes j} \int_{\Delta^{k-j}[0,s]} dx^{\otimes (k-j)} \right\| \\
&\quad \le \sum_{j=1}^{k} \left\| \int_{\Delta^j[s,t]} dx^{\otimes j} \right\| \left\| \int_{\Delta^{k-j}[0,s]} dx^{\otimes (k-j)} \right\| \\
&\quad \le C^k \sum_{j=1}^{k} \frac{1}{\left(\frac{j}{p}\right)!} \omega(s,t)^{j/p} \frac{1}{\left(\frac{k-j}{p}\right)!} \omega(s,t)^{(k-j)/p} \\
&\quad \le C^k \omega(s,t)^{1/p} \sum_{j=1}^{k} \frac{1}{\left(\frac{j}{p}\right)!} \omega(0,T)^{(j-1)/p} \frac{1}{\left(\frac{k-j}{p}\right)!} \omega(0,T)^{(k-j)/p} \\
&\quad \le C^k \omega(s,t)^{1/p} \omega(0,T)^{(k-1)/p} \sum_{j=1}^{k} \frac{1}{\left(\frac{j}{p}\right)!} \frac{1}{\left(\frac{k-j}{p}\right)!},
\end{aligned}$$

and we conclude with the binomial inequality. □

We are now ready for a second major estimate which is the key to define iterated integrals of a path with p-bounded variation when $p \ge 2$.

Theorem 7.21. *Let $p \geq 1$, $K > 0$ and $x, y \in C^{1\text{-var}}([0,T], \mathbb{R}^d)$ such that*

$$\sum_{j=1}^{[p]} \left\| \int dx^{\otimes j} - \int dy^{\otimes j} \right\|_{\frac{p}{j}\text{-var},[0,T]}^{1/j} \leq 1$$

and

$$\left(\sum_{j=1}^{[p]} \left\| \int dx^{\otimes j} \right\|_{\frac{p}{j}\text{-var},[0,T]}^{1/j} \right)^p + \left(\sum_{j=1}^{[p]} \left\| \int dy^{\otimes j} \right\|_{\frac{p}{j}\text{-var},[0,T]}^{1/j} \right)^p \leq K.$$

Then there exists a constant $C \geq 0$ depending only on p and K such that for $0 \leq s \leq t \leq T$ and $k \geq 1$ we have

$$\begin{aligned} &\left\| \int_{\Delta^k[s,t]} dx^{\otimes k} - \int_{\Delta^k[s,t]} dy^{\otimes k} \right\| \\ &\quad \leq \left(\sum_{j=1}^{[p]} \left\| \int dx^{\otimes j} - \int dy^{\otimes j} \right\|_{\frac{p}{j}\text{-var},[0,T]}^{1/j} \right) \frac{C^k}{\left(\frac{k}{p}\right)!} \omega(s,t)^{k/p} \end{aligned}$$

and

$$\left\| \int_{\Delta^k[s,t]} dx^{\otimes k} \right\| + \left\| \int_{\Delta^k[s,t]} dy^{\otimes k} \right\| \leq \frac{C^k}{\left(\frac{k}{p}\right)!} \omega(s,t)^{k/p},$$

where ω is the control

$$\begin{aligned} \omega(s,t) = {} & \frac{\left(\sum_{j=1}^{[p]} \| \int dx^{\otimes j} \|_{\frac{p}{j}\text{-var},[s,t]}^{1/j} \right)^p + \left(\sum_{j=1}^{[p]} \| \int dy^{\otimes j} \|_{\frac{p}{j}\text{-var},[s,t]}^{1/j} \right)^p}{\left(\sum_{j=1}^{[p]} \| \int dx^{\otimes j} \|_{\frac{p}{j}\text{-var},[0,T]}^{1/j} \right)^p + \left(\sum_{j=1}^{[p]} \| \int dy^{\otimes j} \|_{\frac{p}{j}\text{-var},[0,T]}^{1/j} \right)^p} \\ & + \left(\frac{\sum_{j=1}^{[p]} \| \int dx^{\otimes j} - \int dy^{\otimes j} \|_{\frac{p}{j}\text{-var},[s,t]}^{1/j}}{\sum_{j=1}^{[p]} \| \int dx^{\otimes j} - \int dy^{\otimes j} \|_{\frac{p}{j}\text{-var},[0,T]}^{1/j}} \right)^p. \end{aligned}$$

Proof. We prove by induction on k that for some constants C, β,

$$\begin{aligned} &\left\| \int_{\Delta^k[s,t]} dx^{\otimes k} - \int_{\Delta^k[s,t]} dy^{\otimes k} \right\| \\ &\quad \leq \left(\sum_{j=1}^{[p]} \left\| \int dx^{\otimes j} - \int dy^{\otimes j} \right\|_{\frac{p}{j}\text{-var},[0,T]}^{1/j} \right) \frac{C^k}{\beta\left(\frac{k}{p}\right)!} \omega(s,t)^{k/p} \end{aligned}$$

and

$$\left\| \int_{\Delta^k[s,t]} dx^{\otimes k} \right\| + \left\| \int_{\Delta^k[s,t]} dy^{\otimes k} \right\| \leq \frac{C^k}{\beta\left(\frac{k}{p}\right)!} \omega(s,t)^{k/p}.$$

For $k \le p$, we trivially have

$$\begin{aligned}
&\Big\| \int_{\Delta^k[s,t]} dx^{\otimes k} - \int_{\Delta^k[s,t]} dy^{\otimes k} \Big\| \\
&\qquad \le \Big(\sum_{j=1}^{[p]} \Big\| \int dx^{\otimes j} - \int dy^{\otimes j} \Big\|^{1/j}_{\frac{p}{j}\text{-var},[0,T]} \Big)^k \omega(s,t)^{k/p} \\
&\qquad \le \Big(\sum_{j=1}^{[p]} \Big\| \int dx^{\otimes j} - \int dy^{\otimes j} \Big\|^{1/j}_{\frac{p}{j}\text{-var},[0,T]} \Big) \qquad \omega(s,t)^{k/p}
\end{aligned}$$

and

$$\Big\| \int_{\Delta^k[s,t]} dx^{\otimes k} \Big\| + \Big\| \int_{\Delta^k[s,t]} dy^{\otimes k} \Big\| \le K^{k/p} \omega(s,t)^{k/p}.$$

Not let us assume that the result is true for $0 \le j \le k$ with $k > p$. Let

$$\Gamma_{s,t} = \int_{\Delta^k[s,t]} dx^{\otimes (k+1)} - \int_{\Delta^k[s,t]} dy^{\otimes (k+1)}.$$

From the Chen relations, we have for $0 \le s \le t \le u \le T$,

$$\begin{aligned}
\Gamma_{s,u} = \Gamma_{s,t} + \Gamma_{t,u} &+ \sum_{j=1}^{k} \int_{\Delta^j[s,t]} dx^{\otimes j} \int_{\Delta^{k+1-j}[t,u]} dx^{\otimes (k+1-j)} \\
&- \sum_{j=1}^{k} \int_{\Delta^j[s,t]} dy^{\otimes j} \int_{\Delta^{k+1-j}[t,u]} dy^{\otimes (k+1-j)}.
\end{aligned}$$

Therefore, from the binomial inequality

$$\begin{aligned}
&\|\Gamma_{s,u}\| \\
&\ \le \|\Gamma_{s,t}\| + \|\Gamma_{t,u}\| \\
&\quad + \sum_{j=1}^{k} \Big\| \int_{\Delta^j[s,t]} dx^{\otimes j} - \int_{\Delta^j[s,t]} dy^{\otimes j} \Big\| \Big\| \int_{\Delta^{k+1-j}[t,u]} dx^{\otimes (k+1-j)} \Big\| \\
&\quad + \sum_{j=1}^{k} \Big\| \int_{\Delta^j[s,t]} dy^{\otimes j} \Big\| \Big\| \int_{\Delta^{k+1-j}[t,u]} dx^{\otimes (k+1-j)} - \int_{\Delta^{k+1-j}[t,u]} dy^{\otimes (k+1-j)} \Big\|
\end{aligned}$$

$$
\begin{aligned}
&\le \|\Gamma_{s,t}\| + \|\Gamma_{t,u}\| + \frac{1}{\beta^2}\tilde{\omega}(0,T)\sum_{j=1}^{k}\frac{C^j}{\left(\frac{j}{p}\right)!}\omega(s,t)^{j/p}\frac{C^{k+1-j}}{\left(\frac{k+1-j}{p}\right)!}\omega(t,u)^{(k+1-j)/p} \\
&\quad + \frac{1}{\beta^2}\tilde{\omega}(0,T)\sum_{j=1}^{k}\frac{C^j}{\left(\frac{j}{p}\right)!}\omega(s,t)^{j/p}\frac{C^{k+1-j}}{\left(\frac{k+1-j}{p}\right)!}\omega(t,u)^{(k+1-j)/p} \\
&\le \|\Gamma_{s,t}\| + \|\Gamma_{t,u}\| + \frac{2p}{\beta^2}\tilde{\omega}(0,T)C^{k+1}\frac{\omega(s,u)^{(k+1)/p}}{\left(\frac{k+1}{p}\right)!},
\end{aligned}
$$

where

$$
\tilde{\omega}(0,T) = \sum_{j=1}^{[p]}\left\| \int dx^{\otimes j} - \int dy^{\otimes j}\right\|^{1/j}_{\frac{p}{j}\text{-var},[0,T]}.
$$

We deduce that

$$
\|\Gamma_{s,t}\| \le \frac{2p}{\beta^2(1-2^{1-\theta})}\tilde{\omega}(0,T)C^{k+1}\frac{\omega(s,t)^{(k+1)/p}}{\left(\frac{k+1}{p}\right)!}
$$

with $\theta = \frac{k+1}{p}$. A correct choice of β finishes the induction argument. □

The following statement about the continuity of the iterated integrals with respect to a convenient topology can then be proved. The proof uses very similar arguments as in the previous proof, so we leave it as an exercise to the reader.

Theorem 7.22. *Let $p \ge 1$, $K > 0$ and $x, y \in C^{1\text{-var}}([0,T],\mathbb{R}^d)$ such that*

$$
\sum_{j=1}^{[p]}\left\| \int dx^{\otimes j} - \int dy^{\otimes j}\right\|^{1/j}_{\frac{p}{j}\text{-var},[0,T]} \le 1
$$

and

$$
\left(\sum_{j=1}^{[p]}\left\| \int dx^{\otimes j}\right\|^{1/j}_{\frac{p}{j}\text{-var},[0,T]}\right)^p + \left(\sum_{j=1}^{[p]}\left\| \int dy^{\otimes j}\right\|^{1/j}_{\frac{p}{j}\text{-var},[0,T]}\right)^p \le K.
$$

Then there exists a constant $C \ge 0$ depending only on p and K such that for $0 \le s \le t \le T$ and $k \ge 1$,

$$
\begin{aligned}
&\left\| \int_{\Delta^k[0,\cdot]} dx^{\otimes k} - \int_{\Delta^k[0,\cdot]} dy^{\otimes k}\right\|_{p\text{-var},[0,T]} \\
&\quad \le \frac{C^k}{\left(\frac{k}{p}\right)!}\left(\sum_{j=1}^{[p]}\left\| \int dx^{\otimes j} - \int dy^{\otimes j}\right\|^{1/j}_{\frac{p}{j}\text{-var},[0,T]}\right).
\end{aligned}
$$

This continuity result naturally leads to the following definition.

Definition 7.23. Let $p \geq 1$ and $x \in C^{p\text{-var}}([0,T],\mathbb{R}^d)$. We say that x is a *p-rough path* if there exists a sequence $x_n \in C^{1\text{-var}}([0,T],\mathbb{R}^d)$ with $x_n \to x$ in p-variation and such that for every $\varepsilon > 0$ there exists $N \geq 0$ such that

$$\sum_{j=1}^{[p]} \left\| \int dx_n^{\otimes j} - \int dx_m^{\otimes j} \right\|_{\frac{p}{j}\text{-var},[0,T]}^{1/j} \leq \varepsilon$$

for $m, n \geq N$. The space of p-rough paths will be denoted by $\Omega^p([0,T],\mathbb{R}^d)$.

From the very definition, $\Omega^p([0,T],\mathbb{R}^d)$ is the closure of $C^{1\text{-var}}([0,T],\mathbb{R}^d)$ inside $C^{p\text{-var}}([0,T],\mathbb{R}^d)$ for the distance

$$d_{\Omega^p([0,T],\mathbb{R}^d)}(x,y) = \sum_{j=1}^{[p]} \left\| \int dx^{\otimes j} - \int dy^{\otimes j} \right\|_{\frac{p}{j}\text{-var},[0,T]}^{1/j}.$$

If $x \in \Omega^p([0,T],\mathbb{R}^d)$ and if $x_n \in C^{1\text{-var}}([0,T],\mathbb{R}^d)$ is such that $x_n \to x$ in p-variation and such that for every $\varepsilon > 0$ there exists $N \geq 0$ such that for $m, n \geq N$,

$$\sum_{j=1}^{[p]} \left\| \int dx_n^{\otimes j} - \int dx_m^{\otimes j} \right\|_{\frac{p}{j}\text{-var},[0,T]}^{1/j} \leq \varepsilon,$$

then we define $\int_{\Delta^k[s,t]} dx^{\otimes k}$ for $k \leq p$ as the limit of the iterated integrals $\int_{\Delta^k[s,t]} dx_n^{\otimes k}$. However it is important to observe that $\int_{\Delta^k[s,t]} dx^{\otimes k}$ may then depend on the choice of the approximating sequence x_n. Once the integrals $\int_{\Delta^k[s,t]} dx^{\otimes k}$ are defined for $k \leq p$, we can then use the previous theorem to construct all the iterated integrals $\int_{\Delta^k[s,t]} dx^{\otimes k}$ for $k > p$. It is obvious that if $x, y \in \Omega^p([0,T],\mathbb{R}^d)$, then

$$1 + \sum_{k=1}^{[p]} \int_{\Delta^k[s,t]} dx^{\otimes k} = 1 + \sum_{k=1}^{[p]} \int_{\Delta^k[s,t]} dy^{\otimes k}$$

implies that

$$1 + \sum_{k=1}^{+\infty} \int_{\Delta^k[s,t]} dx^{\otimes k} = 1 + \sum_{k=1}^{+\infty} \int_{\Delta^k[s,t]} dy^{\otimes k}.$$

In other words, the signature of a p-rough path is completely characterized by its truncated signature at order $[p]$:

$$\mathfrak{S}_{[p]}(x)_{s,t} = 1 + \sum_{k=1}^{[p]} \int_{\Delta^k[s,t]} dx^{\otimes k}.$$

For this reason, it is natural to present a p-rough path by this truncated signature at order $[p]$ in order to stress that the choice of the approximating sequence to construct the iterated integrals up to order $[p]$ has been made. The following results are straightforward to obtain by a limiting argument.

Lemma 7.24 (Chen relations). *Let $x \in \Omega^p([0,T],\mathbb{R}^d)$, $p \geq 1$. For $0 \leq s \leq t \leq u \leq T$ and $n \geq 1$,*

$$\int_{\Delta^n[s,u]} dx^{\otimes n} = \sum_{k=0}^{n} \int_{\Delta^k[s,t]} dx^{\otimes k} \int_{\Delta^{n-k}[t,u]} dx^{\otimes (n-k)}.$$

Theorem 7.25. *Let $p \geq 1$. There exists a constant $C \geq 0$, depending only on p, such that for every $x \in \Omega^p([0,T],\mathbb{R}^d)$ and $k \geq 1$,*

$$\Big\| \int_{\Delta^k[s,t]} dx^{\otimes k} \Big\| \leq \frac{C^k}{(\frac{k}{p})!} \Big(\sum_{j=1}^{[p]} \Big\| \int dx^{\otimes j} \Big\|^{1/j}_{\frac{p}{j}\text{-var},[s,t]} \Big)^k, \quad 0 \leq s \leq t \leq T.$$

7.4 Rough differential equations

In this section we define solutions of differential equations driven by p-rough paths, $p \geq 1$. We study in detail the case of linear equations for which we prove the Lyons continuity result. At the end of the section we state the full Lyons result.

Let $x \in \Omega^p([0,T],\mathbb{R}^d)$ be a p-rough path with truncated signature

$$\sum_{k=0}^{[p]} \int_{\Delta^k[s,t]} dx^{\otimes k},$$

and let $x_n \in C^{1\text{-var}}([0,T],\mathbb{R}^d)$ be an approximating sequence such that

$$\sum_{j=1}^{[p]} \Big\| \int dx^{\otimes j} - \int dx_n^{\otimes j} \Big\|^{1/j}_{\frac{p}{j}\text{-var},[0,T]} \to 0.$$

Let us consider matrices $M_1,\dots,M_d \in \mathbb{R}^{n\times n}$. We have the following theorem:

Theorem 7.26 (Lyons' continuity theorem, linear case). *Let $y_n\colon [0,T] \to \mathbb{R}^n$ be the solution of the differential equation*

$$y_n(t) = y(0) + \sum_{i=1}^{d} \int_0^t M_i y_n(s) dx_n^i(s).$$

Then, when $n \to \infty$, y_n converges in the p-variation distance to some $y \in C^{p\text{-var}}([0,T],\mathbb{R}^n)$. The path y is called the solution of the rough differential equation

$$y(t) = y(0) + \sum_{i=1}^{d} \int_0^t M_i y(s) dx^i(s).$$

Proof. It is a classical result that the solution of the equation

$$y_n(t) = y(0) + \sum_{i=1}^{d} \int_0^t M_i y_n(s) dx_n^i(s)$$

can be expanded as the convergent Volterra series:

$$y_n(t) = y_n(s) + \sum_{k=1}^{+\infty} \sum_{I=(i_1,\dots,i_k)} M_{i_1} \dots M_{i_k} \Big(\int_{\Delta^k[s,t]} dx_n^I \Big) y_n(s).$$

Therefore, in particular, for $n, m \geq 0$,

$$y_n(t) - y_m(t) = \sum_{k=1}^{+\infty} \sum_{I=(i_1,\dots,i_k)} M_{i_1} \dots M_{i_k} \Big(\int_{\Delta^k[0,t]} dx_n^I - \int_{\Delta^k[0,t]} dx_m^I \Big) y(0),$$

which implies that

$$\|y_n(t) - y_m(t)\| \leq \sum_{k=1}^{+\infty} M^k \Big\| \int_{\Delta^k[0,t]} dx_n^{\otimes k} - \int_{\Delta^k[0,t]} dx_m^{\otimes k} \Big\| \|y(0)\|$$

with $M = \max\{\|M_1\|, \dots, \|M_d\|\}$. From Theorem 7.22, there exists a constant $C \geq 0$ depending only on p and

$$\sup_n \sum_{j=1}^{[p]} \Big\| \int dx_n^{\otimes j} \Big\|_{\frac{p}{j}\text{-var},[0,T]}^{1/j}$$

such that for $k \geq 1$ and n, m big enough one has

$$\begin{aligned} &\Big\| \int_{\Delta^k[0,\cdot]} dx_n^{\otimes k} - \int_{\Delta^k[0,\cdot]} dx_m^{\otimes k} \Big\|_{p\text{-var},[0,T]} \\ &\quad \leq \Big(\sum_{j=1}^{[p]} \Big\| \int dx_n^{\otimes j} - \int dx_m^{\otimes j} \Big\|_{\frac{p}{j}\text{-var},[0,T]}^{1/j} \Big) \frac{C^k}{\big(\frac{k}{p}\big)!}. \end{aligned}$$

As a consequence, there exists a constant $\widetilde{C}$ such that for n, m big enough,

$$\|y_n(t) - y_m(t)\| \le \widetilde{C} \sum_{j=1}^{[p]} \left\| \int dx_n^{\otimes j} - \int dx_m^{\otimes j} \right\|^{1/j}_{\frac{p}{j}\text{-var},[0,T]}.$$

This already proves that y_n converges in the supremum topology to some y. We now have

$$\begin{aligned}&(y_n(t) - y_n(s)) - (y_m(t) - y_m(s)) \\ &\quad = \sum_{k=1}^{+\infty} \sum_{I=(i_1,\dots,i_k)} M_{i_1} \dots M_{i_k} \Big(\int_{\Delta^k[s,t]} dx_n^I y_n(s) - \int_{\Delta^k[s,t]} dx_m^I y_m(s) \Big),\end{aligned}$$

and we can bound

$$\begin{aligned}&\Big\| \int_{\Delta^k[s,t]} dx_n^I y_n(s) - \int_{\Delta^k[s,t]} dx_m^I y_m(s) \Big\| \\ &\le \Big\| \int_{\Delta^k[s,t]} dx_n^I \Big\| \|y_n(s) - y_m(s)\| + \|y_m(s)\| \Big\| \int_{\Delta^k[s,t]} dx_n^I - \int_{\Delta^k[s,t]} dx_m^I \Big\| \\ &\le \Big\| \int_{\Delta^k[s,t]} dx_n^I \Big\| \|y_n - y_m\|_{\infty,[0,T]} + \|y_m\|_{\infty,[0,T]} \Big\| \int_{\Delta^k[s,t]} dx_n^I - \int_{\Delta^k[s,t]} dx_m^I \Big\|.\end{aligned}$$

Now there exists a constant $C \ge 0$ depending only on p and

$$\sup_n \sum_{j=1}^{[p]} \left\| \int dx_n^{\otimes j} \right\|^{1/j}_{\frac{p}{j}\text{-var},[0,T]}$$

such that for $k \ge 1$ and n, m big enough one has

$$\left\| \int_{\Delta^k[s,t]} dx_n^{\otimes k} \right\| \le \frac{C^k}{\left(\frac{k}{p}\right)!} \omega(s,t)^{k/p}, \quad 0 \le s \le t \le T,$$

and

$$\begin{aligned}&\left\| \int_{\Delta^k[s,t]} dx_n^{\otimes k} - \int_{\Delta^k[s,t]} dx_m^{\otimes k} \right\| \\ &\qquad \le \Big(\sum_{j=1}^{[p]} \left\| \int dx_n^{\otimes j} - \int dx_m^{\otimes k} \right\|^{1/j}_{\frac{p}{j}\text{-var},[0,T]} \Big) \frac{C^k}{\left(\frac{k}{p}\right)!} \omega(s,t)^{k/p},\end{aligned}$$

where ω is a control such that $\omega(0,T) = 1$. Consequently, there is a constant $\widetilde{C}$ such that

$$\begin{aligned}&\|(y_n(t) - y_n(s)) - (y_m(t) - y_m(s))\| \\ &\quad \le \widetilde{C} \Big(\|y_n - y_m\|_{\infty,[0,T]} + \sum_{j=1}^{[p]} \left\| \int dx_n^{\otimes j} - \int dx_m^{\otimes k} \right\|^{1/j}_{\frac{p}{j}\text{-var},[0,T]} \Big) \omega(s,t)^{1/p}.\end{aligned}$$

This implies the estimate

$$\|y_n - y_m\|_{p\text{-var},[0,T]} \leq \widetilde{C}\Big(\|y_n - y_m\|_{\infty,[0,T]} + \sum_{j=1}^{[p]} \Big\| \int dx_n^{\otimes j} - \int dx_m^{\otimes k} \Big\|^{1/j}_{\frac{p}{j}\text{-var},[0,T]}\Big)$$

and thus gives the conclusion. □

Exercise 7.27. Let $y_n \colon [0,T] \to \mathbb{R}^n$ be the solution of the differential equation

$$y_n(t) = y(0) + \sum_{i=1}^{d} \int_0^t M_i y_n(s) dx_n^i(s),$$

and y be the solution of the rough differential equation

$$y(t) = y(0) + \sum_{i=1}^{d} \int_0^t M_i y(s) dx^i(s).$$

Show that $y \in \Omega^p([0,T], \mathbb{R}^n)$ and that when $n \to \infty$,

$$\sum_{j=1}^{[p]} \Big\| \int dy^{\otimes j} - \int dy_n^{\otimes j} \Big\|^{1/j}_{\frac{p}{j}\text{-var},[0,T]} \to 0.$$

We can get useful estimates for solutions of rough differential equations. To this end we need the following real analysis lemma:

Proposition 7.28. *For $x \geq 0$ and $p \geq 1$,*

$$\sum_{k=0}^{+\infty} \frac{x^k}{\left(\frac{k}{p}\right)!} \leq p e^{x^p}.$$

Proof. For $\alpha > 0$, we write

$$E_\alpha(x) = \sum_{k=0}^{+\infty} \frac{x^k}{(k\alpha)!}.$$

This is a special function called the Mittag-Leffler function. From the binomial inequality,

$$E_\alpha(x)^2 = \sum_{k=0}^{+\infty} \Big(\sum_{j=0}^{k} \frac{1}{(j\alpha)!((k-j)\alpha)!} \Big) x^k \leq \frac{1}{\alpha} \sum_{k=0}^{+\infty} 2^{\alpha k} \frac{x^k}{(k\alpha)!} = \frac{1}{\alpha} E_\alpha(2^\alpha x).$$

Thus we proved

$$E_\alpha(x) \le \frac{1}{\alpha^{1/2}} E_\alpha(2^\alpha x)^{1/2}.$$

Iterating this inequality k times we obtain

$$E_\alpha(x) \le \frac{1}{\alpha^{\sum_{j=1}^k \frac{1}{2^j}}} E_\alpha(2^{\alpha k} x)^{1/2^k}.$$

It is known (and not difficult to prove, see Exercise 7.29) that

$$E_\alpha(x) \sim_{x\to\infty} \frac{1}{\alpha} e^{x^{1/\alpha}}.$$

By letting $k \to \infty$ we conclude that

$$E_\alpha(x) \le \frac{1}{\alpha} e^{x^{1/\alpha}}.$$

□

Exercise 7.29. For $\alpha \ge 0$, we consider the Mittag-Leffler function

$$E_\alpha(x) = \sum_{k=0}^{+\infty} \frac{x^k}{(k\alpha)!}.$$

Show that

$$E_\alpha(x) \sim_{x\to\infty} \frac{1}{\alpha} e^{x^{1/\alpha}}.$$

Hint. Compute the Laplace transform of $\sum \frac{x^{\alpha k}}{(k\alpha-1)!}$.

This estimate provides the following result:

Proposition 7.30. *Let y be the solution of the rough differential equation*

$$y(t) = y(0) + \sum_{i=1}^d \int_0^t M_i y(s) dx^i(s).$$

Then there exists a constant C depending only on p such that for $0 \le t \le T$,

$$\|y(t)\| \le p\|y(0)\| e^{(CM)^p (\sum_{j=1}^{[p]} \|\int dx^{\otimes j}\|^{1/j}_{\frac{p}{j}\text{-var},[0,t]})^p},$$

where $M = \max\{\|M_1\|, \dots, \|M_d\|\}$.

Proof. We have

$$y(t) = y(0) + \sum_{k=1}^{+\infty} \sum_{I=(i_1,\dots,i_k)} M_{i_1} \dots M_{i_k} \Big(\int_{\Delta^k[0,t]} dx^I \Big) y(0).$$

Thus we obtain

$$\|y(t)\| \le \Big(1 + \sum_{k=1}^{+\infty} \sum_{I=(i_1,\dots,i_k)} M^k \Big\| \int_{\Delta^k[0,t]} dx^I \Big\| \Big) \|y(0)\|,$$

and we conclude by using estimates on iterated integrals of rough paths (see Theorem 7.25) together with Proposition 7.28. □

We now give the Lyons' continuity result for non-linear differential equations. We first introduce the notion of a γ-Lipschitz vector field.

Definition 7.31. A vector field V on $\mathbb{R}^n$ is called *γ-Lipschitz* if it is $[\gamma]$ times continuously differentiable and there exists a constant $M \ge 0$ such that the supremum norm of its k-th derivatives, $k = 0, \dots, [\gamma]$, and the $\gamma - [\gamma]$ Hölder norm of its $[\gamma]$-th derivative are bounded by M. The smallest M that satisfies the above condition is the γ-Lipschitz norm of V and will be denoted by $\|V\|_{\mathrm{Lip}^\gamma}$.

We now formulate the non-linear version of the Lyons continuity theorem. For a proof we refer the interested reader to [29].

Theorem 7.32 (Lyons continuity theorem). *Let $p \ge 1$. Let $x \in \Omega^p([0,T], \mathbb{R}^d)$ be a p-rough path with truncated signature*

$$\sum_{k=0}^{[p]} \int_{\Delta^k[s,t]} dx^{\otimes k},$$

and let $x_n \in C^{1\text{-var}}([0,T], \mathbb{R}^d)$ be an approximating sequence such that

$$\sum_{j=1}^{[p]} \Big\| \int dx^{\otimes j} - \int dx_n^{\otimes j} \Big\|_{\frac{p}{j}\text{-var},[0,T]}^{1/j} \to 0.$$

Assume that $V_1, \dots, V_d$ are γ-Lipschitz vector fields in $\mathbb{R}^n$ with $\gamma > p$. The solution of the equation

$$y_n(t) = y(0) + \sum_{j=1}^{d} \int_0^t V_j(y_n(s)) dx_n^j(s), \quad 0 \le t \le T,$$

converges when $n \to +\infty$ in p-variation to some $y \in C^{p\text{-var}}([0,T], \mathbb{R}^n)$ which we call a solution of the rough differential equation

$$y(t) = y(0) + \sum_{j=1}^{d} \int_0^t V_j(y(s)) dx^j(s), \quad 0 \le t \le T.$$

Moreover, there exists a constant C depending only on p, T and γ such that for every $0 \le s < t \le T$,

$$\|y\|_{p\text{-var},[s,t]} \le C(\|V\|_{\mathrm{Lip}^{\gamma-1}} \|\mathbf{x}\|_{p\text{-var},[s,t]} + \|V\|^p_{\mathrm{Lip}^{\gamma-1}} \|\mathbf{x}\|^p_{p\text{-var},[s,t]}), \tag{7.1}$$

where

$$\|\mathbf{x}\|_{p\text{-var},[s,t]} = \sum_{j=1}^{[p]} \Big\| \int dx^{\otimes j} \Big\|^{1/j}_{\frac{p}{j}\text{-var},[s,t]}.$$

Remark 7.33. The estimate (7.1) is often referred to as the Davie estimate.

7.5 The Brownian motion as a rough path

It is now time to show how rough paths theory can be used to study Brownian motion and stochastic differential equations. We first show in this section that Brownian motion paths are almost surely p-rough paths for $2 < p < 3$. The key estimate is the Garsia–Rodemich–Rumsey inequality.

Theorem 7.34 (Garsia–Rodemich–Rumsey inequality). *Let $T > 0$ and let*

$$\Gamma\colon \{0 \le s, t \le T\} \to \mathbb{R}_{\ge 0}$$

be a continuous symmetric functional ($\Gamma_{s,t} = \Gamma_{t,s}$) that vanishes on the diagonal ($\Gamma_{t,t} = 0$) and such that there exists a constant $C > 0$ such that for every $0 \le t_1 \le \dots \le t_n \le T$,

$$\Gamma_{t_1,t_n} \le C\Big(\sum_{i=1}^{n-1} \Gamma_{t_i,t_{i+1}}\Big). \tag{7.2}$$

For $q > 1$ and $\alpha \in (1/q, 1)$, there exists a constant $K > 0$ such that for all $0 \le s \le t \le T$,

$$\Gamma^q_{s,t} \le K|t-s|^{\alpha q-1} \int_0^T \int_0^T \frac{\Gamma^q_{u,v}}{|u-v|^{1+\alpha q}} du dv.$$

Proof. Step 1. We first assume $T = 1$ and prove that

$$\Gamma^q_{0,1} \le K \int_0^1 \int_0^1 \frac{\Gamma^q_{u,v}}{|u-v|^{1+\alpha q}} du dv.$$

Define $I(v) = \int_0^1 \frac{\Gamma^q_{u,v}}{|u-v|^{1+\alpha q}} du$ so that

$$\int_0^1 I(v) dv = \int_0^1 \int_0^1 \frac{\Gamma^q_{u,v}}{|u-v|^{1+\alpha q}} du dv.$$

We can find $t_0 \in (0,1)$ such that $I(t_0) \le \int_0^1 \int_0^1 \frac{\Gamma^q_{u,v}}{|u-v|^{1+\alpha q}} du dv$. We construct then a decreasing sequence $(t_n)_{n \ge 0}$ by induction as follows. If t_{n-1} has been chosen, then we pick $t_n \in (0, \frac{1}{2} t_{n-1})$ such that

$$I(t_n) \le \frac{4}{t_{n-1}} \int_0^1 \int_0^1 \frac{\Gamma^q_{u,v}}{|u-v|^{1+\alpha q}} du dv, \quad J_{n-1}(t_n) \le \frac{4 I(t_{n-1})}{t_{n-1}},$$

where we have set $J_{n-1}(s) = \frac{\Gamma^q_{s,t_{n-1}}}{|s-t_{n-1}|^{1+\alpha q}}$. We can always find such t_n. Otherwise, we would have $(0, \frac{1}{2} t_{n-1}) = A \cup B$ with

$$A = \Big\{ t \in \Big(0, \frac{1}{2} t_{n-1}\Big), I(t) > \frac{4}{t_{n-1}} \int_0^1 \int_0^1 \frac{\Gamma^q_{u,v}}{|u-v|^{1+\alpha q}} du dv \Big\},$$

$$B = \Big\{ t \in \Big(0, \frac{1}{2} t_{n-1}\Big), J_{n-1}(t) > \frac{4 I(t_{n-1})}{t_{n-1}} \Big\}.$$

But clearly we have

$$\frac{4\mu(A)}{t_{n-1}} \int_0^1 \int_0^1 \frac{\Gamma^q_{u,v}}{|u-v|^{1+\alpha q}} du dv \le \int_A I(t) dt \le \int_0^1 \int_0^1 \frac{\Gamma^q_{u,v}}{|u-v|^{1+\alpha q}} du dv,$$

where μ stands for the Lebesgue measure. Hence, if $\mu(A) > 0$, we have $\mu(A) < \frac{1}{4} t_{n-1}$. Similarly of course, $\mu(B) > 0$ implies $\mu(B) < \frac{1}{4} t_{n-1}$. This contradicts the fact that $\big(0, \frac{1}{2} t_{n-1}\big) = A \cup B$. Hence t_n is well defined if t_{n-1} is. We then have

$$J_{n-1}(t_n) \le \frac{4 I(t_{n-1})}{t_{n-1}} \le \frac{16}{t_{n-1}^2} \int_0^1 \int_0^1 \frac{\Gamma^q_{u,v}}{|u-v|^{1+\alpha q}} du dv.$$

Coming back to the definition of J_{n-1} then yields

$$\frac{\Gamma^q_{t_n, t_{n-1}}}{|t_n - t_{n-1}|^{1+\alpha q}} \le \frac{16}{t_{n-1}^2} \int_0^1 \int_0^1 \frac{\Gamma^q_{u,v}}{|u-v|^{1+\alpha q}} du dv.$$

But it is easily seen that there is a constant $C > 0$ such that

$$\frac{|t_n - t_{n-1}|^{\alpha + 1/q}}{t_{n-1}^{2/q}} \le C (t_n^{\alpha - 1/q} - t_{n+1}^{\alpha - 1/q}).$$

Using then the subadditivity of the functional Γ we end up with

$$\Gamma^q_{0,t_0} \le K \int_0^1 \int_0^1 \frac{\Gamma^q_{u,v}}{|u-v|^{1+\alpha q}} du dv.$$

Similarly, by repeating the above argument on the functional $\Gamma_{1-s,1-t}$ we get that

$$\Gamma^q_{t_0,1} \le K \int_0^1 \int_0^1 \frac{\Gamma^q_{u,v}}{|u-v|^{1+\alpha q}} du dv.$$

Therefore we proved that for some constant K',

$$\Gamma_{0,1}^q \le K' \int_0^1 \int_0^1 \frac{\Gamma_{u,v}^q}{|u-v|^{1+\alpha q}} du dv.$$

Step 2. We now consider a general functional $\Gamma\colon \{0 \le s,t \le T\} \to \mathbb{R}_{\ge 0}$ and fix $0 \le s < t \le T$. Consider then the rescaled functional $\widetilde{\Gamma}_{u,v} = \Gamma_{s+u(t-s),s+v(t-s)}$ which is defined for $0 \le u, v \le 1$. From the first step, we have

$$\widetilde{\Gamma}_{0,1}^q \le K \int_0^1 \int_0^1 \frac{\widetilde{\Gamma}_{u,v}^q}{|u-v|^{1+\alpha q}} du dv,$$

and the result follows by a simple change of variable. □

As we are going to see, for the Brownian motion the natural integral in rough paths theory is not Itô's integral but Stratonovitch's (see Section 6.5 for the definition). If $(B_t)_{t\ge 0}$ is d-dimensional Brownian motion, we can inductively define the iterated Stratonovitch integrals $\int_{0\le t_1\le\cdots\le t_k\le t} \circ dB_{t_1}^{i_1} \circ \cdots \circ dB_{t_k}^{i_k}$, and, as before, it will be convenient to use a concise notation by embedding the set of iterated integrals in the algebra of formal series

$$\int_{\Delta^k[s,t]} \circ dB^{\otimes k} = \sum_{I\in\{1,\dots,d\}^k} \Big(\int_{\Delta^k[s,t]} \circ dB^I \Big) X_{i_1} \dots X_{i_k}.$$

Exercise 7.35. Show that for $m \ge 1$, and $t \ge 0$,

$$\mathbb{E}\Big(\Big| \int_{0\le t_1\le\cdots\le t_k\le t} \circ dB_{t_1}^{i_1} \circ \cdots \circ dB_{t_k}^{i_k} \Big|^m \Big) < +\infty.$$

The following estimate is crucial in order to apply rough paths theory to Brownian motion.

Proposition 7.36. *Let $T \ge 0$, $2 < p < 3$ and $n \ge 1$. There exists a finite positive random variable $C = C(T,p,n)$ such that $\mathbb{E}(C^m) < +\infty$ for every $m \ge 1$, and for every $0 \le s \le t \le T$,*

$$\sum_{k=1}^n \Big\| \int_{\Delta^k[s,t]} \circ dB^{\otimes k} \Big\|^{1/k} \le C|t-s|^{1/p}.$$

Proof. Let us consider the functional

$$\Gamma_{s,t} = \sum_{k=1}^n \Big\| \int_{\Delta^k[s,t]} \circ dB^{\otimes k} \Big\|^{1/k},$$

and prove that it satisfies the chain condition (7.2). Using inductively Chen's relations, we see that

$$\begin{aligned}&\int_{\Delta^k[t_1,t_N]} \circ dB^{\otimes k} \\ &\quad = \sum_{i_1+\dots+i_{N-1}=k} \int_{\Delta^{i_1}[t_1,t_2]} \circ dB^{\otimes i_1} \dots \int_{\Delta^{i_{N-1}}[t_{N-1},t_N]} \circ dB^{\otimes i_{N-1}}.\end{aligned}$$

Therefore we have

$$\begin{aligned}&\Big\| \int_{\Delta^k[t_1,t_N]} \circ dB^{\otimes k} \Big\| \\ &\quad \le \sum_{i_1+\dots+i_{N-1}=k} \Big\| \int_{\Delta^{i_1}[t_1,t_2]} \circ dB^{\otimes i_1} \dots \int_{\Delta^{i_{N-1}}[t_{N-1},t_N]} \circ dB^{\otimes i_{N-1}} \Big\| \\ &\quad \le \sum_{i_1+\dots+i_{N-1}=k} \Big\| \int_{\Delta^{i_1}[t_1,t_2]} \circ dB^{\otimes i_1} \Big\| \dots \Big\| \int_{\Delta^{i_{N-1}}[t_{N-1},t_N]} \circ dB^{\otimes i_{N-1}} \Big\| \\ &\quad \le \sum_{i_1+\dots+i_{N-1}=k} \Gamma^{i_1}_{t_1,t_2} \dots \Gamma^{i_{N-1}}_{t_{N-1},t_N} \\ &\quad \le (\Gamma_{t_1,t_2} + \dots + \Gamma_{t_{N-1},t_N})^k.\end{aligned}$$

This yields

$$\Gamma_{t_1,t_N} \le n(\Gamma_{t_1,t_2} + \dots + \Gamma_{t_{N-1},t_N}).$$

From the Garsia–Rodemich–Rumsey inequality, for $q > 1$ and $\alpha \in (1/q, 1)$, there exists therefore a constant $K > 0$ such that for all $0 \le s \le t \le T$,

$$\Gamma^q_{s,t} \le K|t-s|^{\alpha q - 1} \int_0^T \int_0^T \frac{\Gamma^q_{u,v}}{|u-v|^{1+\alpha q}} du dv.$$

The scaling property of Brownian motion and Exercise 7.35 imply that for some constant $K > 0$

$$\mathbb{E}(\Gamma^q_{u,v}) = K|u-v|^{q/2}.$$

As a consequence

$$\int_0^T \int_0^T \frac{\Gamma^q_{u,v}}{|u-v|^{1+\alpha q}} du dv < +\infty$$

if $\alpha - \frac{1}{q} < \frac{1}{2}$ and the result follows from Fubini's theorem. □

With this estimate in hands we are now in a position to prove that Brownian motion paths are p-rough paths for $2 < p < 3$. We can moreover give an explicit approximating sequence. Let us work on a fixed interval $[0, T]$ and consider a sequence D_n of subdivisions of $[0, T]$ such that $D_{n+1} \subset D_n$ and whose mesh goes to 0 when $n \to +\infty$. An example is given by the sequence of dyadic subdivisions. The family $\mathcal{F}_n = \sigma(B_t, t \in D_n)$ is then a filtration, that is, an increasing family of σ-fields. We denote by B^n the piecewise linear process which is obtained from B by interpolation along the subdivision D_n, that is, for $t_i^n \le t \le t_{i+1}^n$,

$$B_t^n = \frac{t_{i+1}^n - t}{t_{i+1}^n - t_i^n} B_{t_i^n} + \frac{t - t_i^n}{t_{i+1}^n - t_i^n} B_{t_{i+1}^n}.$$

Theorem 7.37. *When $n \to +\infty$, almost surely*

$$\| B^n - B \|_{p\text{-var},[0,T]} + \Big\| \int dB^{n,\otimes 2} - \int \circ dB^{\otimes 2} \Big\|^{1/2}_{\frac{p}{2}\text{-}var,[0,T]} \to 0.$$

Thus, Brownian motion paths are almost surely p-rough paths for $2 < p < 3$.

Proof. We first observe that, due to the Markov property of Brownian motion, we have for $t_i^n \le t \le t_{i+1}^n$,

$$\mathbb{E}(B_t \mid \mathcal{F}_n) = \mathbb{E}(B_t \mid B_{t_i^n}, B_{t_i^{n+1}}).$$

It is then an easy exercise to check that

$$\mathbb{E}(B_t \mid B_{t_i^n}, B_{t_i^{n+1}}) = \frac{t_{i+1}^n - t}{t_{i+1}^n - t_i^n} B_{t_i^n} + \frac{t - t_i^n}{t_{i+1}^n - t_i^n} B_{t_{i+1}^n} = B_t^n.$$

As a conclusion we get

$$\mathbb{E}(B_t \mid \mathcal{F}_n) = B_t^n.$$

It immediately follows that $B_t^n \to B_t$ when $n \to +\infty$. In the same way, for $i \neq j$ we have

$$\mathbb{E}\Big(\int_0^t B_s^i dB_s^j \mid \mathcal{F}_n \Big) = \int_0^t B_s^{n,i} dB_s^{n,j}.$$

Indeed, for $0 < t < T$ and ε small enough, we have by independence of B^i and B^j that

$$\mathbb{E}(B_t^i (B_{t+\varepsilon}^j - B_t^j) \mid \mathcal{F}_n) = \mathbb{E}(B_t^i \mid \mathcal{F}_n) \mathbb{E}(B_{t+\varepsilon}^j - B_t^j \mid \mathcal{F}_n) = B_t^{n,i} (B_{t+\varepsilon}^{n,j} - B_t^{n,j}),$$

and we conclude by using the fact that Itô's integral is a limit in L^2 of Riemann sums. It follows that, almost surely,

$$\lim_{n \to \infty} \int_0^t B_s^{n,i} dB_s^{n,j} = \int_0^t B_s^i dB_s^j.$$

Since B^i is independent from B^j, the quadratic covariation $\langle B^i, B^j \rangle$ is zero. As a consequence we have $\int_0^t B^i_s dB^j_s = \int_0^t B^i_s \circ dB^j_s$. For $i = j$ we have $\int_0^t B^i_s \circ dB^i_s = (B^i_t)^2$. So if we collect the previous results, we established that almost surely

$$\lim_{n \to +\infty} \Big(B^n_t, \int_0^t B^n_s \otimes dB^n_s \Big) = \Big(B_t, \int_0^t B_s \otimes \circ dB_s \Big).$$

To prove the convergence with the variation norms, we need a uniform Hölder estimate.

From Proposition 7.36, we know that there is a finite random variable K_1 (which belongs to L^m for every $m \geq 1$) such that for every $i \neq j, 0 \leq s \leq t \leq T$,

$$\Big| \int_s^t (B^i_u - B^i_s) \circ dB^j_u \Big| \leq K_1 |t - s|^{2/p}.$$

Since

$$\mathbb{E}\Big(\int_s^t (B^i_u - B^i_s) dB^j_u \mid \mathcal{F}_n \Big) = \int_s^t (B^{n,i}_u - B^{n,i}_s) dB^{n,j}_u,$$

we deduce that

$$\Big| \int_s^t (B^{n,i}_u - B^{n,i}_s) dB^{n,j}_u \Big| \leq K_2 |t - s|^{2/p},$$

where K_2 is a finite random variable that belongs to L^m for every $m \geq 1$. Similarly, of course, we have

$$\| B^n_t - B^n_s \| \leq K_3 |t - s|^{1/p}.$$

Combining the pointwise convergence with these uniform Hölder estimates then give the expected result. We let the reader work out the details as an exercise. □

Since a d-dimensional Brownian motion $(B_t)_{t \geq 0}$ is a p-rough path for $2 < p < 3$, we know how to give a sense to the signature of the Brownian motion. In particular, the iterated integrals at any order of the Brownian motion are well defined. It turns out that these iterated integrals, and this comes at no surprise in view of the previous result, coincide with iterated Stratonovitch integrals.

Theorem 7.38. *If $(B_t)_{t \geq 0}$ is a d-dimensional Brownian motion, the signature of B as a rough path is the formal series*

$$\begin{aligned} \mathfrak{S}(B)_t &= 1 + \sum_{k=1}^{+\infty} \int_{\Delta^k[0,t]} \circ dB^{\otimes k} \\ &= 1 + \sum_{k=1}^{+\infty} \sum_{I \in \{1,\dots,d\}^k} \Big(\int_{0 \leq t_1 \leq \dots \leq t_k \leq t} \circ dB^{i_1}_{t_1} \dots \circ dB^{i_k}_{t_k} \Big) X_{i_1} \dots X_{i_k}. \end{aligned}$$

Proof. Let us work on a fixed interval $[0,T]$ and consider a sequence D_n of subdivisions of $[0,T]$ such that $D_{n+1} \subset D_n$ and whose mesh goes to 0 when $n \to +\infty$. As before, by B^n we denote the piecewise linear process which is obtained from B by interpolation along the subdivision D_n, that is, for $t_i^n \le t \le t_{i+1}^n$,

$$B_t^n = \frac{t_{i+1}^n - t}{t_{i+1}^n - t_i^n} B_{t_i^n} + \frac{t - t_i^n}{t_{i+1}^n - t_i^n} B_{t_{i+1}^n}.$$

We know from Theorem 7.37 that B^n converges to B in the p-rough paths topology $2 < p < 3$. In particular all the iterated integrals $\int_{\Delta^k[s,t]} dB^{n,\otimes k}$ converge. We claim that actually

$$\lim_{n\to\infty} \int_{\Delta^k[s,t]} dB^{n,\otimes k} = \int_{\Delta^k[0,t]} \circ\, dB^{\otimes k}.$$

Let us write

$$\int_{\Delta^k[s,t]} \partial B^{\otimes k} = \lim_{n\to\infty} \int_{\Delta^k[s,t]} dB^{n,\otimes k}.$$

We are going to prove by induction on k that $\int_{\Delta^k[s,t]} \partial B^{\otimes k} = \int_{\Delta^k[s,t]} \circ\, dB^{\otimes k}$. We have

$$\begin{aligned}
&\int_0^T B_s^n \otimes dB_s^n \\
&= \sum_{i=0}^{n-1} \int_{t_i^n}^{t_{i+1}^n} B_s^n \otimes dB_s^n \\
&= \sum_{i=0}^{n-1} \int_{t_i^n}^{t_{i+1}^n} \Big(\frac{t_{i+1}^n - s}{t_{i+1}^n - t_i^n} B_{t_i^n} + \frac{s - t_i^n}{t_{i+1}^n - t_i^n} B_{t_{i+1}^n} \Big) ds \otimes \frac{B_{t_{i+1}^n} - B_{t_i^n}}{t_{i+1}^n - t_i^n} \\
&= \frac{1}{2} \sum_{i=0}^{n-1} (B_{t_{i+1}^n} + B_{t_i^n}) \otimes (B_{t_{i+1}^n} - B_{t_i^n}).
\end{aligned}$$

By taking the limit when $n \to \infty$, we therefore deduce that $\int_{\Delta^2[0,T]} \partial B^{\otimes 2} = \int_{\Delta^2[0,T]} \circ\, dB^{\otimes 2}$. In the same way, we have $\int_{\Delta^2[s,t]} \partial B^{\otimes 2} = \int_{\Delta^2[s,t]} \circ dB^{\otimes 2}$ for $0 \le s < t \le T$. Assume now by induction that $\int_{\Delta^k[s,t]} \partial B^{\otimes k} = \int_{\Delta^k[s,t]} \circ dB^{\otimes k}$ for every $0 \le s \le t \le T$ and $1 \le j \le k$. Let us write

$$\Gamma_{s,t} = \int_{\Delta^{k+1}[s,t]} \partial B^{\otimes(k+1)} - \int_{\Delta^{k+1}[s,t]} \circ dB^{\otimes(k+1)}.$$

From the Chen relations, we immediately see that

$$\Gamma_{s,u} = \Gamma_{s,t} + \Gamma_{t,u}.$$

Moreover, it is easy to estimate

$$\|\Gamma_{s,t}\| \le C\omega(s,t)^{\frac{k+1}{p}},$$

where $2 < p < 3$ and $\omega(s,t) = |t-s|$. Indeed, the bound

$$\Big\| \int_{\Delta^{k+1}[s,t]} \partial B^{\otimes(k+1)} \Big\| \le C_1 \omega(s,t)^{\frac{k+1}{p}}$$

comes from Theorem 7.25, and the bound

$$\Big\| \int_{\Delta^{k+1}[s,t]} \circ dB^{\otimes(k+1)} \Big\| \le C_2 \omega(s,t)^{\frac{k+1}{p}}$$

comes from Proposition 7.36. As a consequence of Lemma 7.19, we deduce that $\Gamma_{s,t} = 0$ which proves the induction. □

We finish this section by a very interesting probabilistic object, the expectation of the Brownian signature. If

$$Y = y_0 + \sum_{k=1}^{+\infty} \sum_{I \in \{1,\dots,d\}^k} a_{i_1,\dots,i_k} X_{i_1} \dots X_{i_k}$$

is a random series, that is, if the coefficients are real random variables defined on a probability space, we will write

$$\mathbb{E}(Y) = \mathbb{E}(y_0) + \sum_{k=1}^{+\infty} \sum_{I \in \{1,\dots,d\}^k} \mathbb{E}(a_{i_1,\dots,i_k}) X_{i_1} \dots X_{i_k}$$

as soon as the coefficients of Y are integrable.

Theorem 7.39. *For $t \ge 0$,*

$$\mathbb{E}(\mathfrak{S}(B)_t) = \exp\Big(t\Big(\frac{1}{2}\sum_{i=1}^{d} X_i^2\Big)\Big).$$

Proof. An easy computation shows that if $\mathcal{I}_n$ is the set of words of length n obtained by all the possible concatenations of the words

$$\{(i,i)\}, \quad i \in \{1,\dots,d\}$$

then

(1) if $I \notin \mathcal{I}_n$ then

$$\mathbb{E}\Big(\int_{\Delta^n[0,t]} \circ dB^I\Big) = 0;$$

(2) if $I \in \mathcal{I}_n$ then

$$\mathbb{E}\Big(\int_{\Delta^n[0,t]} \circ\, dB^I\Big) = \frac{t^{\frac{n}{2}}}{2^{\frac{n}{2}}\left(\frac{n}{2}\right)!}.$$

Therefore, we have

$$\mathbb{E}(\mathfrak{S}(B)_t) = 1 + \sum_{k=1}^{+\infty} \sum_{I \in \mathcal{I}_k} \frac{t^{\frac{k}{2}}}{2^{\frac{k}{2}}\left(\frac{k}{2}\right)!} X_{i_1} \dots X_{i_k} = \exp\Big(t\Big(\frac{1}{2}\sum_{i=1}^{d} X_i^2\Big)\Big). \quad \square$$

Based on the previous results, it should come as no surprise that differential equations driven by the Brownian rough path should correspond to Stratonovitch differential equations. This is indeed the case. For linear equations, the result is easy to prove as a consequence of the previous proposition.

Theorem 7.40. *Let us consider matrices $M_1, \dots, M_d \in \mathbb{R}^{n \times n}$. For $y_0 \in \mathbb{R}^n$, the solution of the rough differential equation*

$$y(t) = y(0) + \sum_{i=1}^{d} \int_0^t M_i y(s) dB^i(s)$$

coincides with the solution of the Stratonovitch differential equation

$$y(t) = y(0) + \sum_{i=1}^{d} \int_0^t M_i y(s) \circ dB^i(s).$$

Proof. The solution of the rough differential equation can be expanded as the Volterra series:

$$y(t) = y(0) + \sum_{k=1}^{+\infty} \sum_{I=(i_1,\dots,i_k)} M_{i_1} \dots M_{i_k} \Big(\int_{\Delta^k[0,t]} \partial B^I\Big) y(0),$$

where $\int_{\Delta^k[0,t]} \partial B^I$ is the iterated integral in the rough path sense. This iterated integral coincides with the iterated Stratonovitch integral, so we have

$$y(t) = y(0) + \sum_{k=1}^{+\infty} \sum_{I=(i_1,\dots,i_k)} M_{i_1} \dots M_{i_k} \Big(\int_{\Delta^k[0,t]} \circ dB^I\Big) y(0),$$

which is the Volterra expansion of the Stratonovitch stochastic differential equation. $\square$

The previous theorem extends to the non-linear case. We refer the interested reader to [29] for a proof of the following general result.

Theorem 7.41. *Let $\gamma > 2$ and let $V_1, \dots, V_d$ be γ-Lipschitz vector fields on $\mathbb{R}^n$. Let $x_0 \in \mathbb{R}^n$. The solution of the rough differential equation*

$$X_t = x_0 + \sum_{i=1}^{d} \int_0^t V_i(X_s)\, dB_s^i$$

is the solution of the Stratonovitch differential equation

$$X_t = x_0 + \sum_{i=1}^{d} \int_0^t V_i(X_s) \circ dB_s^i.$$

To finish we mention that rough paths theory provides a perfect framework to study differential equations driven by Gaussian processes. Indeed, let B be a Gaussian process with covariance function R. A lot of the information concerning the Gaussian process B is encoded in the rectangular increments of R, which are defined by

$$R_{uv}^{st} = \mathbb{E}((B_t - B_s) \otimes (B_v - B_u)).$$

We then call two-dimensional ρ-variation of R the quantity

$$V_\rho(R) \equiv \sup \Big\{ \Big(\sum_{i,j} \| R_{s_i s_{i+1}}^{t_j t_{j+1}} \|^\rho \Big)^{1/\rho} ; (s_i), (t_j) \in \Pi \Big\},$$

where Π stands for the set of partitions of $[0, 1]$. One has the following fundamental Friz–Victoir theorem concerning the Gaussian rough path existence:

Proposition 7.42 (Friz–Victoir). *If there exists $1 \le \rho < 2$ such that R has finite ρ-variation, the process B is a p-rough path for any $p > 2\rho$.*

As an illustration, the previous result applies in particular to fractional Brownian motion. Let $B = (B^1, \dots, B^d)$ be a d-dimensional fractional Brownian motion defined on a complete probability space $(\Omega, \mathcal{F}, \mathbb{P})$, with Hurst parameter $H \in (0, 1)$. It means that B is a centered Gaussian process whose coordinates are independent and satisfy

$$\mathbb{E}((B_t^j - B_s^j)^2) = |t - s|^{2H} \quad \text{for } s, t \ge 0.$$

For instance when $H = 1/2$, then a fractional Brownian motion is a Brownian motion. It can then be proved that the hypothesis of Proposition 7.42 is satisfied for a fractional Brownian motion with Hurst parameter $H \in (\frac{1}{4}, 1)$, and in that case $\rho = 1/(2H)$. Thus, by using the rough paths theory, we may define stochastic differential equations driven by fractional Brownian motions with Hurst parameter $H \in (\frac{1}{4}, 1)$.

Notes and comments

Section 7.2. The signature of the path is a fascinating object at the intersection between algebraic geometry, topology, Lie group theory and sub-Riemannian geometry. Several sub-Riemannian aspects of the signature are studied in the book [6].

Sections 7.3, 7.4. The rough paths theory is a relatively new theory that was built in the paper [55] by T. Lyons. There are now several comprehensive books about the theory: we mention in particular [29], [57] and [56]. At the time of writing, the theory is still a very active domain of research. Recent important applications have been found in the theory of ill-posed stochastic partial differential equations (see [35]).

Section 7.5. The expectation of the signature of the Brownian motion as a rough path can be used to construct parametrices for semigroups and finds applications to index theory (see [7]). The rough paths point of view on stochastic differential equations has several advantages. It comes with previously unknown deterministic estimates like Davie's and often provides straightforward proofs and improvements of classical results like the Stroock–Varadhan support theorem. We refer to the book [29] for an overview of these applications. Besides stochastic differential equations driven by Brownian motions, rough paths theory also applies to stochastic differential equations driven by general semimartingales because it may be proved that such processes always are p-rough paths for $2 < p < 3$. Of course, the advantage is that the theory even applies to equations driven by very irregular processes that are not semimartingales like rough Gaussian processes (for instance the fractional Brownian motion, see [15]). In this case Itô's theory does not apply and Lyons' theory becomes the only way to define and study such equations.

Appendix A
Unbounded operators

It is a fact that many interesting linear operators are not bounded and only defined on a dense subset of a Banach space (think of differential operators in L^p). We collect here some general definitions and basic results about such operators that are used at some places in this book. The details and the proofs can be found in the reference book [62].

Let $(\mathcal{B}_1, \|\cdot\|_1)$ and $(\mathcal{B}_2, \|\cdot\|_2)$ be two Banach spaces. Let $T: V \to \mathcal{B}_2$ be a linear operator defined on a linear subspace $V \subset \mathcal{B}_1$. The space V on which T is defined is called the *domain* of T and is usually denoted by $\mathcal{D}(T)$. If $\mathcal{D}(T)$ is dense in $\mathcal{B}_1$, then T is said to be *densely defined.*

In the study of unbounded operators like T, it is often useful to consider the graph

$$\boldsymbol{G}_T = \{(v, Tv), v \in \mathcal{D}(T)\},$$

which is a linear subspace of the Banach space $\mathcal{B}_1 \oplus \mathcal{B}_2$.

Definition A.1. The operator T is said to be a *closed operator* if its graph $\boldsymbol{G}_T$ is a closed linear subspace of $\mathcal{B}_1 \oplus \mathcal{B}_2$.

The following theorem is known as the closed graph theorem.

Theorem A.2. *Let us assume that $\mathcal{D}(T) = \mathcal{B}_1$. The operator T is bounded if and only if it is closed.*

In general, the closure $\bar{\boldsymbol{G}}_T$ of $\boldsymbol{G}_T$ does not need to be the graph of an operator.

Definition A.3. The operator T is said to be *closable* if there is an operator $\bar{T}$ such that

$$\bar{\boldsymbol{G}}_T = \boldsymbol{G}_{\bar{T}}.$$

The operator $\bar{T}$ is then called the *closure* of T.

If T is densely defined, we define the *adjoint* of T as the linear operator

$$T' \colon \mathcal{D}(T') \subset \mathcal{B}_2^* \to \mathcal{B}_1^*,$$

defined on

$$\mathcal{D}(T') = \{u \in \mathcal{B}_2^*, \ \exists c(u) \geq 0, \ \forall v \in \mathcal{D}(T), \ |\langle u, Tv\rangle| \leq c(u)\|v\|_1\}$$

and characterized by the duality formula

$$\langle u, Tv\rangle = \langle T'u, v\rangle, \quad u \in \mathcal{D}(T'), \ v \in \mathcal{D}(T).$$

We then have the following result.

Proposition A.4. *Assume that the Banach spaces $\mathcal{B}_1$ and $\mathcal{B}_2$ are reflexive. A densely defined operator*

$$T : \mathcal{D}(T) \subset \mathcal{B}_1 \to \mathcal{B}_2$$

is closable if and only if T' is densely defined.

Let $(\mathcal{H}, \langle \cdot, \cdot \rangle_{\mathcal{H}}$ be a Hilbert space and let A be a densely defined operator on a domain $\mathcal{D}(A)$. We have the following basic definitions.

- The operator A is said to be *symmetric* if for $f, g \in \mathcal{D}(A)$,

$$\langle f, Ag \rangle_{\mathcal{H}} = \langle Af, g \rangle_{\mathcal{H}}.$$

- The operator A is said to be a *non-negative symmetric* operator if it is symmetric and if for $f \in \mathcal{D}(A)$,

$$\langle f, Af \rangle_{\mathcal{H}} \geq 0.$$

- The *adjoint* A^* of A is the operator defined on the domain

$$\mathcal{D}(A^*) = \{ f \in \mathcal{H}, \; \exists\, c(f) \geq 0, \; \forall\, g \in \mathcal{D}(A), \; |\langle f, Ag \rangle_{\mathcal{H}}| \leq c(f) \|g\|_{\mathcal{H}} \}$$

  and given through the Riesz representation theorem by the formula

$$\langle A^* f, g \rangle_{\mathcal{H}} = \langle f, Ag \rangle_{\mathcal{H}},$$

  where $g \in \mathcal{D}(A)$, $f \in \mathcal{D}(A^*)$.
- The operator A is said to be *self-adjoint* if it is symmetric and if $\mathcal{D}(A^*) = \mathcal{D}(A)$.

Let us observe that, in general, the adjoint A^* is not necessarily densely defined, however it is readily checked that if A is a symmetric operator then, from the Cauchy–Schwarz inequality, $\mathcal{D}(A) \subset \mathcal{D}(A^*)$. We have the following first criterion for self-adjointness which may be useful.

Lemma A.5. *Let $A : \mathcal{D}(A) \subset \mathcal{H} \to \mathcal{H}$ be a densely defined operator. Consider the graph of A,*

$$\mathbf{G}_A = \{ (v, Av), v \in \mathcal{D}(A) \} \subset \mathcal{H} \oplus \mathcal{H},$$

and the complex structure

$$\mathcal{J} : \mathcal{H} \oplus \mathcal{H} \to \mathcal{H} \oplus \mathcal{H}, \quad (v, w) \to (-w, v).$$

Then the operator A is self-adjoint if and only if

$$\mathbf{G}_A^{\perp} = \mathcal{J}(\mathbf{G}_A).$$

Proof. It is checked that for any densely defined operator A

$$G_{A^*} = \mathcal{J}\left(G_A^{\perp}\right),$$

and the conclusion follows from routine computations. □

The following result is useful in the proof of Friedrichs' extension theorem (Theorem 4.6):

Lemma A.6. *Let $A\colon \mathcal{D}(A) \subset \mathcal{H} \to \mathcal{H}$ be an injective densely defined self-adjoint operator. Let us denote by $\mathcal{R}(A)$ the range of A. The inverse operator*

$$A^{-1}\colon \mathcal{R}(A) \to \mathcal{H}$$

is a densely defined self-adjoint operator.

Proof. First, observe that

$$\mathcal{R}(A)^{\perp} = \operatorname{Ker}(A^*) = \operatorname{Ker}(A) = \{0\}.$$

Therefore $\mathcal{R}(A)$ is dense in $\mathcal{H}$ and A^{-1} is densely defined. Now,

$$G_{A^{-1}}^{\perp} = \mathcal{J}(G_{-A})^{\perp} = \mathcal{J}(G_{-A}^{\perp}) = \mathcal{J}\mathcal{J}(G_{-A}) = \mathcal{J}(G_{A^{-1}}).$$ □

Definition A.7. Let X and Y be Banach spaces. A bounded operator $T: X \to Y$ is called *compact* if T transforms bounded sets into relatively compact sets. That is, T is compact if and only if for every bounded sequence $(x_n)_{n\in\mathbb{N}}$ in X the sequence $(Tx_n)_{n\in\mathbb{N}}$ has a subsequence convergent in Y.

For instance, an operator whose range is finite-dimensional is necessarily a compact operator.

We have the following results about compact operators. Let X and Y be Banach spaces and let $T\colon X \to Y$ be a bounded operator.

- If $(T_n)_{n\in\mathbb{N}}$ is a sequence of compact operators and $T_n \to T$ in the norm topology, then T is compact.
- If S is a bounded operator from Y to a Banach space Z and if T or S is compact, then ST is compact.
- T is compact if and only if its adjoint T' is a compact operator.

For self-adjoint compact operators, the spectral theorem takes a particularly nice form:

Theorem A.8 (Hilbert–Schmidt theorem). *Let T be a compact and self-adjoint operator defined on a separable Hilbert space $\mathcal{H}$. Then there is a complete orthonormal basis $(\phi_n)_{n\in\mathbb{N}}$ for $\mathcal{H}$ and a sequence $(\lambda_n)_{n\in\mathbb{N}}$ of real numbers such that*

$$T\phi_n = \lambda_n\phi_n,$$

and $\lambda_n \to 0$ as $n \to \infty$.

There are two interesting ideals in the class of compact operators: The trace-class operators and the Hilbert–Schmidt operators.

Definition A.9. Let $\mathcal{H}$ be a separable Hilbert space, $(\phi_n)_{n\in\mathbb{N}}$ an orthonormal basis. Then for every bounded and positive operator T, we define

$$\operatorname{Tr}(T) = \sum_{n\in\mathbb{N}} \langle \phi_n, T\phi_n \rangle.$$

This (possibly infinite) number is called the *trace* of T and is independent of the orthonormal basis chosen.

With obvious notations, we have the following properties:

- $\operatorname{Tr}(A + B) = \operatorname{Tr}(A) + \operatorname{Tr}(B)$.
- $\operatorname{Tr}(\lambda A) = \lambda \operatorname{Tr}(A)$.
- If $0 \le A \le B$, $\operatorname{Tr}(A) \le \operatorname{Tr}(B)$.

Definition A.10. A bounded operator T defined on a separable Hilbert space $\mathcal{H}$ is said to be a *trace class operator* if

$$\operatorname{Tr}(\sqrt{T^*T}) < \infty.$$

It turns out that trace class operators necessarily are compact operators. Moreover the following holds:

- If S is a bounded operator if T is a trace class operator, then ST and TS are trace class operators.
- T is a trace class operator if and only if its adjoint T' is a trace class operator.
- If T is a trace class operator and if $(\phi_n)_{n\in\mathbb{N}}$ is an orthonormal basis, the series

  $$\sum_{n\in\mathbb{N}} \langle \phi_n, T\phi_n \rangle$$

  converges absolutely and the limit is independent of the choice of the basis. This limit is called the *trace* of T and is denoted as above by $\operatorname{Tr}(T)$.

Definition A.11. A bounded operator T defined on a separable Hilbert space $\mathcal{H}$ is said to be a *Hilbert–Schmidt operator* if

$$\operatorname{Tr}(T^*T) < \infty.$$

Hilbert–Schmidt operators necessarily are compact operators. Moreover, we have the following properties:

- If S is a bounded operator if T is a Hilbert–Schmidt operator, then ST and TS are Hilbert–Schmidt operators.

- T is a Hilbert–Schmidt operator if and only if its adjoint T' is a Hilbert–Schmidt operator.
- The set of Hilbert–Schmidt operators endowed with the inner product

$$\langle T_1, T_2 \rangle_2 = \mathrm{Tr}(T_1^* T_2)$$

  is a Hilbert space.
- A bounded operator is a trace class operator if and only if it is a product of two Hilbert–Schmidt operators.

The following theorem completely describes the Hilbert–Schmidt operators of the L^2 space of some measure μ.

Theorem A.12. *Let (Ω, μ) be a measure space and $\mathcal{H} = L^2_\mu(\Omega, \mathbb{R})$. A bounded operator K on $\mathcal{H}$ is a Hilbert–Schmidt operator if and only if there is a kernel*

$$K \in L^2_{\mu\otimes\mu}(\Omega \times \Omega, \mathbb{R}),$$

such that

$$Kf(x) = \int_\Omega K(x, y) f(y) \mu(dy).$$

Moreover,

$$\|K\|_2^2 = \int_\Omega K(x, y)^2 \mu(dx)\mu(dy).$$

Appendix B
Regularity theory

We present some basic facts of the theory of distributions and Sobolev spaces and applications to elliptic differential operators. Most of these facts will be stated without proof. The material sketched here is covered in more detail in the books [27] or [67] to which we refer the interested reader for the proofs.

Let Ω be a non-empty open set in $\mathbb{R}^n$. We denote by $\mathcal{C}_c(\Omega, \mathbb{R})$ the set of smooth and compactly supported functions on Ω and by $\mathcal{C}^k(\Omega, \mathbb{R})$, $k \geq 0$, the set of functions on Ω that are k times continuously differentiable.

It is convenient to use the multi-index notation, that is, if $\alpha = (\alpha_1, \dots, \alpha_m) \in \{1, \dots, n\}^m$ and $f \in \mathcal{C}_c^k(\Omega, \mathbb{R})$, for $|\alpha| = \alpha_1 + \dots + \alpha_m \leq k$, we write

$$\partial^\alpha f = \frac{\partial^{|\alpha|}}{\partial x_{\alpha_1} \dots \partial x_{\alpha_n}}.$$

As a first step, we define the notion of sequential convergence on $\mathcal{C}_c(\Omega, \mathbb{R})$. A sequence $\phi_n \in \mathcal{C}_c(\Omega, \mathbb{R})$ is said to *sequentially converge* to $\phi \in \mathcal{C}_c(\Omega, \mathbb{R})$,

$$\phi_n \overset{\text{s.c.}}{\longrightarrow} \phi,$$

if the $\phi_n' s$ are supported in a common compact subset of Ω and $\partial^\alpha \phi_n \to \partial^\alpha \phi$ uniformly for every multi-index α.

If u is linear form on $\mathcal{C}_c(\Omega, \mathbb{R})$, for $\phi \in \mathcal{C}_c(\Omega, \mathbb{R})$ we write

$$\langle u, \phi \rangle = u(\phi).$$

Definition B.1. A distribution on Ω is a linear form u on $\mathcal{C}_c(\Omega, \mathbb{R})$ which is continuous in the sense that if $\phi_n \overset{\text{s.c.}}{\longrightarrow} \phi$, then $\langle u, \phi_n \rangle \to \langle u, \phi \rangle$. The space of distributions on Ω is denoted by $\mathcal{D}'(\Omega)$. The space $\mathcal{D}'(\Omega)$ is naturally endowed with the weak topology: a sequence $u_n \in \mathcal{D}'(\Omega)$ converges to $u \in \mathcal{D}'(\Omega)$ if and only $\langle u_n, \phi \rangle \to \langle u, \phi \rangle$ for every $\phi \in \mathcal{C}_c(\Omega, \mathbb{R})$.

Distributions may be viewed as generalized functions. Indeed, if f is a locally integrable on Ω, then $u(\phi) = \int_\Omega f \phi dx$ defines a distribution on Ω, this correspondence being one to one (if we regard two functions that are almost surely equal as equal). That is why it is very common to use the notation

$$u(\phi) = \int_\Omega u \phi dx,$$

even when u is not a function. More generally, any Borel measure on Ω defines a distribution: $u(\phi) = \int_\Omega \phi d\mu$.

One very nice feature of distributions that make them extremely useful is that we may differentiate them as many times as we wish. Indeed, let α be a multi-index. The operator ∂^α defined on $\mathcal{C}_c(\Omega, \mathbb{R})$ is sequentially continuous, that is, if $\phi_n \overset{\text{s.c.}}{\longrightarrow} \phi$ then $\partial^\alpha \phi_n \overset{\text{s.c.}}{\longrightarrow} \partial^\alpha \phi$. We can then define ∂^α on $\mathcal{D}'(\Omega)$, through the integration by parts formula:

$$\langle \partial^\alpha u, \phi \rangle = (-1)^{|\alpha|} \langle u, \partial^\alpha \phi \rangle.$$

In the same spirit we may multiply distributions by smooth functions. Let $f \in \mathcal{C}^\infty(\Omega, \mathbb{R})$. The operator $\phi \to f\phi$ defined on $\mathcal{C}_c(\Omega, \mathbb{R})$ is sequentially continuous. For $u \in \mathcal{D}'(\Omega)$ we can therefore define $fu \in \mathcal{D}'(\Omega)$ by the formula

$$\langle fu, \phi \rangle = \langle u, f\phi \rangle.$$

In particular, if

$$L = \sum_{i,j=1}^n \sigma_{ij}(x) \frac{\partial^2}{\partial x_i \partial x_j} + \sum_{i=1}^n b_i(x) \frac{\partial}{\partial x_i}$$

is a diffusion operator and $u \in \mathcal{D}'(\Omega)$, then we have

$$\langle Lu, \phi \rangle = \langle u, L'\phi \rangle,$$

where

$$L'\phi = \sum_{i,j=1}^n \frac{\partial^2}{\partial x_i \partial x_j}(\sigma_{ij}\phi) - \sum_{i=1}^n \frac{\partial}{\partial x_i}(b_i \phi).$$

More generally, of course, any differential operator on Ω can be defined on distributions. It is then possible to try to solve partial differential equations in the space $\mathcal{D}'(\Omega)$. We then speak of weak solutions, by opposition to strong solutions where the unknown is a function.

Clearly, if $u \in \mathcal{C}^k(\Omega, \mathbb{R})$, the distribution derivatives of u of order $\leq k$ are just the pointwise derivatives but the converse is also true; namely, if $u \in \mathcal{C}^0(\Omega, \mathbb{R})$ and if its distribution derivatives are also in $\mathcal{C}^0(\Omega, \mathbb{R})$, for $|\alpha| \leq k$, then $u \in \mathcal{C}^k(\Omega, \mathbb{R})$.

There is a special class of distributions that are of great interest: the tempered distributions. These are the distributions that have a Fourier transform.

To start with, let us first observe that we defined distributions as linear forms on the space of real functions $\mathcal{C}_c(\Omega, \mathbb{R})$, but of course we may define in a similar way complex distributions by using the set of complex functions $\mathcal{C}_c(\Omega, \mathbb{C})$ as a set of test functions. It is readily checked that what we claimed so far for distributions may be extended in a trivial manner to complex distributions.

Let us now recall that the Schwartz space $\mathcal{S}$ is the space of smooth rapidly decreasing complex-valued functions. This space can be endowed with the following topology: $\phi_n \to \phi$ in $\mathcal{S}$ if and only if

$$\sup_{x\in\mathbb{R}^n} |x^\alpha(\partial^\beta \phi_n - \partial^\beta \phi)| \to 0,$$

for every multi indices α and β (the notation x^α means $x_1^{\alpha_1}\dots x_n^{\alpha_n}$).

Definition B.2. A *tempered distribution* on $\mathbb{R}^n$ is a linear form u on $\mathcal{S}$ which is continuous in the sense that if $\phi_n \to \phi$ in $\mathcal{S}$, then $\langle u, \phi_n\rangle \to \langle u, \phi\rangle$. The set of tempered distributions is denoted by $\mathcal{S}'$.

The space $\mathcal{C}_c(\mathbb{R}^n, \mathbb{C})$ is dense in $\mathcal{S}$ for the topology of $\mathcal{S}$. Therefore every tempered distribution is eventually a distribution. The following two facts may be checked:

- If $u \in \mathcal{S}'$, then for every multi index α, $\partial^\alpha u \in \mathcal{S}'$.
- If $u \in \mathcal{S}'$ and if $f \in \mathcal{C}^\infty(\mathbb{R}^n, \mathbb{C})$ is such that for every multi index $\partial^\alpha f$ grows at most polynomially at infinity, then $fu \in \mathcal{S}'$.

The importance of tempered distributions lies in the fact that they admit a Fourier transform. We recall that if $f : \mathbb{R}^n \to \mathbb{R}$ is an integrable function then its Fourier transform is the bounded function on $\mathbb{R}^n$ defined by

$$\hat{f}(\xi) = \int_{\mathbb{R}^n} e^{-2i\pi\langle \xi, x\rangle} f(x)dx.$$

The Fourier transform maps $\mathcal{S}$ onto itself, is continuous (for the topology on $\mathcal{S}$ described above), and moreover satisfies for $f, g \in \mathcal{S}$,

$$\int_{\mathbb{R}^n} \hat{f} g dx = \int_{\mathbb{R}^n} f \hat{g} dx.$$

It is therefore consistent to define the Fourier transform of a tempered distribution $u \in \mathcal{S}'$ by the requirement

$$\langle \hat{u}, \phi\rangle = \langle u, \hat{\phi}\rangle, \quad \phi \in \mathcal{S}.$$

Definition B.3. Let $s \in \mathbb{R}$. We define the *Sobolev space of order* s:

$$\mathcal{H}_s(\mathbb{R}^n) = \Big\{ f \in \mathcal{S}',\ \hat{f} \text{ is a function and } \|f\|_s^2 = \int_{\mathbb{R}^n} |\hat{f}(\xi)|^2 (1 + \|\xi\|^2)^s d\xi < +\infty \Big\}.$$

The Sobolev space $\mathcal{H}_s(\mathbb{R}^n)$ is a Hilbert space with Hermitian inner product

$$\langle f, \bar{g}\rangle_s = \int_{\mathbb{R}^n} \hat{f}(\xi)\hat{g}(\xi)(1 + \|\xi\|^2)^s d\xi,$$

and the Fourier transform is a unitary isomorphism from $\mathcal{H}_s(\mathbb{R}^n)$ to $L^2_\mu(\mathbb{R}^n, \mathbb{C})$, where

$$\mu(d\xi) = (1 + \|\xi\|^2)^s d\xi.$$

If $s \leq t$, we have $\mathcal{H}_s(\mathbb{R}^n) \subset \mathcal{H}_t(\mathbb{R}^n)$. In particular, for $s \geq 0$, $\mathcal{H}_s(\mathbb{R}^n) \subset \mathcal{H}_0(\mathbb{R}^n) = L^2(\mathbb{R}^n, \mathbb{C})$.

One of the most useful results of the theory of Sobolev spaces is the following theorem, which is sometimes called the Sobolev lemma. It quantifies the simple idea that if f is a tempered distribution whose Fourier transform decreases fast enough at infinity, then f is actually a function that satisfies some regularity properties. More precisely:

Theorem B.4 (Sobolev lemma). *If $s > k + \frac{n}{2}$, then $\mathcal{H}_s(\mathbb{R}^n) \subset \mathcal{C}^k(\mathbb{R}^n, \mathbb{C})$.*

Proof. The proof is very simple and explains where the $k + \frac{n}{2}$ comes from.

Let $f \in \mathcal{H}_s(\mathbb{R}^n)$, with $s > k + \frac{n}{2}$. We have for $|\alpha| \leq k$,

$$\begin{aligned}
\int_{\mathbb{R}^n} |\xi^\alpha \hat{f}(\xi)| d\xi &\leq C \int_{\mathbb{R}^n} (1 + |\xi|^2)^{k/2} |\hat{f}(\xi)| d\xi \\
&\leq C \int_{\mathbb{R}^n} (1 + |\xi|^2)^{\frac{k-s}{2}} |\hat{f}(\xi)| (1 + \|\xi\|^2)^{\frac{s}{2}} d\xi \\
&\leq C \|f\|_s \Big(\int_{\mathbb{R}^n} (1 + |\xi|^2)^{k-s} d\xi \Big)^{1/2}.
\end{aligned}$$

Since $s > k + \frac{n}{2}$, the integral $\int_{\mathbb{R}^n} (1 + |\xi|^2)^{k-s} d\xi$ is finite. We deduce therefore that if $|\alpha| \leq k$, then $|\xi^\alpha \hat{f}(\xi)|$ is integrable, which immediately implies from the inverse Fourier transform formula that $f \in \mathcal{C}^k(\mathbb{R}^n, \mathbb{C})$. □

As a corollary of the Sobolev lemma, we obtain

Corollary B.5. *If $f \in \mathcal{H}_s(\mathbb{R}^n)$, for all $s \in \mathbb{R}$, then $f \in \mathcal{C}^\infty(\mathbb{R}^n, \mathbb{C})$.*

An important feature of Sobolev spaces is that they may be localized. To be precise, if for an open set $\Omega \subset \mathbb{R}^n$, we define $\mathcal{H}_s^{\mathrm{loc}}(\Omega)$ as the set of distributions f such that for every $\phi \in \mathcal{C}_c(\Omega, \mathbb{C})$, $f\phi \in \mathcal{H}_s(\mathbb{R}^n)$, then it can be shown that

$$\mathcal{H}_s(\mathbb{R}^n) \subset \mathcal{H}_s^{\mathrm{loc}}(\Omega).$$

If Ω is an open set in $\mathbb{R}^n$, we define $\mathcal{H}_s^0(\Omega)$ as the closure of $\mathcal{C}_c(\Omega, \mathbb{C})$ in $\mathcal{H}_s(\mathbb{R}^n)$.

The following compactness result theorem is then often extremely useful.

Theorem B.6 (Rellich theorem). *If $\Omega \subset \mathbb{R}^n$ is bounded and $s > t$, the inclusion map*

$$\iota\colon \mathcal{H}_s^0(\Omega) \to \mathcal{H}_t^0(\Omega)$$

is compact.

A diffusion operator

$$L = \sum_{i,j=1}^{n} \sigma_{ij}(x) \frac{\partial^2}{\partial x_i \partial x_j} + \sum_{i=1}^{n} b_i(x) \frac{\partial}{\partial x_i},$$

where b_i and σ_{ij} are continuous functions on $\mathbb{R}^n$ such that for every $x \in \mathbb{R}^n$ the matrix $(\sigma_{ij}(x))_{1 \le i,j \le n}$ is symmetric and positive definite, is said to be an *elliptic diffusion operator*. The first and canonical example of an elliptic diffusion operator is the Laplace operator on $\mathbb{R}^n$:

$$\Delta = \sum_{i=1}^{n} \frac{\partial^2}{\partial x_i^2}.$$

The Sobolev embedding theorem for elliptic diffusion operators is the following theorem.

Theorem B.7 (Local Sobolev embedding theorem). *Let L be an elliptic diffusion operator with smooth coefficients. Suppose that Ω is a bounded open set in $\mathbb{R}^n$. Then, for any $s \in \mathbb{R}$, there is a positive constant C such that for every $u \in \mathcal{H}_s^0(\Omega)$,*

$$\|u\|_s \le C(\|Lu\|_{s-2} + \|u\|_{s-1}).$$

As a corollary one has the following *regularization property* for L.

Corollary B.8. *Let L be an elliptic diffusion operator with smooth coefficients. Suppose that Ω is an open set in $\mathbb{R}^n$. If $u \in \mathcal{H}_s^{\mathrm{loc}}(\Omega)$ and $Lu \in \mathcal{H}_{s-1}^{\mathrm{loc}}(\Omega)$, then $u \in \mathcal{H}_{s+1}^{\mathrm{loc}}(\Omega)$.*

The Sobolev embedding theorem is related to the notion of hypoellipticity:

Definition B.9. Let A be a differential operator on $\mathbb{R}^n$. Then A is said to be *hypoelliptic* if for every open set $\Omega \subset \mathbb{R}^n$ and every distribution u the following holds: If $Au \in \mathcal{C}^\infty(\Omega, \mathbb{C})$, then $u \in \mathcal{C}^\infty(\Omega, \mathbb{C})$.

A fundamental consequence of the Sobolev embedding theorem is the following theorem which, in this form, is due to Hermann Weyl.

Theorem B.10. *Any elliptic diffusion operator with smooth coefficients is hypoelliptic.*

Another consequence of the theory of Sobolev spaces is the following regularization property of elliptic diffusion operators.

Proposition B.11. *Let L be an elliptic diffusion operator with smooth coefficients on $\mathbb{R}^n$ which is symmetric with respect to a Borel measure μ. Let $u \in L^2_\mu(\mathbb{R}^n, \mathbb{R})$ be such that*

$$Lu, L^2u, \dots, L^k u \in L^2_\mu(\mathbb{R}^n, \mathbb{R})$$

for some positive integer k. If $k > \frac{n}{4}$, then u is a continuous function. Moreover, for any bounded open set $\Omega \subset \mathbb{R}^n$ and any compact set $K \subset \Omega$, there exists a positive constant C (independent of u) such that

$$(\sup_{x \in K} |u(x)|)^2 \leq C\Big(\sum_{j=0}^{k} \|L^j u\|^2_{L^2_\mu(\Omega,\mathbb{R})}\Big).$$

More generally, if $k > \frac{m}{2} + \frac{n}{4}$ for some non-negative integer m, then $u \in \mathcal{C}^m(\mathbb{R}^n, \mathbb{R})$, and for any bounded open set $\Omega \subset \mathbb{R}^n$ and any compact set $K \subset \Omega$ there exists a positive constant C (independent of u) such that

$$(\sup_{|\alpha| \leq m} \sup_{x \in K} |\partial^\alpha u(x)|)^2 \leq C\Big(\sum_{j=0}^{k} \|L^j u\|^2_{L^2_\mu(\Omega,\mathbb{R})}\Big).$$

Bibliography

[1] D. Applebaum, *Lévy processes and stochastic calculus*. Cambridge Stud. Adv. Math. 93, Cambridge University Press, Cambridge 2004. 88, 96

[2] L. Bachelier, Théorie de la spéculation. *Ann. Sci. École Norm. Sup.* (3) **17** (1900), 21–86. 1

[3] D. Bakry, Un critère de non-explosion pour certaines diffusions sur une variété riemannienne complète. *C. R. Acad. Sci. Paris Sér. I Math.* **303** (1986), 23–26. 136

[4] R. Bañuelos, The foundational inequalities of D. L. Burkholder and some of their ramifications. *Illinois J. Math.* **54** (2010), 789–868. 184

[5] R. F. Bass, *Diffusions and elliptic operators*. Springer-Verlag, New York 1998. 96

[6] F. Baudoin, *An introduction to the geometry of stochastic flows*. Imperial College Press, London 2004. 255

[7] F. Baudoin, Brownian Chen series and Atiyah–Singer theorem. *J. Funct. Anal.* **254** (2008), 301–317. 255

[8] D. R. Bell, *The Malliavin calculus*. Pitman Monographs Surveys Pure Appl. Math. 34, Longman Scientific & Technical, Harlow 1987. 220

[9] J. Bertoin, *Lévy processes*. Cambridge Tracts in Math. 121, Cambridge University Press, Cambridge 1996. 96

[10] A. Beurling and J. Deny, Espaces de Dirichlet. I. Le cas élémentaire. *Acta Math.* **99** (1958), 203–224. 136

[11] K. Bichteler, *Stochastic integration with jumps*. Encyclopedia Math. Appl. 89, Cambridge University Press, Cambridge 2002. 183

[12] B. Bru and M. Yor, Comments on the life and mathematical legacy of Wolfgang Doeblin. *Finance Stoch.* **6** (2002), 3–47. 184, 220

[13] Z.-Q. Chen and M. Fukushima, *Symmetric Markov processes, time change, and boundary theory*. London Math. Soc. Monogr. 35, Princeton University Press, Princeton, NJ, 2012. 137

[14] K. L. Chung, *Lectures from Markov processes to Brownian motion*. Grundlehren Math. Wiss. 249, Springer-Verlag, Berlin 1982. 96, 184

[15] L. Coutin and Z. Qian, Stochastic analysis, rough path analysis and fractional Brownian motions. *Probab. Theory Related Fields* **122** (2002), 108–140. 255

[16] E. B. Davies, *Heat kernels and spectral theory*. Cambridge Tracts in Math. 92, Cambridge University Press, Cambridge 1989. 107, 136

[17] C. Dellacherie and P.-A. Meyer, *Probabilités et potentiel. Chapitres V à VIII*. Revised ed., Actualités Scientifiques et Industrielles 1385, Hermann, Paris 1980. 34, 183

[18] W. Doeblin, Sur l'équation de Kolmogoroff. *C. R. Acad. Sci. Paris Sér. I Math.* **331** (2000), 1059–1128. Avec notes de lecture du pli cacheté par Bernard Bru. 184

[19] J. L. Doob, *Stochastic processes*. John Wiley & Sons, New York 1953. 1, 34

[20] J. L. Doob, *Classical potential theory and its probabilistic counterpart*. Grundlehren Math. Wiss. 262, Springer-Verlag, Berlin 1984. 184

[21] R. Durrett, *Brownian motion and martingales in analysis*. Wadsworth Mathematics Series, Wadsworth International Group, Belmont, CA, 1984. 184

[22] R. Durrett, *Probability: theory and examples*. 4th ed., Camb. Ser. Stat. Probab. Math., Cambridge University Press, Cambridge 2010. 62

[23] E. B. Dynkin, *Markov processes*. Vols. I, II, Grundlehren Math. Wiss. 121, 122, Springer-Verlag, Berlin 1965. 96

[24] S. N. Ethier and T. G. Kurtz, *Markov processes*. Wiley Ser. Probab. Math. Statist. Probab. Math. Statist., John Wiley & Sons, New York 1986. 96

[25] W. Feller, The parabolic differential equations and the associated semi-groups of transformations. *Ann. of Math.* (2) **55** (1952), 468–519. 136

[26] G. B. Folland, *Introduction to partial differential equations*. 2nd ed., Princeton University Press, Princeton, NJ, 1995. 127

[27] G. B. Folland, *Real analysis*. 2nd ed., Pure Appl. Math., John Wiley & Sons, New York 1999. 262

[28] A. Friedman, *Partial differential equations of parabolic type*. Prentice-Hall, Englewood Cliffs, N.J., 1964. 136

[29] P. K. Friz and N. B. Victoir, *Multidimensional stochastic processes as rough paths*. Cambridge Stud. Adv. Math. 120, Cambridge University Press, Cambridge 2010. 183, 220, 244, 253, 255

[30] M. Fukushima, Y. Oshima, and M. Takeda, *Dirichlet forms and symmetric Markov processes*. de Gruyter Stud. Math. 19, Walter de Gruyter & Co., Berlin 1994. 136, 137

[31] B. Gaveau, Principe de moindre action, propagation de la chaleur et estimées sous elliptiques sur certains groupes nilpotents. *Acta Math.* **139** (1977), 95–153. 184

[32] J. Hadamard, Sur un problème mixte aux dérivées partielles. *Bull. Soc. Math. France* **31** (1903), 208–224. 136

[33] J. Hadamard, Principe de Huygens er prolongement analytique. *Bull. Soc. Math. France* **52** (1924), 241–278. 136

[34] J. Hadamard, *Lectures on Cauchy's problem in linear partial differential equations*. Dover Publications, New York 1953. 136

[35] M. Hairer, Solving the KPZ equation. *Ann. of Math.* (2) **178** (2013), 559–664. 255

[36] E. Hille, Sur les semi-groupes analytiques. *C. R. Acad. Sci. Paris* **225** (1947), 445–447. 136

[37] E. Hille, *Functional Analysis and Semi-Groups*. Amer. Math. Soc. Colloq. Publ.. 31, Amer. Math. Soc., New York 1948. 136

[38] N. Ikeda and S. Watanabe, *Stochastic differential equations and diffusion processes*. 2nd ed., North-Holland Math. Library 24, North-Holland Publishing Co., Amsterdam 1989. 34, 183

[39] K. Itô, Stochastic integral. *Proc. Imp. Acad. Tokyo* **20** (1944), 519–524. 183

[40] K. Itô, On a stochastic integral equation. *Proc. Japan Acad.* **22** (1946), 32–35. 219

[41] K. Ito, *Lectures on stochastic processes*. 2nd ed., Tata Inst. Fund. Res. Lectures on Math. and Phys. 24, Springer-Verlag, Berlin 1984. 62

[42] N. Jacob, *Pseudo differential operators and Markov processes*. Vol. I: Fourier analysis and semigroups, Imperial College Press, London 2001. 86, 96

[43] N. Jacob, *Pseudo differential operators & Markov processes*. Vol. II: Generators and their potential theory, Imperial College Press, London 2002. 96

[44] J. Jacod and A. N. Shiryaev, *Limit theorems for stochastic processes*. 2nd ed., Grundlehren Math. Wiss. 288, 2nd ed., Springer-Verlag, Berlin 2003. 34, 62

[45] I. Karatzas and S. E. Shreve, *Methods of mathematical finance*. Appl. Math. 39, Springer-Verlag, New York 1998. 184

[46] A. N. Kolmogorov, *Foundations of the theory of probability*. Chelsea Publishing Company, New York, N.Y., 1950. 1

[47] H. Kunita, *Stochastic flows and stochastic differential equations*. Cambridge Stud. Adv. Math. 24, Cambridge University Press, Cambridge 1990. 192, 220

[48] S. Kusuoka and D. Stroock, Applications of the Malliavin calculus. I. In *Stochastic analysis* (Katata/Kyoto, 1982), North-Holland Math. Library 32, North-Holland, Amsterdam 1984, 271–306. 220

[49] S. Kusuoka and D. Stroock, Applications of the Malliavin calculus. II. *J. Fac. Sci. Univ. Tokyo Sect. IA Math.* **32** (1985), 1–76. 220

[50] S. Kusuoka and D. Stroock, Applications of the Malliavin calculus. III. *J. Fac. Sci. Univ. Tokyo Sect. IA Math.* **34** (1987), 391–442. 220

[51] G. F. Lawler, *Conformally invariant processes in the plane*. Math. Surveys. Monogr. 114, Amer. Math. Soc., Providence, RI, 2005. 184

[52] G. F. Lawler and V. Limic, *Random walk: a modern introduction*. Cambridge Stud. Adv. Math. 123, Cambridge University Press, Cambridge 2010. 62

[53] P. Lévy, Wiener's random function, and other Laplacian random functions. In *Proceedings of the second Berkeley symposium on mathematical statistics and probability* (July 31 – August 12, 1950), University of California Press, Berkeley 1951, 171–187. 184

[54] P. Lévy, *Processus stochastiques et mouvement brownien*. Suivi d'une note de M. Loève. Deuxième édition revue et augmentée, Gauthier-Villars, Paris 1965. 62

[55] T. J. Lyons, Differential equations driven by rough signals. *Rev. Mat. Iberoamericana* **14** (1998), 215–310. 255

[56] T. J. Lyons, M. Caruana, and T. Lévy, *Differential equations driven by rough paths*. Lecture Notes in Math. 1908, Springer, Berlin 2007. 255

[57] T. Lyons and Z. Qian, *System control and rough paths*. Oxford Math. Monogr., Oxford University Press, Oxford 2002. 255

[58] P. Malliavin, Stochastic calculus of variation and hypoelliptic operators. In *Proceedings of the international symposium on stochastic differential equations* (Res. Inst. Math. Sci., Kyoto Univ., Kyoto, 1976), Wiley, New York 1978, 195–263. 206, 220

[59] P. Malliavin, *Stochastic analysis*. Grundlehren Math. Wiss. 313, Springer-Verlag, Berlin 1997. 220

[60] D. Nualart, *The Malliavin calculus and related topics*. 2nd ed., Probab. Appl., Springer-Verlag, Berlin 2006. 206, 213, 220

[61] P. E. Protter, *Stochastic integration and differential equations*. 2nd ed., Appl. Math. 21, Springer-Verlag, Berlin 2004. 34, 158, 161, 183, 220

[62] M. Reed and B. Simon, *Methods of modern mathematical physics*. Vol I: Functional analysis, 2nd ed., Academic Press, New York 1980. 103, 136, 257

[63] M. Reed and B. Simon, *Methods of modern mathematical physics*. Vol. II: Fourier analysis, self-adjointness, Academic Press, New York 1975. 112, 136

[64] D. Revuz and M. Yor, *Continuous martingales and Brownian motion*. 3rd ed., Grundlehren m Math. Wiss. 293, Springer-Verlag, Berlin 1999. 34

[65] M. Riesz, Sur les maxima des formes bilinéaires et sur les fonctionnelles linéaires. *Acta Math.* **49** (1927), 465–497. 136

[66] L. C. G. Rogers and D. Williams, *Diffusions, Markov processes and martingales*. Vol. 1: Foundations, Reprint of the second (1994) edition, Cambridge Math. Lib., Cambridge University Press, Cambridge 2000. 62

[67] W. Rudin, *Functional analysis*. 2nd ed., McGraw-Hill, New York 1991. 262

[68] K.-i. Sato, *Lévy processes and infinitely divisible distributions*. Cambridge Stud. Adv. Math. 68, Cambridge University Press, Cambridge 1999. 96

[69] R. L. Schilling, R. Song, and Z. Vondraček, *Bernstein functions*. de Gruyter Stud. Math. 37, Walter de Gruyter & Co., Berlin 2010. 93

[70] I. Shigekawa, Derivatives of Wiener functionals and absolute continuity of induced measures. *J. Math. Kyoto Univ.* **20** (1980), 263–289. 220

[71] I. Shigekawa, *Stochastic analysis*. Transl. Math. Monogr. 224, Amer. Math. Soc., Providence, RI, 2004. 220

[72] B. Simon, An abstract Kato's inequality for generators of positivity preserving semigroups. *Indiana Univ. Math. J.* **26** (1977), 1067–1073. 136

[73] F. Spitzer, *Principles of random walk*. 2nd ed., Grad. Texts in Math. 34, Springer-Verlag, New York 1976. 62

[74] E. M. Stein, Interpolation of linear operators. *Trans. Amer. Math. Soc.* **83** (1956), 482–492. 136

[75] R. S. Strichartz, Analysis of the Laplacian on the complete Riemannian manifold. *J. Funct. Anal.* **52** (1983), 48–79. 136

[76] D. W. Stroock, Some applications of stochastic calculus to partial differential equations. In *Ecole d'Été de Probabilités de Saint-Flour XI – 1981*, Lecture Notes in Math. 976, Springer-Verlag, Berlin 1983, 267–382. 220

[77] D. W. Stroock and S. R. S. Varadhan, *Multidimensional diffusion processes*. Classics Math., Springer-Verlag, Berlin 2006. 220

[78] K. T. Sturm, Diffusion processes and heat kernels on metric spaces. *Ann. Probab.* **26** (1998), 1–55. 136

[79] N. Wiener, Differential-space. *J. Math. Phys. of the Massachusetts Institute of Technology* **2** (1923), 131–174. 62

[80] M. Yor, *Some aspects of Brownian motion*. Part I: Some special functionals, Lectures Math. ETH Zürich, Birkhäuser, Basel 1992. 184

[81] K. Yosida, On the differentiability and the representation of one-parameter semi-group of linear operators. *J. Math. Soc. Japan* **1** (1948), 15–21. 136

[82] K. Yosida, *Functional analysis*. 6th ed., Grundlehren Math. Wiss. 123, Springer-Verlag, Berlin 1980. 136

Index

Adapted process, 10
Arcsine law, 59, 198
Ascoli theorem, 56

Bernstein function, 93
Bessel process, 66, 172, 199
Binomial inequality, 230
Black–Scholes process, 66, 189
Brownian bridge, 38, 189
Brownian motion, 1, 35
Burkholder–Davis–Gundy inequalities, 176

Canonical process, 8
Carathéodory extension theorem, 12
Carré du champ, 98
Cauchy's semigroup, 106
Chapman–Kolmogorov relation, 63, 108
Chen relations, 227, 239
Clark–Ocone formula, 209
Continuous process, 8
Control, 222
Courrège theorem, 86
Cylindrical set, 6

Döblin–Itô formula, 161, 163–165
Dambis, Dubins–Schwarz theorem, 172
Daniell–Kolmogorov theorem, 11
Diffusion operator, 97
Diffusion process, 82
Diffusion semigroup, 81, 104
Dirichlet form, 99
Dirichlet process, 133
Dirichlet semigroup, 128
Donsker theorem, 58
Doob convergence theorem, 24
Doob maximal inequalities, 32
Doob regularization theorem, 28
Doob stopping theorem, 20
Dynkin theorem, 83

Essentially self-adjoint, 101

Feller–Dynkin process, 76
Feller–Dynkin semigroup, 76
Feynman–Kac formula, 197
Filtration, 10
Fractional Brownian motion, 37, 254
Friedrichs extension, 99

Garsia–Rodemich–Rumsey inequality, 245
Gaussian process, 15
Generator, 80
Girsanov theorem, 178

Hölder function, 16
Heat kernel, 108
Hilbert–Schmidt operator, 260
Hilbert–Schmidt theorem, 259
Hille–Yosida theorem, 116

Integration by parts formula, 162, 207
Itô integral, 141, 153
Itô representation theorem, 170

Kato inequality, 109
Killed Markov process, 72
Kolmogorov continuity theorem, 16

Lévy area formula, 182
Lévy characterization theorem, 171
Lévy measure, 89
Lévy process, 87
Lévy–Khinchin theorem, 89
Law of iterated logarithm, 42, 52
Lenglart inequality, 175

Local martingale, 156
Local time, 55, 75
Lyons' continuity theorem, 239, 244

Malliavin derivative, 206
Malliavin matrix, 214
Markov process, 48, 63, 64, 194
Martingale, 20
McDonald's function, 105
Measurable process, 8
Modification of a process, 16

Neumann process, 133
Neumann semigroup, 131
Novikov condition, 181

Ornstein–Uhlenbeck process, 38, 66, 82, 188

Poisson process, 87
Progressively measurable process, 10
Prokhorov theorem, 56

Quadratic covariation, 152
Quadratic variation, 147, 157, 159

Random walk, 45
Recurrence property, 49
Recurrence property of Brownian motion, 40
Recurrent, 167
Reflection principle, 54, 74
Rellich theorem, 265
Reproducing kernel, 131
Rough path, 238

Semigroup, 64
Semimartingale, 158
Signature of a path, 227
Skohorod integral, 209
Sobolev embedding theorem, 266
Sobolev lemma, 265
Spectral theorem, 103
Spitzer theorem, 174
Stochastic differential equation, 188
Stochastic flow, 192
Stochastic process, 8
Stopping time, 18, 48
Stratonovitch integral, 160, 201
Strong Markov process, 49, 72, 200
Sub-Markov process, 71, 72
Sub-Markov semigroup, 71, 109
Submartingale, 20
Subordinator, 92
Supermartingale, 20

Trace class operator, 260
Trace of an operator, 260
Transition function, 63

Uniformly integrable family, 23
Usual completion, 27
Usual conditions, 27

Vector field, 201, 226

Wiener chaos expansion, 211
Wiener measure, 35

Young integral, 140

DATE DUE

			PRINTED IN U.S.A.

QA 274 .B397 2014

Baudoin, Fabrice,

Diffusion processes and stochastic calculus